T0390615

Breaking Tolerance to Anti-Cancer Cell-Mediated Immunotherapy

γδ T CELL CANCER IMMUNOTHERAPY

VOLUME 7

Breaking Tolerance to Anti-Cancer Cell-Mediated Immunotherapy

Series Editor: Benjamin Bonavida, PhD

Volume 1: Autophagy in Immune Response: Impact on Cancer Immunotherapy
Edited by Salem Chouaib

Volume 2: Immunotherapy in Resistant Cancer: From the Lab Bench Work to Its Clinical Perspectives
Edited by Jorge Morales-Monto and Mariana Segovia-Mendoza

Volume 3: Immunotherapeutic Strategies for the Treatment of Glioma
Edited by Michael Lim and Christopher Jackson

Volume 4: NK cells in cancer immunotherapy: Successes and Challenges
Edited by Anahid Jewett and Yuman Fong

Volume 5: Immune Landscape of Pancreatic Cancer Development and Drug Resistance
Edited by Batoul Farran and Ganji Purnachandra Nagaraju

Volume 6: Principles of Immunotherapy in Breast and Gastrointestinal Cancers
Edited by Michele Ghidini

Volume 7: γδ T Cell Cancer Immunotherapy
Edited by Marta Barisa

Upcoming Volumes in Series:

Breaking Tolerance to Pancreatic Cancer Unresponsive to Immunotherapy
Edited by Kasuya Hideki and Itzel Bustos Villalobos

Gammadelta T Cells in Cancer Immunotherapy
Edited by Ilan Bank

T Cell Metabolism and Cancer Immunotherapy
Edited by Jianxun Song

Breaking Tolerance to Anti-Cancer Cell-Mediated Immunotherapy

γδ T CELL CANCER IMMUNOTHERAPY

Evidence-Based Perspectives for Clinical Translation

VOLUME 7

Edited by

MARTA BARISA

University College London, Experimental Paediatric Oncology Research Group and Allogeneic Immunotherapy Research Group Zayed Centre for Research, Great Ormond Street Hospital, University College London Great Ormond Street Institute of Child Health, London, UK

Academic Press is an imprint of Elsevier
125 London Wall, London EC2Y 5AS, United Kingdom
525 B Street, Suite 1650, San Diego, CA 92101, United States
50 Hampshire Street, 5th Floor, Cambridge, MA 02139, United States

ISBN 978-0-443-21766-1

For information on all Academic Press publications
visit our website at https://www.elsevier.com/books-and-journals

Publisher: Stacy Masucci
Senior Acquisition Editor: Linda Versteeg-Buschman
Editorial Project Manager: Samantha Allard
Production Project Manager: Selvaraj Raviraj
Cover Designer: Vicky Pearson Esser

Typeset by STRAIVE, India

Cover Image Insert

White T cells approach a large, red, and adhered cancer cell. Set on a blue background.

Aims and scope of series "Breaking Tolerance to Anti-Cancer Cell-Mediated Immunotherapy"

The role of the immune system in the eradication of cancers has been investigated for several decades with controversial findings. The controversy was the result of a poor understanding of the underlying mechanisms that govern responsiveness and unresponsiveness. Hence, significant advances have been made with respect to the regulation of the host immune response against cancer and several immunotherapeutics have been recently introduced and used clinically. These include both antibody and cell-mediated immunity targeting the cancer cells. Such immunotherapies led to significant clinical responses in various cancer types that were unresponsive to conventional therapies. However, only a subset of cancer patients responds to such immunotherapeutics and also there is a responding subset that develops resistance to further treatment. Various studies have examined potential underlying mechanisms involved in resistance and identified a variety of gene products that play pivotal roles in maintaining the resistant phenotype of the cancer cells to cell-mediated immunotherapy.

The main objective of the proposed series "Breaking Cancer Resistance to Cell-Mediated Immunotherapies" is the development of individual volumes that are focused on the application of particular sensitizing agents that, when used in combination with cell-mediated immunotherapy, result in the reversal of resistance.

A variety of different classes of immunosensitizing agents has been reported. Each individual volume will focus on one class of immunosensitizing agents and their effects on the reversal of cell-mediated immune resistance in different cancers. Emphasis will be on biochemical, molecular, and genetic mechanisms by which the sensitizing agents mediate their effects individually and/or in combination with immunotherapy. Each editor will compile nonoverlapping review chapters on the therapeutic role of specific sensitizing agents used in combination with conventional immunotherapy and the reversal of resistance. There will also be an emphasis on discrimination of responses obtained in various cancer types.

The scope of the series is to provide updated information to scientists and clinicians that is valuable in their quest to gather information, carry out new investigations, and develop novel immunosensitizing agents that are both more potent and also that might be active whereby the existing ones were not active.

Benjamin Bonavida, PhD (Series Editor)

About the series editor

Dr. Benjamin Bonavida, PhD (series editor), is currently distinguished research professor at the University of California, Los Angeles (UCLA). His research career, thus far, has focused on basic immunochemistry and cancer immunobiology. His research investigations have ranged from the mechanisms of cell-mediated killing and sensitization of resistant tumor cells to chemo-/immunotherapies, characterization of resistant factors in cancer cells, cell-signaling pathways mediated by therapeutic anticancer antibodies, and characterization of a dysregulated NF-κB/Snail/YY1/RKIP/PTEN loop in many cancers that regulate cell survival, proliferation, invasion, metastasis, and resistance. He has also investigated the role of nitric oxide in cancer and its potential antitumor activity. Many of the above studies are centered on the clinical challenging features of cancer patients' failure to respond to both conventional and targeted therapies. The development and activity of various chemosensitizing agents and their modes of action to reverse chemo- and immunoresistance are highlighted in many refereed publications.

Acknowledgments

The series editor acknowledges the Department of Microbiology, Immunology and Molecular Genetics and the UCLA David Geffen School of Medicine for their continuous support. The series editor also acknowledges the assistance of Mr. Rafael Teixeira, acquisitions editor for Elsevier/Academic Press, and the excellent assistance and support of Ms. Samantha Allard, editorial project manager for Elsevier/Academic Press, for their continuous cooperation throughout the development of this book.

Volume editor biography

Dr. Marta Barisa is a Senior Fellow in Experimental Oncology at University College London (UCL). She is based at the Cancer Section of UCL's Zayed Centre of the Great Ormond Street Institute of Child Health.

She completed her undergraduate degree at St. George's Hospital Medical School in London, United Kingdom, followed by training in molecular immunology and cellular immunotherapy at UCL, as well as the Massachusetts Institute of Technology in Boston, United States. Most of her career has focused on immunotherapy drug development for various oncology indications—alone or in combination with additional immunotherapeutic, radio-, and chemotherapeutic interventions.

For much of her research career, she has been based at UCL, which is a world-leading center for cellular immunotherapy development. It is one of a few institutions dominating global innovation in the space; more than 400 CAR-T patents and 40 patent families have been registered by UCL alone. The Institute of Child Health, where Dr. Barisa is based, is closely associated with Great Ormond Street Hospital for Children. The hospital, with its associated research institute, is Europe's biggest and highest-ranked center for pediatric health, excelling particularly with the quality and quantity of its translational research that is coupled with clinical trials.

Dr. Barisa's specialism rests with the design, development, and translation of genetically-modified adoptive cellular immunotherapies for solid tumors. In this capacity, she is based across two different immunotherapy groups at UCL: (i) the Innate Immune Engineering Lab, which is focused on allogeneic, gene-modified $\gamma\delta$ T cell therapeutic development for adolescent and adult solid cancers, including carcinoma and sarcoma, and (ii) the Experimental Paediatric Oncology Lab, where she evaluates a range of autologous as well as allogeneic chimeric antigen receptor (CAR-T) $\alpha\beta$ and $\gamma\delta$ T cell interventions for their ability to target pediatric neuroendocrine tumors.

Throughout her career, she has woven together academic and commercial cell therapy development. She has held roles in both private biotech companies and academic institutions—often building novel cell therapy concepts in the academic space, and then funding the late-stage development of these with commercial partners. She serves on scientific advisory boards and holds patents pertaining to novel cellular immunotherapy designs and manufacturing methodologies. She lectures on cell therapy development regularly, and is a co-lead of a postgraduate program for Cell & Gene Therapy at UCL.

Dr. Barisa is regularly invited to speak at and chair discussions at international immunotherapy meetings and publishes in the space frequently. Most importantly in the context of this collection, her research has contributed to a number of world-first clinical trial developments for diverse solid tumor oncology indications. These range from the evaluation of novel checkpoint blocker antibodies as boosters to traditional chemoimmunotherapy, to autologous, drug-switchable αβ CAR-T, to allogeneic, armored, gene-modified γδ T cells. For many of these drug candidates, she has been involved with their journey to a patient from inception: the ideation of a new interventional product, its design, the subsequent product development, the manufacturing process development, and ultimately—translation to the clinic. The clinical collaborators she and her partners work with are most often closely integrated with the UCL translational research network, but can also be found at other United Kingdom locations, as well as various sites in the European Union and the United States. In addition to being coupled with Europe's busiest pediatric CAR-T trial center, the Zayed Centre also houses its own cell therapy and lentiviral manufacturing facilities with whom Dr. Barisa works closely. It is this broad bench-to-bedside viewpoint and translational perspective that she hopes to bring to the herein enclosed collection of reviews on the current state of γδ T cell immunotherapy.

Preface

In a therapeutic landscape that would have been difficult to conceive of even a few decades ago, immunotherapy now stands solidly alongside chemotherapy, surgery, and radiotherapy as a pillar of cancer treatment. Antibody-based therapies dominate novel oncology drug development pipelines, from antibody-drug conjugates to checkpoint inhibitors and antibody-dependent cellular cytotoxicity drugs, from bispecific T cell engagers to single chain variable fragment (scFv)-directed chimeric antigen receptor (CAR) cellular therapies. Unlocking the potential of modulating immune responses in an scFv-directed manner has transformed not only how cancer is treated, but also how it is diagnosed and monitored.

To date, cellular CAR therapies based on canonical αβ T cells have had a particularly profound effect on how we treat hematological malignancies. Six CAR-T therapies have been FDA approved to date: Kymriah for B cell acute lymphoblastic leukemia from Novartis (FMC63 anti-CD19-BBζ CAR), Yescarta for large B cell leukemia, and Tecartus for mantle cell lymphoma from Kite (FMC63 anti-CD19-28ζ CAR), Breyanzi for large B cell leukemia from Juno (FMC63 anti-CD19-BBζ CAR), Abecma for multiple myeloma from Bluebird Bio (BB2121 anti-BCMA-BBζ CAR) and Carvykti for multiple myeloma from Johnson & Johnson and Legend Therapeutics (dual camelid single-domain antibody anti-BCMA-BCMA-BBζ CAR). Along with these clinical successes, however, have come soaring treatment costs, complex manufacturing logistics, and strained supply chains.

All of the currently approved CAR-T therapies are autologous products manufactured from patient leukapheresis. A reduction in the cost and complexity of product manufacture almost certainly necessitates a shift to allogeneic product sourcing, and "off-the-shelf" supply. To this end, substantial public and private investments have been directed toward developing allogeneic CAR-T drug products. Since this is technically difficult to deliver using αβ T cells due to their alloreactive T cell receptors (TCRs), alternative immune chassis are being sought. Natural killer (NK) cells are the most widely studied alternative cell therapy delivery vehicle in this context, followed by a noncanonical T cell subset that expresses a TCR composed of a gamma and delta chain: γδ T cells.

γδ T cells possess elements of both αβ T cell and NK cell biology. They depend developmentally on the signaling of their TCR, but also express a range of NK cell-like receptors that determine their tumor-directed responses. γδ T cells are a particularly attractive cellular immunotherapy vehicle in the solid tumor context. In contrast to αβ T cells and NK cells, γδ T cells reside mostly in tissues and are adapted to tissue penetration and residence. Indeed, the presence of γδ T cells inside the tumor has been linked to conferring a significant survival advantage in a range of solid tumor indications.

Nonetheless, despite much promising preclinical data that have been generated for tumor immunotherapy with γδ T cells, substantial challenges to therapeutic success remain. First and foremost, their unique and primate-restricted biology—which may be their most attractive translational feature—presents many challenges to our understanding of the mechanisms that determine their behavior. This lack of mechanistic insight substantially complicates evidence-based therapeutic design. The herein enclosed collection of reviews—the chapters of this book—aim to address the most pressing translational questions that remain outstanding in the context of utilizing γδ T cells for the treatment of cancer.

The authors of each chapter were chosen based on their world-leading expertise. It was my distinct honor to work with this diverse group of scientists and clinicians, each a key thought leader in their respective fields. Above all, it was my pleasure to read their thoughts on the knowns and unknowns of γδ T cell cancer immunotherapy, their opinions about the field's direction of travel, and their perspectives on how therapeutic designs can be improved.

Chapter 1, titled "Exploiting fundamental γδ T cell immunobiology in cancer immunotherapy: Seven steps toward a Great Leap Forward," was authored by the formidable pair that is Carrie and Ben Willcox. It lays the groundwork for our journey by delving into the fundamental biology of human γδ T cells, and what is—and what is not—known about it. Are some γδ T cell subsets more suitable as a basis for therapeutic development than others? What defines tissue versus blood-resident γδ T cells? What evidence is there for the contribution of their TCRs to tumor responses? It considers these questions and many more in the complex evidence landscape that we are presented with regarding this enigmatic immune cell subset.

Chapter 2 is titled "Examining γδ T cell receptor (γδ-TCR) structure and signaling in the context of cellular immunotherapy design", and was written by Marta Barisa and John Anderson, with illustrations from Gaya Nair. John is a world-leading expert in pediatric experimental oncology and CAR-T cell development, with a particular interest in ways to optimize T cell signaling using synthetic biology approaches. Canonical CAR design and improvement is based around signaling of the αβ-TCR. This chapter discusses the synthetic immunotherapy learnings we can take away from what is known about γδ-TCR. It also examines just how much remains yet to be understood about this enigmatic immune receptor.

Chapter 3 was penned by Francesco Dieli, Serena Meraviglia, and the team, and is titled "γδ T cell immunotherapy: Requirement for combinations?" This chapter explores whether γδ T cell immunotherapy stands as a solitary therapeutic product, or whether it may benefit from therapeutic combinations. It then examines the evidence basis for the lead therapeutic candidates that warrant exploration in combination with γδ T cells. In order to do that, the chapter explores the data that suggests direct and indirect γδ T cell suppression by tumors and their microenvironment. Are γδ T cells susceptible to hypoxia-driven anergy? Do they express checkpoint receptors? Are tumor microenvironment myeloid cells potential suppressors of γδ T cell function? For answers to these questions and more, join Professors Dieli and Meraviglia on their exploration of the interplay between γδ T cells with other elements of the immune system, and how additional therapies can be used to capitalize on these interactions to mediate greater anticancer efficacy.

In Chapter 4, "Appraising γδ T cell exhaustion and differentiation in the context of synthetic engineering for cancer immunotherapy," John Anderson takes us on a journey into the realm of γδ T cell functional exhaustion and differentiation. What is the evidence for γδ T cell exhaustion? What are the latest debates on how "exhaustion" is defined in an anticancer cell therapy context? How is exhaustion combated in canonical CAR-T therapy? What are the synthetic engineering strategies that could be applied to γδ T cell immunotherapy? Professor Anderson summarizes the cutting-edge evidence from the αβ T cell field, and provides a piercing look at how this may apply (or not) to γδ T cell therapeutic development.

Chapter 5 was written by me. In "γδ T cells for cancer immunotherapy: A 2024 comprehensive systematic review of clinical trials" I summarize all the γδ T cell-directed oncology clinical trials that we were able to identify. I examine the trials by the type of γδ T cell intervention they aimed to utilize, the cancer indications they targeted, and, ultimately, to synthesize what the learnings from these trials were. I discuss why γδ T cell clinical trials to date have largely been disappointing, and how future trials can be improved.

Finally, in Chapter 6, Daniel Fowler considers "Allograft persistence: The next frontier for allogeneic γδ T cell therapy." This chapter examines the available evidence on T cell immunotherapeutic persistence in the autologous as well as allogeneic contexts. What is the evidence for the contribution of various preinfusion lymphodepleting chemotherapies to improving engraftment? What about radiotherapy? What relevant learnings regarding stealth synthetic engineering can be adapted from αβ T cell immunotherapy to γδ T cells? Lymphodepleting chemotherapy and stealth engineering are likely to become an increasingly relevant topic of discussion for the entire field, if CAR- (and otherwise-modified) γδ T cell interventions begin to yield a more persistent efficacy signal in upcoming trials.

Taken together, these chapters cover the most translationally relevant questions facing the γδ T cell cancer immunotherapy field. Wherever possible, the authors and I considered data from human γδ T cell studies. Where murine or other nonhuman studies were included, it is clearly spelled out in the narrative. Intended to be informative as well as provocative, I hope you find these pages as educational, compelling, and forward-looking as I did.

This work, like any scientific endeavor, was the product of the collaborative efforts of many individuals. I am grateful to the authors who have dedicated their expertise to the creation of this collection. I also extend my appreciation to my colleagues, mentors, and students, as well as the patients who inspire all of us to push the boundaries of γδ T cell immunotherapy onwards and upwards. My special thanks to John Anderson and Jonathan Fisher, whose distinct and inspiring scientific as well as clinical leadership continue to inform much of my translational thinking. I thank my Assistant Editor at Elsevier, Samantha Allard, and overall Series Editor, University of California Los Angeles Professor Emeritus Benjamin Bonavida, for suggesting I take on designing and editing this collection. Finally, I thank my family—Kass, you have been so patient and supportive throughout the many evenings and weekends this took to complete.

I hope this book serves as a guide for researchers, graduate students, and postdoctoral fellows entering the field. I hope also to provide insights to clinicians and commercial cell therapy developers working on γδ T cell immunotherapies for cancer. Ultimately, we are all on the same journey—to advance the treatment of cancer with therapies that are more effective, safer, and affordable.

Sincerely,

Marta Barisa
University College London,
London, United Kingdom

Contents

5. γδ T cells for cancer immunotherapy: A 2024 comprehensive systematic review of clinical trials

Marta Barisa, Callum Nattress, Daniel Fowler, John Anderson, and Jonathan Fisher

6. Allograft persistence: The next frontier for allogeneic γδ T cell therapy

Daniel Fowler and Jonathan Fisher

Contributors

John Anderson UCL Great Ormond Street Institute of Child Health, University College London, London, United Kingdom

Marta Barisa UCL Great Ormond Street Institute of Child Health, University College London, London, United Kingdom

Anna Maria Corsale Central Laboratory of Advanced Diagnosis and Biomedical Research (CLADIBIOR); Department of Health Promotion, Mother and Child Care, Internal Medicine and Medical Specialties, University of Palermo, Palermo, Italy

Marta Di Simone Central Laboratory of Advanced Diagnosis and Biomedical Research (CLADIBIOR); Department of Biomedicine, Neurosciences and Advanced Diagnosis, University of Palermo, Palermo, Italy

Francesco Dieli Central Laboratory of Advanced Diagnosis and Biomedical Research (CLADIBIOR); Department of Biomedicine, Neurosciences and Advanced Diagnosis, University of Palermo, Palermo, Italy

Jonathan Fisher UCL Great Ormond Street Institute of Child Health, University College London, London, United Kingdom

Daniel Fowler UCL Great Ormond Street Institute of Child Health, University College London, London, United Kingdom

Serena Meraviglia Central Laboratory of Advanced Diagnosis and Biomedical Research (CLADIBIOR); Department of Biomedicine, Neurosciences and Advanced Diagnosis, University of Palermo, Palermo, Italy

Gaya Nair Faculty of Mathematical and Physical Sciences, University College London, London, United Kingdom

Callum Nattress UCL Cancer Institute, University College London, London, United Kingdom

Benjamin E. Willcox Institute of Immunology and Immunotherapy; Cancer Immunology and Immunotherapy Centre, University of Birmingham, Birmingham, United Kingdom

Carrie R. Willcox Institute of Immunology and Immunotherapy; Cancer Immunology and Immunotherapy Centre, University of Birmingham, Birmingham, United Kingdom

CHAPTER

1

Exploiting fundamental γδ T cell immunobiology in cancer immunotherapy

Benjamin E. Willcox[a,b] *and Carrie R. Willcox*[a,b]

[a]Institute of Immunology and Immunotherapy, University of Birmingham, Birmingham, United Kingdom [b]Cancer Immunology and Immunotherapy Centre, University of Birmingham, Birmingham, United Kingdom

Abstract

Immunotherapies, such as immune checkpoint blockade (ICB) and chimeric antigen receptor-T (CAR-T) cells, have transformed the therapeutic landscape of cancer but have predominantly impacted certain malignancies/patient groups, with major areas of unmet clinical need remaining. γδ T cells, a third lymphocyte lineage bearing a recombined antigen receptor, are of intense interest in cancer immunotherapy, but currently unexploited. Here we focus on fundamental γδ T cell immunobiology in the context of emerging cancer immunotherapeutic strategies. Unique major histocompatibility complex (MHC)-unrestricted T cell receptor (TCR) recognition modes are a defining feature of γδ T cells, distinguishing them from αβ T cell and B cells, and arguably reflect their status as "Nature's CAR-Ts." Well-established stress recognition and cancer immunosurveillance capabilities, combined with potent effector functions, pharmacological manipulability, and potential to coordinate downstream immune responses, suggest strong anticancer potential, tempered by limited γδ T cell frequencies in blood/tissues. Repurposing γδ T cell stress signature recognition mechanisms, both TCR-intrinsic and extrinsic, to enhance antitumor responses is an attractive aim. Diverse γδ T cell-focused immunotherapy approaches are currently in development. Some aim to boost γδ T cell levels using ex vivo or in vivo expansion/activation, others aim to engineer supra-physiological tumor targeting via cellular therapy and antibody-based and/or T cell engager approaches. Collectively, these may expand the clinical footprint of cancer immunotherapy, catalyzing development of allogeneic "off-the-shelf" cellular therapy platforms, tumor microenvironment adjuvantization strategies, and approaches for indications poorly served by current immunotherapy treatments, such as MHC-negative and/or low mutational burden tumor settings. Finally, we raise seven aspects of γδ T cell immunobiology that may impact and direct future γδ T cell cancer immunotherapy development.

γδ T Cell Cancer Immunotherapy
https://doi.org/10.1016/B978-0-443-21766-1.00004-7

Abbreviations

Ab	antibody
ADCC	antibody-dependent cellular cytotoxicity
ADCI	antibody-dependent cell-mediated inhibition
ADCP	antibody-dependent cellular phagocytosis
Ag	antigen
AML	acute myeloid leukemia
AREG	amphiregulin
BTN	butyrophilin
BTNL	butyrophilin-like
CAR	chimeric antigen receptor
CCR	C–C chemokine receptor
CD	cluster of differentiation
CDR	complementarity determining region
CMV	cytomegalovirus
CRS	cytokine release syndrome
DETC	dendritic epidermal T cell
DOT	Delta One T
EBV	Epstein–Barr virus
EGFR	epidermal growth factor receptor
EMRA	effector memory cells re-expressing CD45RA
EPCR	endothelial protein C receptor
EphA2	ephrin receptor A2
FPPS	farnesyl pyrophosphate synthase
GAB	gamma delta TCR anti-CD3 bispecific molecules
GD2	disialoganglioside 2
GF	growth factor
GvHD	graft versus host disease
GvL	graft versus leukemia
HER2	human epidermal growth factor receptor 2
HHV	human herpesvirus
HIV	human immunodeficiency virus
HMBPP	(*E*)-4-hydroxy-3-methyl-but-2-enyl pyrophosphate
IBD	inflammatory bowel disease
ICANS	immune effector cell-associated neurotoxicity syndrome
ICB	immune checkpoint blockade
IEL	intraepithelial lymphocyte
IFN	interferon
IL	interleukin
ITAM	immunoreceptor tyrosine-based activatory motif
ITIM	immunoreceptor tyrosine-based inhibitory motif
KIR	killer immunoglobulin-like receptor
LILR	leukocyte immunoglobulin-like receptor
mAb	monoclonal antibody
MAIT	mucosal-associated invariant T cells
MHC	major histocompatibility complex
N-BP	amino bisphosphonate
NHL	non-Hodgkin lymphoma
NKR	natural killer cell receptor
P-Ag	phosphoantigen
PAMP	pathogen-associated molecular pattern
RM	resident memory
RNA	ribonucleic acid

SCT stem cell transplantation
TCR T cell receptor
TEG T cells engineered with a defined gamma delta TCR

Conflict of interest statement

B.E.W. provides consultancy regarding the development of γδ T cell immunotherapy approaches for Ferring Ventures SA, linked to Ferring Pharmaceuticals.

Introduction

Immunotherapy-based approaches have drastically reconfigured the therapeutic landscape for certain cancers over the past 15 years, ushering in a new era of therapy development in this field. In particular, immune checkpoint blockade (ICB) antibodies have transformed treatment of some solid cancers, including in traditionally highly challenging advanced disease cohorts, where they have achieved remarkable durable responses in substantial subsets of patients [1,2]. For melanoma, this has resulted in ICB therapies, administered either as single agent or two-agent combinations, displacing chemotherapy as a frontline therapy for most patients. These treatments have also been highly successful for lung cancer, previously viewed as largely immunologically inert. In parallel, application of autologous chimeric antigen receptor engineered (CAR) T cells (CAR-T cells) to certain hematological cancers has also been transformative, and entered standard of care for some B cell malignancies, with proven capability to establish durable, complete responses in cancers previously resistant to traditional therapeutics such as chemo- and radiotherapy [3].

Both ICB and CAR-T therapies have arguably been built on decades of fundamental immunology research. Such research has progressed our understanding of two of the three compartments of vertebrate lymphocytes that bear somatically rearranged antigen receptors, namely, B cells and αβ T cells, establishing a knowledge base regarding B cell and antibody functionality, and also defining critical activatory and inhibitory pathways regulating αβ T cells. In contrast, a third compartment, comprising γδ T cells [4,5], which has been retained alongside B cells and αβ T cells throughout the ~500 million of vertebrate evolution, has been seemingly excluded from this wave of new cancer immunotherapies. In part, this has reflected greater focus on understanding the biology of B cell and αβ T cells relative to their γδ T cell counterparts. Nevertheless, γδ T cells are a focus of considerable interest for exploitation in cancer immunotherapy [6], due significantly to their potent antitumor effector functions, lack of major histocompatibility complex (MHC) restriction, and novel T cell receptor (TCR)-mediated recognition modalities. Moreover, there is considerable hope that exploitation of γδ T cell-focused approaches may at least partially address some of the limitations of current ICB- and CAR-T-based strategies, including the fact that their success, while striking, has been restricted to a limited set of cancers, and to subsets of patients [1,2]. Specifically, ICB therapies have generally been less successful in low mutational burden cancers; conversely, the identification of safe targets for CAR-T approaches covering a range of cancers is demanding [3], and both strategies ultimately confront the dual challenge of highly mutable cancer targets and typically a decidedly immunosuppressive tumor microenvironment. To realize this potential, an appreciation of γδ T cell immunobiology, how this both overlaps and differs from that of αβ T cells and B cells, and how it might feasibly be exploited in the cancer setting, is required.

In this article, we review and summarize key aspects of γδ T cell immunobiology, focusing predominantly on human studies, explore how these elements link with emerging immunotherapeutic strategies, and raise seven questions that may be important to address to help catalyze transformative advances.

γδ T cells: A third lineage of adaptive lymphocytes retained in vertebrates

γδ T cells were discovered unexpectedly in the 1980s when a TCRγ chain was identified during the search for the gene encoding the elusive TCRα chain [7]. Prior to that, a second type of T cell was not even predicted by functional studies.

The γδ TCR is structurally homologous to the αβ TCR and associates with the CD3 complex on the T cell surface. The TCRγ and TCRδ chains are generated by somatic recombination of Variable (V), Diversity (D), and Joining (J) regions, as is also the case with both the αβ TCR, and the B cell receptor. γδ T cells therefore represent a third lineage of lymphocytes bearing antigen receptors that undergo somatic recombination alongside B cells and αβ T cells. However, while αβ T cells use their TCR complementarity determining regions (CDRs) to recognize peptides presented by MHC class I or class II molecules, γδ T cells are generally thought to be MHC unrestricted and develop in MHC-deficient mice [8]; only very limited evidence is suggestive of MHC recognition in humans [9,10]. However, the antigens recognized by γδ T cells remain enigmatic [5], and although some progress has been made in identifying ligands for γδ TCRs (reviewed in Willcox and Willcox 2019 [11]), a clear understanding of the full range of γδ TCR ligands, or indeed, underlying principles of recognition, is currently lacking. Furthermore, the relative contribution of the TCR versus other surface receptors to γδ T cell activation is poorly understood and may depend significantly on the γδ T cell subset and location.

γδ T cell subsets exhibit distinct immunobiology

γδ T cells are often regarded as "innate-like" effectors, and this may be the case for many murine γδ T cell populations. While αβ T cell populations use V(*D*)J recombination including nontemplated (N) nucleotide addition by terminal deoxynucleotide transferase (TdT), to generate the diverse TCRs able collectively to recognize virtually any antigenic peptide presented by self-MHC molecules, many mouse γδ T cell populations express TCR repertoires of limited diversity, often restricted to expression of a single Vγ gene segment at particular tissue sites and exhibiting little to no CDR3 diversity [12,13]. This has led to the hypothesis that such populations may recognize a limited set of host ligands [14]. In addition, several murine γδ T cell subsets generally display an effector phenotype from early in life, consistent with an "innate-effector" status [15–17].

In this review, our focus is predominantly on human γδ T cells, for which in recent years, a combination of phenotypic, TCR repertoire, and functional studies has revealed two broad classes in the human, innate-like and adaptive-like, each of which is being explored for therapeutic use. Importantly, human γδ T cells can be delineated into functionally distinct subsets by expression of particular TCR Vγ and Vδ genes, which relates both to development and function (Fig. 1).

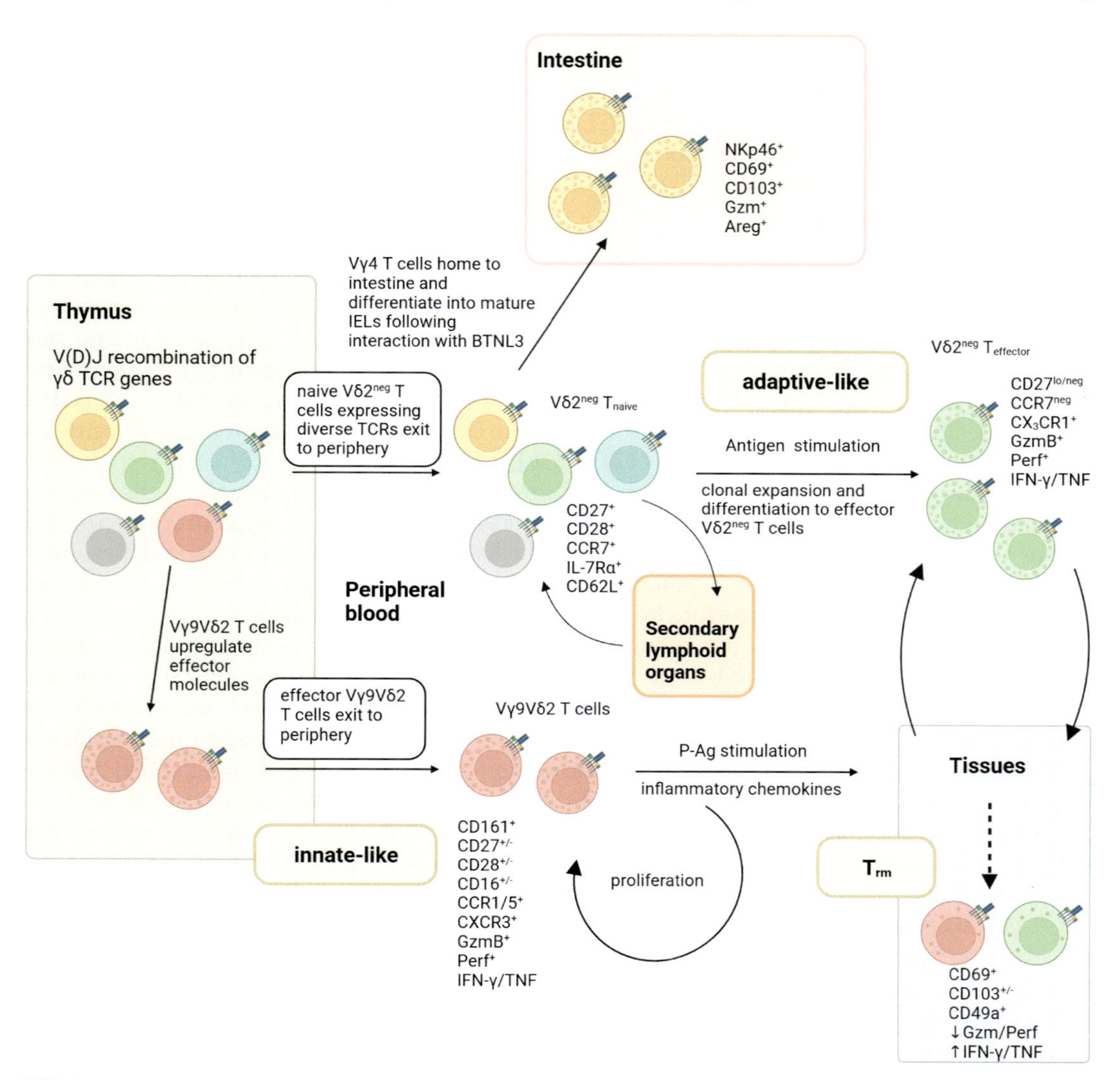

FIG. 1 Current understanding of the development, differentiation, and homing of human γδ T cell subsets. Human γδ T cells develop in the thymus. Innate-like Vγ9Vδ2 T cells mature in the thymus and exit to the periphery as differentiated effector cells. Upon P-Ag stimulation, Vγ9Vδ2 T cells proliferate and home to tissues. Vδ2^{neg} T cells exit the thymus as naïve adaptive-like T cells which recirculate through the blood and secondary lymphoid tissues. Following antigen stimulation, Vδ2^{neg} T cells proliferate, differentiate into effectors and home instead to the tissues, where, alongside tissue-associated Vγ9Vδ2 T cells, they may become resident memory T cells (T_{rm}). Vγ9^{neg}Vδ2 T cells likely follow a similar differentiation pathway to Vδ2^{neg} T cells. Vγ4 T cells seed the intestine and become intraepithelial lymphocytes, retained by interactions with BTNL3. *Created with BioRender.com.*

Vγ9Vδ2 T cells

Vγ9Vδ2 T cells are the best candidate for innate-like γδ T cell effectors in humans and express Vγ9 chains with semi-invariant CDR3 regions, alongside Vδ2 chains with diverse CDR3 regions. Vγ9Vδ2 T cells are among the first T cells generated during fetal development, initially in the fetal liver from week 5 of gestation [18] and later from the fetal thymus [19] and subsequently differentiate into effector cells in the periphery during the first months of life [20–22]. This occurs concurrently with expansion of the subset, such that Vγ9Vδ2 T cells soon become the most numerous γδ T cells in the peripheral blood of the majority of healthy individuals. Consistent with subtle TCR repertoire changes between neonatal and adult Vγ9Vδ2 T cells [23–25], recent data suggest that most adult Vγ9Vδ2 T cells develop and mature in the postnatal thymus [26]. Of note, Vγ9Vδ2 T cells bear transcriptional and phenotypic similarities to mucosal-associated invariant T (MAIT) cells, a semi-invariant αβ T cell population [27]. Both Vγ9Vδ2 T cell and MAIT cells express the surface receptor CD161 and can proliferate in response to cytokines IL-12/18 [28].

Consistent with an important function in antimicrobial immunity, Vγ9Vδ2 T cells respond *en masse* to target cells exposed to small phosphorylated antigens (P-Ags) produced in the nonmevalonate pathway of isoprenoid biosynthesis by some bacteria and parasites, such as (*E*)-4-hydroxy-3-methyl-but-2-enyl pyrophosphate (HMBPP), and synthetic derivatives thereof [29]. HMBPP could therefore be considered a pathogen-associated molecular pattern (PAMP). Dysregulation in the mevalonate pathway of isoprenoid biosynthesis, for example, in human tumor cells, can also lead to the accumulation of the endogenous P-Ag isoprenyl pyrophosphate (IPP), a host P-Ag that also activates Vγ9Vδ2 T cells, albeit substantially less potently than HMBPP [29]. The presence of Vγ9Vδ2 T cells in all patients, and their ability to be modulated by both P-Ag (directly) and (indirectly) by pharmacological agents that perturb P-Ag levels, makes these cells a prime candidate for immunotherapeutic approaches.

The detailed mechanism of P-Ag recognition by Vγ9Vδ2 T cells is still being explored but importantly, from the T cell perspective, is encoded by the Vγ9Vδ2 TCR itself. From the target cell perspective, the process involves members of the butyrophilin (BTN) family of proteins (BTN2A1, BTN3A1, BTN3A2, and BTN3A3) [30–33], which are expressed on all human cells and are structurally related to B7 costimulatory molecules and the PD-L1 coinhibitory ligand. Although some molecular details are uncertain, it is clear that P-Ag must enter the target cell and bind to the intracellular B30.2 domain of BTN3A1 [34–36], which arguably acts as the critical initiating P-Ag sensor and subsequently associates with the B30.2 domain of BTN2A1 [37,38]. This is thought to result in close association and/or topological rearrangement of the extracellular domains of these molecules on the cell surface, creating a recognition complex for Vγ9Vδ2 TCR binding. While the nature of this recognition complex is unclear, recent data support the concept that the Vγ9Vδ2 TCR is able to make low-affinity interactions with both BTN2A1 [31,32] and possibly also BTN3A, which following P-Ag exposure are closely juxtaposed at the cell surface, forming a "composite ligand" capable of triggering TCR signaling [39,40].

Upon P-Ag recognition, Vγ9Vδ2 T cells proliferate, secrete IFN-γ, TNFα, and other cytokines and chemokines, and release cytotoxic molecules. Infections such as TB, malaria, ehrlichiosis, toxoplasmosis, and Epstein–Barr virus (EBV) lead to expansion of the Vγ9Vδ2 population in peripheral blood [4,29].

Non-Vγ9Vδ2 T cells

The remaining non-Vγ9Vδ2 γδ T cells in human peripheral blood are thought to be more adaptive-like based on several criteria. This subset expresses diverse TCR chains: Vδ1, Vδ3, and, less commonly, other Vα gene segments can be rearranged to Dδ and Jδ gene segments to generate TCRδ chains, paired with any of the functional Vγ chains (known collectively as $V\delta2^{neg}$ T cells), and even $V\gamma9^{neg}V\delta2$ cells are part of this subset (featuring Vδ2 chains paired with any non-Vγ9 chain) [23,41] (Fig. 1). $V\delta2^{neg}$ and $V\gamma9^{neg}V\delta2$ T cells share many features such as expression of TCRs with highly diverse CDR3 regions despite using only a few functional Vγ and Vδ regions. Diversity is increased by the addition of many nontemplated (N) nucleotides (>20 N nucleotides is not uncommon in CDR3δ) added by TdT during V(*D*)J recombination [23,41]. $V\delta2^{neg}$ and $V\gamma9^{neg}V\delta2$ clonotypes can undergo clonal expansion following cytomegalovirus (CMV) infection [22,23,42,43]. However, other infections may also drive $V\delta2^{neg}$ T cell clonal expansion as many CMV-seronegative healthy donors have clonally focused $V\delta2^{neg}$ T cell compartments [41]. Other infections that are thought to induce clonal expansion within this compartment are EBV, human immunodeficiency virus (HIV), malaria, and human herpes virus-6/7 (HHV-6/7) [44–50]. Because of the highly diverse nature of the $V\delta2^{neg}$ TCR rearrangements, clonal expansions are incredibly private, i.e., not shared between individuals, even those with the same infection history and genetic makeup [51].

TCR repertoire, phenotypic, and transcriptional studies [20,23,41] have been important in delineating this compartment into two differentiation states in the peripheral blood: a $CD27^{hi}$, $TCF7^{+}$, LEF-1^{+}, $CCR7^{+}$, IL-$7R\alpha^{+}$, $Granzyme^{neg}$ TCR diverse $T_{naïve}$ state seemingly largely devoid of effector functionality; and a $CD27^{lo/neg}$, $Tbet^{+}$, $EOMES^{+}$, $Granzyme^{+}$, $Perforin^{+}$ $T_{effector}$ state with a highly focused TCR repertoire, which is both TCR-sensitive and capable of both cytokine secretion and cytotoxicity [41]. These two states are also strongly delineated for expression of activatory and inhibitory receptors, many of which are upregulated in $T_{effector}$ versus T_{naive} subsets (see below) [20]. Current data support the idea that $CD27^{lo}$ $V\delta2^{neg}$ $T_{effector}$ cells are relatively long-lived effector cells [23,41] similar to CD8 T_{EMRA} cells that help to control chronic viral infections such as CMV. In fact, in a transcriptional comparison of Vδ1 $T_{effector}$ cells with CD8 T_{EMRA}, very few genes were found to be differentially regulated [20]. Importantly, pathogen infection drives both clonotypic expansion from the T_{naive} repertoire, but also phenotypic transition to the $T_{effector}$ state, including for both CMV [22,23] and malaria [50].

By analogy with conventional αβ T cell populations, the transition from Vδ1 T_{naive} to $T_{effector}$ status is hypothesized to be driven significantly by γδ TCR triggering and hence cognate ligand interactions, most likely involving CDR3 regions, explaining the strong clonotypic focusing in the $T_{effector}$ subset [5,11,52,53]. Not surprisingly, the ligands recognized by such clonally expanded TCR clonotypes are largely unclear but are the subject of considerable interest [5,11], in the context of both infection and antitumor immunity. Of relevance, the LES $V\delta2^{neg}$ γδ TCR clonotype expanded following CMV infection was found to directly recognize endothelial protein C receptor [54], a protein expressed on target cells infected with CMV, with strong dependence on both LES TCR CDR3γ and CDR3δ [54,55]. Whether such $T_{effector}$ populations formally fulfill criteria for immunological memory such as responding more quickly to secondary antigenic challenge is also a subject of ongoing investigation, although notably mouse studies on CMV support such potential [56,57].

Emerging principles of human γδ T cell tissue homing

As discussed earlier, Vγ9Vδ2 [29], Vδ2^{neg} [52], and Vγ9^{neg}Vδ2 T cells [53] are found in the peripheral blood, where together they constitute collectively ~1%–10% of T cells, with the remainder αβ T cells. Vγ9Vδ2 T cells make up on average ~65% of PB γδ T cells [58], although this can vary between individuals. Of the remaining PB γδ T cells, the majority express Vδ1 TCRs and may be $T_{naïve}$ or $T_{effector}$ cells (see above). In an analysis of 20 UK adult blood donors aged 18–30, approximately two-thirds had a Vδ1 compartment with substantial clonotypically focused $T_{effector}$ subsets, with the remainder dominated by TCR diverse T_{naive} cells [52]. It is thought that γδ T cells are not "resident" in the peripheral blood, but instead recirculate through blood, lymph, and some peripheral tissues [53], similar to αβ T cells. Naïve Vδ2^{neg} and Vγ9^{neg}Vδ2 T cells express CCR7, which is hypothesized to allow them to recirculate through lymphoid tissues as do naïve αβ T cells [52,53]. Most effector Vδ2^{neg} and Vγ9^{neg}Vδ2 T cells display decreased expression of CCR7, suggesting they would be excluded from lymphoid tissue [52,53], and conversely have elevated expression of CX_3CR1 [23,41], consistent with homing to peripheral sites. Vγ9Vδ2 T cells generally express low levels of CCR7 but instead can respond to inflammatory chemokines through CCR1, CCR5, and CXCR3 and migrate into inflamed tissues [59,60].

γδ T cells are found in secondary lymphoid tissues such as tonsil and spleen. In the tonsil, they are present at low frequencies (~1–2%) [61], but they are reportedly present at increased frequency in the spleen (up to 17% of lymphocytes) [62], and unlike αβ T cells, are predominantly localized to the red pulp. It is currently unclear which γδ T cell subsets are present in the spleen and tonsil (Vγ9Vδ2 or Vδ2^{neg}, or $T_{naïve}$ or $T_{effector}$ subsets), although further TCR repertoire and phenotypic analysis would shed light on this question. While γδ T cells are present in the peripheral blood at low frequencies compared to αβ T cells, Vδ2^{neg} γδ T cells in particular are enriched in many tissues, including the intestine, liver, lung, and skin [52], where it is thought they play a role in immune surveillance against infection and cancer and may also play a role in tissue homeostasis. In such peripheral tissues, most Vδ2^{neg} cells are hypothesized to be enriched for differentiated CD27^{lo} $T_{effector}$ subsets that in blood display upregulated CX_3CR1 and decreased CCR7 and CD62L and express granzymes [52].

Studies have confirmed enrichment of γδ T cells in healthy liver relative to blood (>10% of T cells), with Vδ2^{neg} γδ T cells noted as more prevalent than Vγ9Vδ2 T cells [63]. TCR repertoire and phenotypic analysis of liver γδ T cells indicated the Vδ2^{neg} and Vγ9^{neg}Vδ2 T cells have a $T_{effector}$ phenotype and are clonally expanded, with many hepatic clonotypes also present in the peripheral blood of individual donors. However, some liver-specific clonotypes were also identified, and these cells had a distinct phenotype and effector profile when compared to the shared liver/blood clonotypes, suggesting adaptation to the hepatic microenvironment [63]. Liver-specific Vδ2^{neg} T cells expressed resident memory (T_{rm}) markers such as CD69 (though not CD103), and liver homing receptors CXCR6. In contrast, cells bearing clonotypes shared with PB expressed markers such as CX_3CR1 and were more cytotoxic than liver-specific cells, which instead made more IFN-γ and TNFα upon activation [63]. These findings were extended by Zakeri et al., who demonstrated unequivocally the presence of T_{rm} populations in both the Vγ9Vδ2 and Vδ2^{neg} hepatic subsets [64].

γδ T cells are also found in the dermal layer of human skin [65,66]. These cells have an effector phenotype and feature clonal expansions in both the Vδ1^{+} and Vδ2 compartments, though the latter may reflect clonal expansion in the $V\gamma9^{neg}V\delta2$ subset. Moreover, a proportion of peripheral blood Vγ9Vδ2 T cells express cutaneous lymphocyte antigen (CLA), potentially permitting homing to the skin [67].

In recent years, characterization of γδ T cells present in the human intestine has shed light on their role in immunosurveillance. Vγ9Vδ2 T cells can be found to varying degrees in the lamina propria [67]; however, $V\delta2^{neg}$ T cells are more prevalent in the intraepithelial layer (representing ~25% of intraepithelial T cells in the colon) [68] (Fig. 1). Intestinal $V\delta2^{neg}$ T cells adopt a T_{rm}-like phenotype ($CD103^+$ and $CD69^+$), are in direct contact with multiple epithelial cells, and may monitor the epithelial layer for signs of infection or epithelial cell integrity, in alignment with data supporting such a role in mice [69]. Many human intraepithelial lymphocytes (IELs) express Vγ4 TCRs [70], which has been shown to bind to the butyrophilin-like (BTNL) molecule BTNL3 expressed on epithelial cells [55], and which induces subsequent TCR downregulation [55,70,71]. Human Vγ4 is closely related to mouse Vγ7, and the mouse Vγ7 IEL compartment is also dependent on BTNL molecules Btnl1/6 for their maturation and function [70]. The functional consequence of BTNL/Btnl binding by IELs remains somewhat unclear, as well as whether γδ IELs use their TCR diversity to also recognize diverse ligands in infection and tumorigenesis. However, defective BTNL3/8-induced TCR downregulation has been noted by γδ IELs from inflammatory bowel disease (IBD) patients versus those from healthy controls [72], although whether this contributes to or results from gut inflammation is unclear.

Expression of activating and inhibitory receptors

Although the γδ TCR is not only the defining feature of γδ T cells but is also thought to be the dominant receptor involved in γδ T cell activation, γδ T cell subsets express a number of other activating and inhibitory receptors that may modulate overall γδ T cell activation and downstream responses (Fig. 2). These include the natural killer receptors (NKRs) NKG2D, NKG2A, NKG2C, NKp30, NKp46, NKp44, DNAM-1, CD16, killer immunoglobulin-like receptors (KIRs), leukocyte immunoglobulin-like transcripts (LILRs), costimulatory molecules such as CD28, CD27, 4-1BB/CD137, and coinhibitory receptors TIGIT, CTLA-4, PD-1, and TIM-3. We will consider the expression of these molecules by different γδ T cell subsets.

Vγ9Vδ2 T cells

Vγ9Vδ2 T cells in peripheral blood adopt a $T_{effector}$ status but are characterized by a range of phenotypic states that frequently co-exist within an individual, a profile analyzed most comprehensively by Ryan et al. [73]. Subsets expressing the costimulatory molecules CD27 and CD28 or alternatively just CD27 are common, though in some individuals Vγ9Vδ2 T cells may also exhibit differentiation into late effectors which are $CD27^{neg}CD28^{neg}$ [73]. In addition, terminally differentiated Vγ9Vδ2 T cells may express CD16 [73], an F_c receptor that can facilitate antibody-dependent cell-mediated cytotoxicity (ADCC) [74].

Unlike most $V\delta2^{neg}$ T cells in the peripheral blood, Vγ9Vδ2 T cells typically express the inhibitory NK receptor NKG2A [20]. NKG2A dimerizes with CD94 and binds to the nonclassical MHC class I molecule HLA-E, providing an inhibitory signal that may modulate responses, potentially setting a higher threshold for later activation by P-Ag and/or cytokine

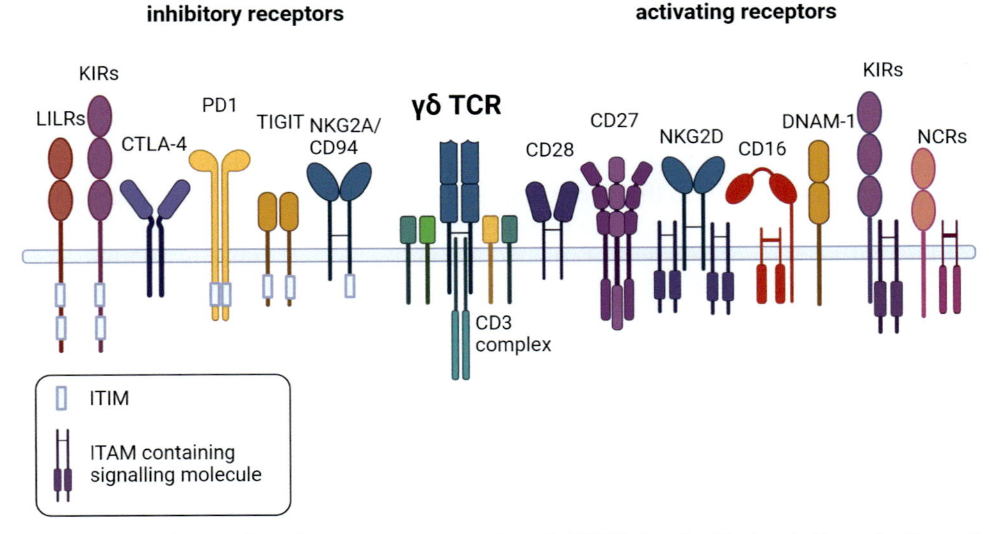

FIG. 2 Activation of γδ T cells is dependent on integration of γδ TCR signal with signals through other activating and inhibitory receptors. Costimulatory signals can be provided by CD28 and CD27, which bind to B7 molecules and CD70, respectively. NKG2DL binding to NKG2D can provide costimulatory signals, or in some situations, full activation signals. Binding of the low-affinity Fc receptor CD16 to antibody–antigen complexes can induce ADCC, ADCI, or ADCP. NK receptors DNAM-1 and the NCRs NKp46, NKp30, NKp44 bind to self or pathogen-encoded ligands. MHC class I receptors of the KIR or LILR families can provide activating or inhibitory signals, while CD94/NKG2A recognizes MHC class I leader peptides presented by HLA-E molecules. Classical coinhibitory receptors CTLA-4, PD1, and TIGIT can provide inhibitory signals to γδ T cells following recognition of B7 molecules, PD-L1/2 or CD155/CD112, respectively. Many activating receptors are associated with signaling complexes which contain an immunoreceptor tyrosine-based activatory motif (ITAM), while many inhibitory molecules contain an immunoreceptor tyrosine-based inhibitory motif (ITIM). *Created with BioRender.com.*

activation based on the presence of an intact class I MHC pathway in target cells. Vγ9Vδ2 T cells may also express NKG2C, an activating receptor related to NKG2A, which also dimerizes with CD94 but instead provides an activating signal in response to HLA-E binding.

Vγ9Vδ2 T cells also typically express the activating NKG2D receptor [75]. NKG2D is a homodimer that binds to a variety of NKG2D ligands (NKG2DL) expressed on stressed or transformed cells. It is thought that NKG2D can propagate a costimulatory signal in peripheral blood Vγ9Vδ2 T cells. NKG2D signaling is not thought to be sufficient for Vγ9Vδ2 T cell activation in the absence of a TCR signal such as P-Ag [75,76], though this is controversial [77]. In addition, other activating NK receptors such as DNAM-1, and the NCRs (NKp30, NKp44, NKp46) are also expressed on Vγ9Vδ2 T cells and may enable recognition of ligands expressed on infected or tumor targets [20].

Vδ2^{neg} T cells in the peripheral blood

Vδ2^{neg} and also Vγ9^{+}Vδ2^{neg} T cells in the PB are generally found in one of two differentiation states. Naïve Vδ2^{neg}/Vγ9^{+}Vδ2^{neg} cells express CD27 and CD28 costimulatory molecules, but are largely devoid of effector molecules such as granzymes and perforin, and do not express checkpoint receptors [20,23,41]. They generally do not express activating NK receptors, such as NKG2D and DNAM, but may express low levels of NKp30 [20]. Effector $V\delta2^{neg}/V\gamma9^{+}V\delta2^{neg}$ T cells, on the other hand, express many NKRs and checkpoint receptors and exhibit downregulated expression of CD27 and CD28 [20,23,41]. Effector $V\delta2^{neg}/V\gamma9^{+}V\delta2^{neg}$ cells express NKG2D, DNAM-1, and may express NKp30, NKp46, and NKp44 [20]. In addition, effector $V\delta2^{neg}/V\gamma9^{+}V\delta2^{neg}$ cells may express CD16, which in addition to enabling ADCC, can also facilitate antibody-dependent cell-mediated inhibition (ADCI) of viral infection by inducing IFN-γ production [78]. They may also express inhibitory receptors NKG2A, KIRs, and LILRs [20], which likely function to regulate T cell activation. Effector $V\delta2^{neg}/V\gamma9^{+}V\delta2^{neg}$ T cells also express the coinhibitory receptors TIGIT and PD-1 and may express CTLA-4 [20]. However, expression of many checkpoint receptors does not necessarily indicate that these cells are exhausted, as they can be activated ex vivo by anti-CD3 and can proliferate in response to IL-2/15 (as can CD8 T_{EMRA}) [20]. Of note, at the RNA level, effector Vδ1 T cells were nearly indistinguishable from CD8 T_{EMRA} cells [41].

Current understanding of how inhibitory and activatory cell axes regulate $V\delta2^{neg}/V\gamma9^{+}V\delta2^{neg}$ T cells in solid tissues is somewhat limited. Nevertheless, recent data suggest that while NK receptors such as NKG2D appear to play a costimulatory role in peripheral blood γδ T cells activated through the TCR [75,76], γδ T cells in tissues may be able to respond solely to NKR ligands in the absence of TCR signal and in some cases exhibit altered NKR expression. Analysis of intratumoral Vδ1 T cells in breast cancer indicated that unlike their peripheral blood equivalents, they may be responsive to NKG2DL on tumor cells [79], and a substantial subset of Vδ1 IELs in the intestine expresses the NK-activating receptors NKp46 and NKp44 [68,80], although the role these receptors play in IEL function is still unclear. Of relevance, the NKp46 ligand ecto-calreticulin can be upregulated in ER stress or infection [81], and therefore may conceivably lead to activation of Vδ1 IELs; the potential role of other NKp44 and NKp46 ligands in intestinal immunosurveillance is unclear. Relative to T_{naive} subsets, $V\delta2^{neg}/V\gamma9^{+}Vd2^{neg}$ $T_{effector}$ cells display both upregulation of diverse inhibitory receptors and an altered homing receptor expression profile that closely matches that of CD8 T_{EMRA} subset [20] and allows access to diverse peripheral tissues. Consistent with this dual observation, the presence of $V\delta2^{neg}$ $T_{effector}$ subsets expressing checkpoint receptors PD-1, TIGIT, and TIM3 [82,83] as well as the costimulatory molecule 4-1BB (CD137) [82], has been noted in solid tumors. Although the degree to which inhibitory checkpoint receptors directly regulate γδ T cells is unclear, following PD-1 therapy, $V\delta2^{neg}$ T cells are proposed to contribute to responses against MHC class I negative colorectal tumors [82]. Although the mechanistic basis of this response is still a subject of ongoing interest, it may rely partly on release of functional suppression by class I MHC-binding inhibitory receptors, of which there are several candidates (KIRs, LILRs, NKG2A), with KIRs highlighted as most significant to date [82]. These findings raise the interesting question of whether for γδ T cells, rather than merely an irrelevant target, class I MHC may represent an important inhibitory checkpoint, the presence and absence of which might skew immunity to conventional αβ T cell and conversely γδ T cell responses, respectively.

A 360° review of γδ T cell effector functions

An attractive feature of γδ T cells with respect to immunotherapy development is the diverse effector functions associated with their activation and transition to effector status (Fig. 3). These include well-recognized activation-triggered cytolytic capabilities (aligning with expression of diverse granzymes and perforin), and cytokine production (e.g., IFN-γ, TNF) [84], underpinned by widespread expression of both Tbet and Eomes in γδ $T_{effector}$ cells, including both Vγ9Vδ2 "innate-effectors" and $V\delta2^{neg}/V\gamma9^{+}V\delta2^{neg}$ more adaptive-like

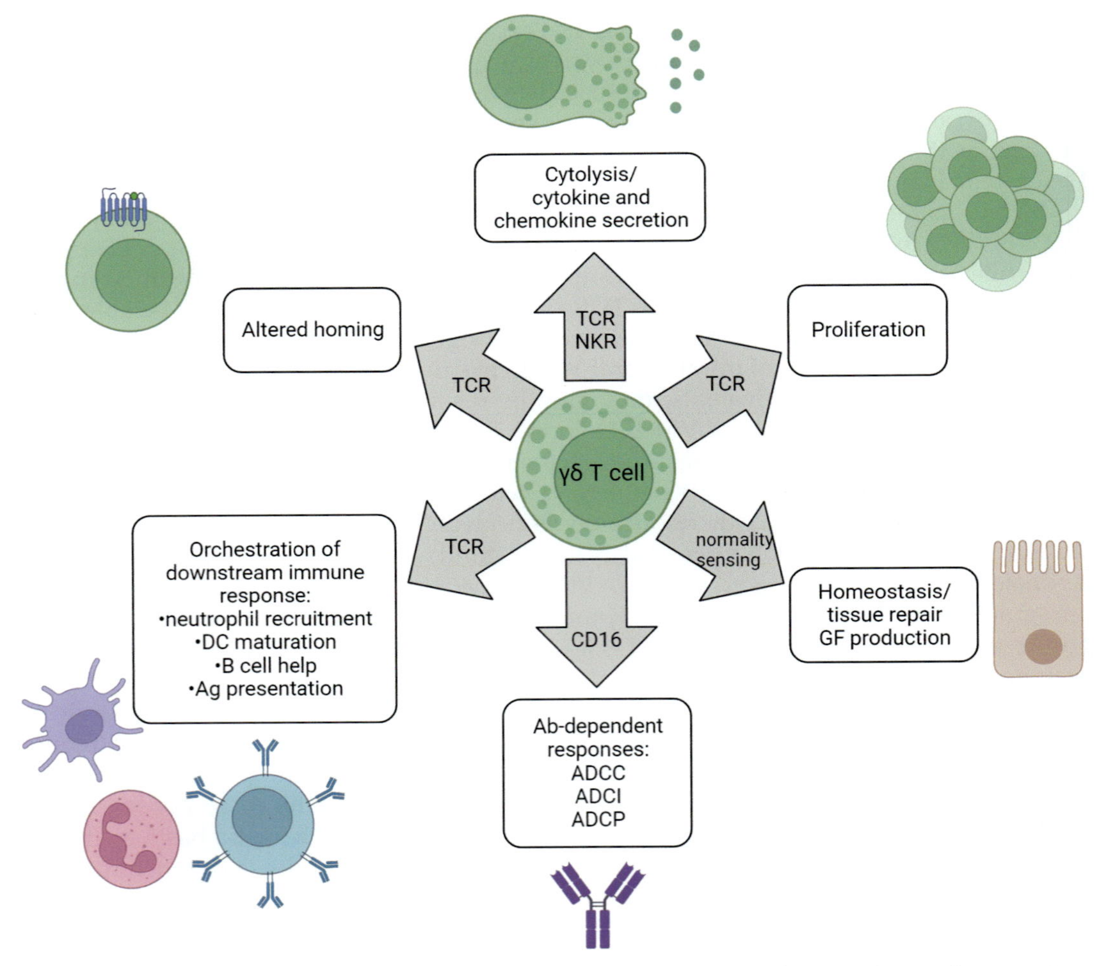

FIG. 3 Activation-induced responses of γδ T cells. TCR signal can induce proliferation of γδ T cells, cytotoxic responses, cytokine secretion, altered homing, and orchestration of downstream immune responses. Tonic TCR signals received following TCR recognition of BTN family molecules may result in a homeostatic or tissue repair program, for example by growth factor (GF) production. Ligation of the Fc receptor CD16 by target cell-bound antibody can result in ADCC, ADCI, or ADCP. *Created with BioRender.com.*

effectors [20]. In the mouse γδ T cell compartment, the existence of a semi-invariant thymically programmed $CD27^{neg}$ IL-17-producing subset is well established and can promote tumorigenesis in some models [85]. In contrast, there is little evidence of IL-17 production in human γδ T cells, with the exception of some inflammatory and pathological scenarios [85–87]. In addition to IFN-γ and TNFα, a number of inflammatory chemokines have been ascribed to γδ T cells, including MIP-1a (CCL3), MIP-1b (CCL4), and RANTES (CCL5) [84,88].

Following activation, γδ T cells also exhibit profound proliferation. This occurs in vivo typically after relevant infections, whereby they can expand up to ~50% of circulating T cells [4]. Relevant examples include mycobacterial infections that drive Vγ9Vδ2 T cells, and viral infections such as CMV that trigger both clonal expansion and differentiation of $V\delta2^{neg}/V\gamma9^{+}V\delta2^{neg}$ subsets [4,5,29]. Both innate-like and adaptive-like subsets can also be readily expanded in vitro, following obligatory TCR stimulation, combined with appropriate cytokine stimulation. Despite this, the question of the degree of persistence of expanded γδ T cell populations remains important from both fundamental and immunotherapeutic perspectives. Analyses of Vδ1 T cells have demonstrated clonotypic persistence for several years in human blood [41], and studies on Vγ9Vδ2 T cells have linked more highly differentiated status with decreased proliferative capacity [73].

Evidence suggests activation of both Vγ9Vδ2 and $V\delta2^{neg}/V\gamma9^{+}V\delta2^{neg}$ T cells drives altered homing. Transition from T_{naive} to $T_{effector}$ status, which is triggered by in vivo infection [22,23,50], drastically alters homing marker expression of $V\delta2^{neg}/V\gamma9^{+}V\delta2^{neg}$ T cells away from lymphoid homing markers to peripheral homing capability. For Vγ9Vδ2 T cells, P-Ag-mediated activation leads to temporary induction of a putative lymphoid homing program comprising increased CCR7 and decreased CCR5 and CCR2 [60], and acute activation in vivo in nonhuman primates is associated with a dramatic exodus of Vγ9Vδ2 T cells from the blood [89]. Once in tissues, at least some γδ T cell subsets are likely to contribute to tissue homeostasis and repair. In murine skin, the dendritic epidermal T cell (DETC) subset has been associated with growth factor production and promotion of epidermal barrier integrity [90,91]. In both mice and humans, γδ IELs may have a unique phenotype (Fig. 1), capable of cytokine production and cytotoxicity as well as involvement in epithelial maintenance. Intestinal intraepithelial Vγ4Vδ1 T cells from healthy human donors express granzymes, perforin, TNF, and IFN-γ, but also regulatory molecules such as amphiregulin (AREG), a growth factor that promotes epithelial repair [80,92]. Normality sensing involving Vγ4 γδ TCR binding to the BTNL molecule BTNL3 expressed on intestinal epithelial cells may conceivably regulate AREG production and epithelial homeostatic functionality. Finally, mouse imaging studies on intestinal IELs support a role for such subsets in the elimination of dysregulated intestinal epithelial cells from the mucosal barrier [69,93].

Orchestration of downstream immunity has long been highlighted as an important feature of γδ T cell immunobiology, and likely operates in diverse scenarios. Various mouse models support critical roles for γδ T cells either in activation of neutrophils [94,95], or in maturation of dendritic cells (DCs) [96–100]. In addition, P-Ag-mediated activation of peripheral blood Vγ9Vδ2 T cells was reported to upregulate potent antigen presentation functions, enabling antigen presentation to and activation of downstream αβ T cells [101]. Finally, evidence in mice supports a role for γδ T cells in B cell help, helping to facilitate IgE production [102].

Also consistent with synergism with B cell responses, expression of the Fc receptor CD16 on Vγ9Vδ2 T cells is proposed to enable synergy with humoral immunity in the context of pathogen infection [74]. In blood-stage malaria infection, this was proposed to enable antibody-dependent cellular phagocytosis, directed at parasite infected erythrocytes [103]. Moreover, CD16 has also been shown to efficiently trigger antibody-dependent cell-mediated cytotoxicity (ADCC) against opsonized target cells [104], as well as ADCI of viral infection by inducing IFN-γ production [78].

γδ T cell contributions to tumor immunosurveillance and progression: Evidence from mouse models

One of the major challenges with the development of human γδ T cell therapy for cancer is the lack of a good preclinical model in which to study γδ T cell responses. Several human γδ T cell subsets do not have an analogous subset in the mouse. For example, the most abundant γδ T cell subset in human blood, Vγ9Vδ2 T cells, has no murine equivalent. Not only are these TCR chains not conserved in mouse, but mice lack BTN3 molecules and mouse γδ T cells do not respond to P-Ag [105]. This does not preclude the use of xenograft models exploiting either human tumor cell lines or patient-derived material and subsequent adoptive transfer of Vγ9Vδ2 T cells from donors/patients. However, although transferred Vγ9Vδ2 T cells can mount strong responses against human-derived tumor cells, these models do not permit analysis of off-target effects because the mouse tissue lacks requisite BTN3A and BTN2A molecules [105]. One potential exception to this is intestinal human Vγ4Vδ1 T cell subset that recognizes BTNL3/BTNL8, for which the mouse Vγ7 IEL γδ subset reactive to Btnl1/6 may be functionally analogous [55,70,71]. Despite this, spontaneous and induced mouse models of cancer have shed light on potential antitumor immunosurveillance and protumoral roles that may be of important generic relevance to human γδ immunobiology.

Some of the best evidence of γδ tumor immunosurveillance comes from the DETC population present in the mouse skin. In two separate models of chemical-induced carcinogenesis, TCRδ−/− mice lacking DETC in the skin developed more tumors than wild-type animals [106,107]. Furthermore, the tumors in these mice grew larger, suggesting the γδ T cells play a role in later stages of tumor growth as well as immunosurveillance of early stage tumors. Tumor recognition was at least partially mediated through NKG2D recognition of NKG2DL upregulated on tumor cells.

Since these results, γδ T cells have been shown to help control tumor development and growth in several spontaneous tumorigenesis models. In a spontaneous prostate cancer model, mice deficient for γδ T cells developed more extensive tumors than wild-type mice, and adoptive transfer of syngeneic splenic γδ T cells significantly reduced tumor growth [108]. Likewise, in a model of spontaneous lymphomagenesis, γδ T cells along with NK cells could reject MHC class I-deficient B cell lymphomas [109]. The γδ T cells proliferated and clustered around the tumor cells in the spleen, suggesting these cells have a direct role in tumor suppression.

Finally, in a chemical carcinogenesis model of colorectal cancer, TCRδ−/− mice also developed more tumors than wild-type mice [110]. However, later studies revealed that there are several different populations of γδ T cells in the mouse intestine, including the cytotoxic Vγ7 Btnl-dependent IEL compartment [70], but also pro-tumor IL-17 producing γδ T cells, which express Vγ6 or Vγ4. Reis et al. (2022) generated mice deficient in Vγ7 or other Vγ chains and showed that the cytotoxic Vγ7 IELs and Vγ1 IELs are important for immune surveillance in a chemically induced and a spontaneous model of colorectal carcinoma, while other IL-17 producing γδ subsets contributed to cancer progression [111]. It is interesting to speculate whether human BTNL-reactive intestinal intraepithelial Vγ4Vδ1 T cells may act similarly to suppress colorectal carcinogenesis. Of note, mouse IL-17 producing γδ T cells have been shown to be tumor-promoting in a number of other mouse tumor models, such as breast [112], lung [113,114], ovarian [115], and hepatocellular carcinoma [116]. To what extent this pro-tumoral biology of mouse IL-17-producing γδ T cells is replicated in human tumors is far less clear. Whereas IL-17-secreting γδ T cells have been reported in several tumors [6], other studies have failed to observe substantial IL-17 signatures in tumor-infiltrating γδ T cells [92,111].

Emerging immunotherapeutic strategies

Expansion of Vγ9Vδ2 T cells

Vγ9Vδ2 T cells have been the focus for the majority of γδ T cell-centric immunotherapy development to date (Fig. 4). This is perhaps not surprising given this subset is universally present in humans, is highly prevalent in blood, can be expanded readily in vitro and in vivo, and demonstrates both cytotoxicity toward tumor targets and cytokine production in response to activation, via a mode of reactivity to target cells that can be manipulated pharmacologically [29]. Two chief immunotherapy modalities have been explored (reviewed in more detail by Hoeres et al. [117]). The first involves expansion of Vγ9Vδ2 populations in vivo in patients, typically achieved with IL-2 combined with aminobisphosphonate (N-BP) drugs, which increase intracellular concentrations of the host P-Ag IPP, based on inhibition of farnesyl pyrophosphate synthase (FPPS), the enzyme responsible for its catabolism within the mammalian mevalonate pathway. This strategy was applied initially in clinical trials for both lymphoid malignancies [118] and prostate cancer [119], with in vivo Vγ9Vδ2 T cell proliferation and Vγ9Vδ2 effector T cell numbers correlating with objective clinical responses, respectively, and has now been applied to diverse solid tumors. Alternatively, ex vivo expansion of Vγ9Vδ2 T cells with N-BP drugs has been used in conjunction with adoptive transfer in clinical trials for a range of hematological and solid tumors [117]. For both modalities, a low toxicity profile has been observed, but only modest therapeutic efficacy, despite objective responses being observed in some patients. Various reasons may account for this, including limited potency of N-BPs relative to pathogen-derived P-Ags, a lack of cancer specificity, intracellular access that is dependent upon active transport pathways that may be compromised in cancer cells, and N-BP accumulation in bone [117].

At least some pharmacodynamic limitations of N-BPs might be addressed by the development of novel Vγ9Vδ2-immunostimulatory compounds, which it is hoped will have enhanced clinical efficacy [117,120] by direct or indirect enhancement of P-Ags. These include P-Ag/P-Ag phosphonate prodrugs [36,121–125], and alternatively "indirect," bisphosphonate prodrugs [126,127]. An advantage is they are likely to enable passive uptake into cells [124], and can enable highly potent activation of Vγ9Vδ2 γδ T cells and sensitization to attack [123,124,126,127], including in preclinical cancer mouse models [126,127]. Although not inherently targeted toward cancer cells, the high potency of some such compounds [123] could form the basis of conjugation approaches to enhance their cancer target specificity.

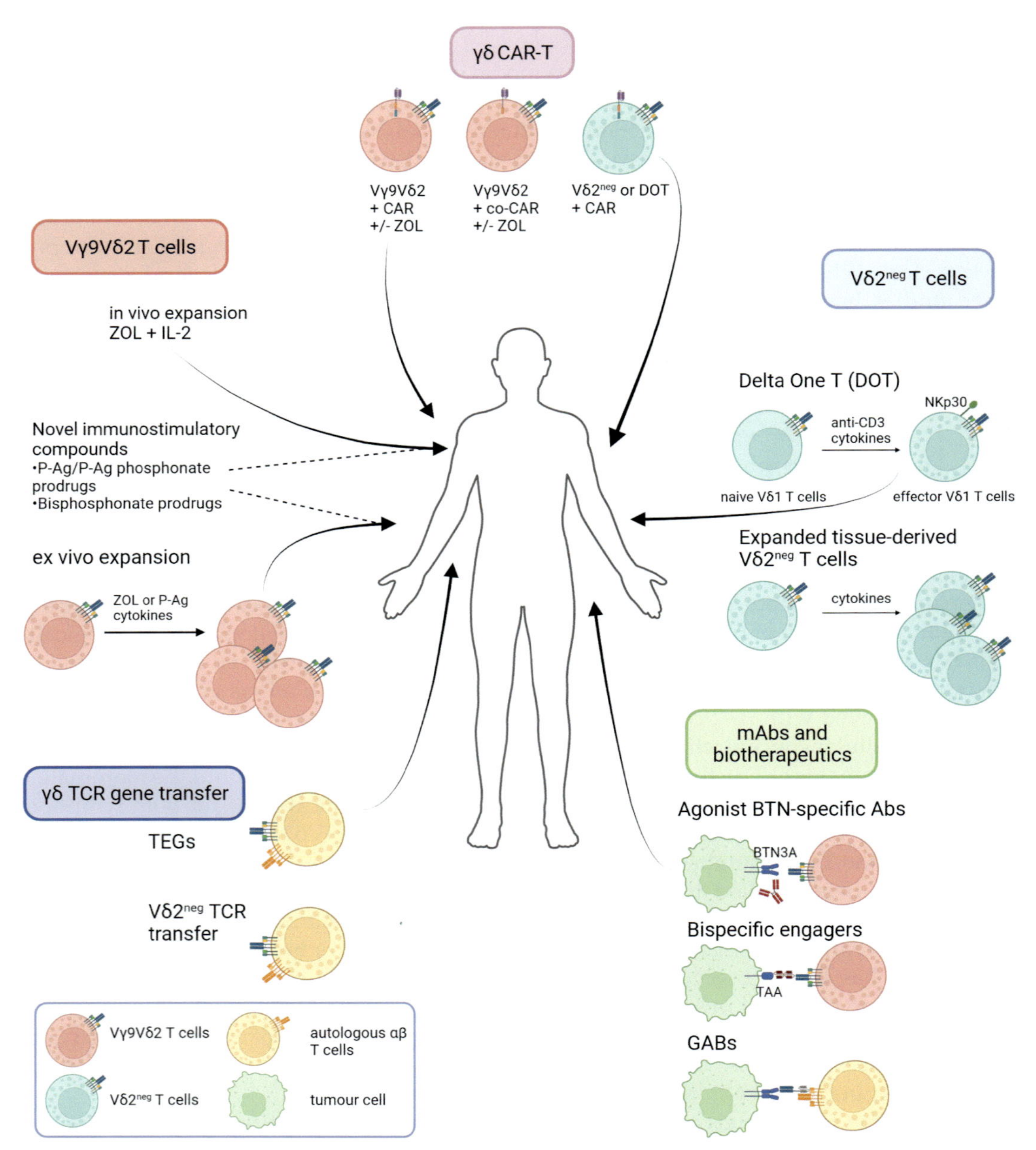

FIG. 4 γδ T cell immunotherapeutic strategies. Vγ9Vδ2-focused strategies (*upper left*) include in vivo or in vitro expansion via zoledronate and IL-2 (the latter combined with adoptive transfer), with novel P-Ag immunostimulatory drugs (including bisphosphonate prodrugs and P-Ag/P-Ag-phosphonate prodrugs) potentially contributing to either in vivo or in vitro approach. $V\delta2^{neg}$ focused strategies (*upper right*) include expansion of DOTs, or expansion of tissue-derived $V\delta2^{neg}$ T cells. γδ CAR-T approaches (*upper center*) leverage both Vγ9Vδ2 and Vδ1 T cells, and may involve second-/third-generation CAR approaches, or co-stimulatory CAR designs. γδ TCR gene transfer approaches (*lower left*) include transduction of Vγ9Vδ2 TCR into autologous αβ T cells to create T cells engineered with a defined gamma delta TCR (TEGs), and in future could leverage $V\delta2^{neg}$ TCRs. Antibody and biotherapeutic approaches (*lower right*) include use of agonist BTN3A-targeted antibodies to activate Vγ9Vδ2 T cells, bispecific antibodies targeting both a tumor-associated antigen (TAA) and TCR γ or δ chain on patient γδ T cells, or gamma delta TCR anti-CD3 bispecific molecules (GABs), Vγ9Vδ2 TCR-anti-CD3 fusion approaches that reprogram patient αβ T cells to become responsive to P-Ag exposed target cells. *Created with BioRender.com.*

Expanded Vδ2^{neg} T cells

Ex vivo expansion of Vδ2^{neg} cells, either from peripheral blood [128] or based on cells extracted from a nonhematopoietic source (e.g., skin) [129], has been used to generate candidate cell therapy products. One such cell therapy product, Delta One T (DOT) cells, is expanded from enriched peripheral blood Vδ1 $\gamma\delta$ T cells cultured with a cocktail of cytokines and anti-CD3 for 3 weeks [128]. The protocol appears to generate Vδ1 $T_{effector}$-like cells from polyclonal Vδ1 T_{naive} cells typically present in peripheral blood and may mimic conditions (i.e., TCR triggering, and cytokine stimulation) that drive "natural" in vivo adaptive $\gamma\delta$ T cell differentiation in response to infectious challenge [22,23,41,52]. Consistent with anti-CD3-driven polyclonal expansion, DOT cells exhibit a diverse TCR repertoire [130], unlike physiological Vδ1 $T_{effector}$ cells, which are highly clonotypically focused [23,41,50], most likely reflecting cognate antigen interactions that are highly selective in triggering specific $\gamma\delta$ TCR clonotypes. Nevertheless, like physiological Vδ1 $T_{effector}$ cells [20], DOT cells display upregulated NK receptors (including the activatory NKRs NKG2D, NKp30, NKp44, and DNAM-1) and cytotoxic molecules relative to T_{naive} counterparts and are potent effectors that can kill transformed cell lines [128,130]. Consistent with use of anti-CD3 to drive in vitro polyclonal expansion, such tumor cell killing was not affected by anti-TCR antibodies but was instead partially dependent on NKp30, the ligand for which, B7-H6, is expressed on many tumor cells [130]. Therefore, DOT cells may rely significantly on TCR-extrinsic stress sensing pathways to trigger tumor cell recognition and killing, themselves dependent on the complex signatures of stress ligands present on a given tumor cell surface, which could exhibit considerable heterogeneity between and within tumor types, requiring empirical investigations to identify suitable indications. Also, whether such NK-like stress sensing pathways are sufficient to drive strong and durable antitumor responses in vivo is unclear. However, DOT cells have shown promising results in preclinical xenograft mouse models and are currently being tested in a Phase 1 trial for acute myeloid leukemia (AML), by Takeda Pharmaceuticals [6].

CAR-armed $\gamma\delta$ T cells

The idea of "arming" $\gamma\delta$ T cells with CAR constructs, which have been successfully employed in $\alpha\beta$-based T cell therapy platforms to treat B cell malignancies [3], to generate CAR-$\gamma\delta$ T cells, is attractive for several reasons. First, it provides a means to enhance $\gamma\delta$ T cell reactivity against the tumor cell and do so in a rational way dependent on expression of a predetermined antigenic target, as opposed to relying on TCR-extrinsic stress sensing pathways, whereby the stress signatures that sensitize tumor cells for attack may be less well defined, and subject to immune escape mechanisms. Second, it provides a potential route for development of broadly applied allogeneic, "off-the-shelf" CAR-T cell therapy products, whereas current approaches reliant on engineering of MHC-restricted $\alpha\beta$ T cells necessitate patient-personalized autologous (and therefore MHC-matched) products, incurring time and cost penalties, and significant failure rates due to patient T cell dysfunction [131].

Both Vγ9Vδ2 T cells and Vδ2^{neg} cells are a focus of ongoing interest in this setting [6], typically involving second- or third-generation CAR-T designs, whereby CD3zeta (signal 1) and costimulatory (signal 2) motif(s) are incorporated in a single construct. These include modification of Vγ9Vδ2 T cells to express CARs targeting a range of antigens, including CD19 [132,133], a well-established antigen for CAR-T targeting of B cell tumors; GD2 [133,134], an established CAR/mAb antigen expressed in neuroectodermal tumors including neuroblastoma; Muc1-Tn [135], a tumor-associated glycoform of mucin-1 overexpressed in various solid tumors; and prostate stem cell antigen [136], an established prostate cancer antigen. Such approaches have typically resulted in in vitro cytotoxicity toward CAR target antigen-expressing tumor lines, and regression of tumors in xenograft models, in some cases involving combination with Zoledronate [136], which resulted in enhanced cytokine secretion and antitumor efficacy. An alternative Vγ9Vδ2-focused approach involved transduction of NKG2D-based CARs, permitting cytotoxicity to several solid tumor lines and extended survival in a xenograft mouse model of ovarian cancer, albeit requiring multiple γδ CAR-T infusions [137].

In parallel, Vδ1 T cells have been modified with CARs targeting glypican-3 [138], a solid tumor antigen expressed in HCC; CD123 [139], a target antigen for AML; and CD20 [140], a target for B cell malignancies. These have resulted in encouraging results in in vitro and in vivo preclinical models. CD123-directed γδ CAR-T cells displayed enhanced cytotoxicity toward primary AML samples and regression of tumor outgrowth in patient-derived xenograft mouse models [139]. In addition, positive preclinical mouse model data in B cell lymphoma xenograft models [140] have led to an ongoing trial of allogeneic CD20-targeting Vδ1 γδ-CAR-T cells (Adicet Bio, ADI-001) in patients with relapsed/refractory B cell malignancies, which has reported promising interim results [141] (ongoing complete responses were observed in four of six patients), albeit with limited time for patient follow-up to date (<9 months).

A major question going forward is the long-term persistence of such allogeneic populations and hence the durability of γδ T cell-mediated antitumor responses [131]. An interesting modification of such approaches involves use of "co-stimulatory CAR" constructs employed preclinically for Vγ9Vδ2 T cells. This utilized a "co-CAR" construct targeted toward GD2 lacking a CD3zeta signaling domain and containing only a NKG2D-derived DAP10 "Signal 2" costimulatory endodomain, resulting in a cell product exhibiting enhanced tumor selectivity. GD2-directed co-CAR recognition synergized with naturally enhanced Vγ9Vδ2 TCR-mediated "Signal 1" detection on tumor cells presumably due to elevated mevalonate-pathway flux in GD2-expressing tumor cells relative to GD2 expressing nontumor cells. This strategy, which could conceivably be combined with pharmacological stimulation of P-Ags in the tumor, has shown promise in in vitro models [142].

γδ TCR gene transfer approaches

An alternative gene modification approach to CAR-T therapy involves use of gene transfer to confer TCR-mediated reactivity to a recipient cell type. Although both approaches involve redirecting recipient cells to recognize a preselected tumor-associated antigen, TCR gene transfer can offer certain advantages relative to CAR-T approaches. Whereas CAR-T therapy is constrained to targeting of intact cell surface antigens, use of αβ TCRs enables targeting of intracellular proteins presented on MHC molecules, expanding the range of targetable tumor-specific targets, and potentially allows targeting of the cancer mutagenome [143]. Conversely, use of γδ TCRs is attractive in that it might permit exploitation of the unique antigen recognition capabilities associated with specific γδ T cell subsets or TCR clonotypes, including circumvention of the requirement for MHC-matching, allowing broad applicability across patient cohorts.

This approach has been developed furthest in relation to Vγ9Vδ2 T cell P-Ag sensing, which is encoded within the γδ TCR itself. Kuball and colleagues have exploited gene transfer of Vγ9Vδ2 TCRs into αβ T cells, to create so-called "T cells engineered with a defined γδ TCR" or TEGs [144]. This approach aims to leverage a well-established knowledge base for modification and clinical use of αβ T cells combined with the potential for broad TCR-mediated tumor targeting elicited by the Vγ9Vδ2 T cell subset. To date, TEGs have been investigated preclinically in vitro, where they permit P-Ag dependent killing of tumor targets and primary tumor targets, and in vivo, where they have shown promise in xenograft cancer models [144–146]. Currently, they are in clinical trials (developed by Gadeta) for multiple myeloma and ovarian cancer, with the use of αβ T cells as a vehicle necessitating an autologous approach [6]. Although exploitation of the Vγ9Vδ2 TCR as a natural "stress sensor" is attractive, the applicability of the approach to different tumors may be determined by the mevalonate pathway flux and accumulation of IPP, and the same mechanisms that may permit tumor evasion of endogenous Vγ9Vδ2 T cells may also apply to TEGs. Moreover, the TCR-centric nature of the approach means contributions to tumor cell recognition from TCR-extrinsic pathways such as NKRs on γδ T cells are lost. Conceivably combining the strategy with P-Ag-stimulating drugs might enhance antitumor responses.

In principle, γδ TCR gene transfer could be applied to the $V\delta2^{neg}$ subset, and this is a potential area for future investigation. A number of $V\delta2^{neg}$ clonotypes have been shown to recognize antigenic targets upregulated on tumors, including endothelial protein C receptor (EPCR) [54], EphA2 [147], and Annexin A2 [148]. Moreover, the diversity of antigens recognized by the adaptive-like $V\delta2^{neg}$ compartment [11] suggests potential for exploring specificities to a broad range of antigenic targets.

Therapeutic antibodies and other biotherapeutics

The dependence of Vγ9Vδ2 T cell recognition upon target cell expression of members of the BTN family, specifically BTN3A [30] and BTN2A [31,32], suggests routes for therapeutic antibody development. Building on the development of an anti-BTN3A agonist monoclonal antibody (mAb) (20.1) that is a potent activator of Vγ9Vδ2 T cells in research settings [30], anti-BTN antibodies are in development for therapeutic application in cancer. A humanized BTN3A-specific mAb, ICT01 (developed by ImCheck Therapeutics), was found to elicit efficient tumor cell targeting in vitro and in hematological and solid tumor xenograft models in mice adoptively transferred for Vγ9Vδ2 T cells and was well tolerated in cynologous monkeys [89]. Interestingly, although the mechanism of action was dependent upon both BTN3A and BTN2A, which are widely expressed in diverse human cells, ICT01 appeared to selectively sensitize tumor but not normal cells for attack by Vγ9Vδ2 T cells [89], although the mechanistic basis of this tumor selectivity is not understood. A first-in-human phase 1/2a clinical study of ICT01 with patients with advanced solid tumors confirmed the drug was well tolerated [89]. Importantly, it appeared to be pharmacodynamically active in patients, leading to Vγ9Vδ2 activation and exodus from the blood, and a notable increase in intratumoral immune infiltration was observed in a patient with melanoma, suggestive of adjuvantization of the tumor microenvironment. Despite the predominant hematological localization of Vγ9Vδ2 T cells, their low frequency in solid tissues, and potential to traffic to lymphoid tissue following activation [60], this represents a promising development, albeit with further patient follow-up required.

In addition to BTN3A-based mAb approaches, a diversification of antibody and biotherapeutic-based approaches is evident and will no doubt continue over coming years. Given the importance of BTN2A alongside BTN3A in Vγ9Vδ2 activation, parallel development of agonistic anti-BTN2A antibodies may also be a subject of investigation. Moreover, the field of bispecific engagers is of interest and has seen development of T cell and NK cell-focused bi/tri-specifics [149], which, although predominantly focused on other immune compartments, could be relevant in the γδ T cell space. In this context, developing agents that specifically exploit critical γδ T cell recognition pathways and/or functional activation axes will be important considerations.

A number of studies have developed specific γδ-T cell engagers comprising bispecific antibodies that fuse a tumor antigen-targeting domain with a γδ T cell-binding domain, the latter directed against the γδ TCR. These have been directed to the Vγ9Vδ2 subset, either by targeting the Vγ9 chain [150,151] or the Vδ2 chain [152–154], with the tumor targeting domain specific for a range of different tumor-associated antigens, including HER2 (in the context of pancreatic cancer) [151], CD40 (in chronic lymphocytic leukemia (CLL)/multiple myeloma) [154], CD1d (in CLL) [153], CD123 (in AML) [150], and EGFR (in colorectal cancer) [152]. These strategies therefore leverage the central role the γδ TCR plays in activating the effector functions of γδ T cells, and consistent with this, generally these agents have triggered robust in vitro γδ T cell activation and tumor killing, and generated encouraging results in preclinical immunocompromised mouse models [6]. Such agents are being developed toward clinical testing, with a critical comparator being the performance of such γδ T cell lineage-specific formats relative to other bispecific T cell engagers that engage the entire $CD3^+$ T cell compartment [149].

Another interesting recent development is novel bispecifics termed gamma delta TCR anti-CD3 bispecific molecules (GABs), comprising the extracellular domains of the Vγ9Vδ2 TCR linked to an anti-CD3 antibody targeting moiety [155]. By incorporating the unique specificity of the Vγ9Vδ2 TCR for P-Ag-exposed target cells, GABs were able to redirect αβ T cells toward diverse hematopoietic and solid tumor targets, including primary AML cells, as shown in a mouse model of multiple myeloma [155]. As well as exploiting the increased mevalonate pathway flux in tumor cells via Vγ9Vδ2 TCR-mediated P-Ag recognition, the approach also leverages the αβ T cell compartment, which represents a greater proportion of the tumor microenvironment than γδ T cells.

Human γδ T cells in immunotherapy

Here we highlight seven translationally aligned questions, each representing both a challenge and an opportunity, which specifically relate to fundamental aspects of γδ T cell immunobiology, answers to which could impact therapy development.

i. Is there an adaptive γδ T cell response to cancer?

In recent years, an apparently adaptive-like immunobiology has emerged for $V\delta2^{neg}$ T cells [22,23,41,50,52,53,63]. Logical sequelae of current observations, many analogous to the CD8 T cell compartment, include the likelihood of γδ TCR-mediated and MHC-independent cognate ligand interactions triggering transition from T_{naive} to long-lived $T_{effector}$ status [5,11], potentially enabling antigen-specific memory, commensurate with changes in homing receptor expression facilitating trafficking of antigen-experienced $T_{effector}$ subsets to peripheral tissues [5,11,63]. Although understanding of the antigenic targets underlying this paradigm is incomplete, it is clear that infections including CMV [22,23] and *Plasmodium* sp. [50] can trigger such adaptive responses. Nevertheless, such pathogens have highly evolved immune evasion capabilities, and γδ TCR responses to host-encoded proteins dysregulated by infectious stress may be as likely as direct recognition of pathogenic targets.

How this paradigm might apply to cancer is currently very poorly understood. The highly diverse TCR repertoire and ligand recognition capabilities of adaptive-like γδ T cell subsets [5,11] combined with the myriad of antigen changes in cancer certainly suggests the potential for an adaptive response to cancer, and previously we have noted that by surveilling for changes in intact cell surface proteins, γδ T cells may act as "Nature's CAR-Ts" [5]. Moreover, the correlation of γδ T cell intratumoral enrichment with favorable prognosis across several cancers (e.g., pancreatic ductal adenocarcinoma, colorectal cancer, hepatocellular carcinoma, gastric, and Merkel cell cancer) [6,85,156] could be interpreted in this context. However, some caution should be exercised in such conclusions. First, alongside uncertainties of how such adaptive-like responses are initiated, a key question is whether de novo adaptive-like responses may be initiated to non-MHC cell surface neoantigens on cancer cells. Alternatively, γδ TCR-mediated recognition of cancer cells by such subsets could reflect the presence of stress antigens common to infection and cancer. In support of this, CMV-reactive $V\delta2^{neg}$ clones have been shown to cross-react with cancer cells [157], and antigen targets defined for individual $V\delta2^{neg}$ TCR clonotypes expanded during infection (EPCR [54,158], EphA2 [147]) have been shown to be upregulated in cancer. Moreover, studies on human lung cancer have revealed that enrichment of $V\delta2^{neg}$ γδ T cells in normal tissue correlated with improved prognosis [159]. Collectively, these findings suggest the possibility that as opposed to de novo unconventional adaptive immunity to cancer, pathogen-driven adaptive-like responses may drive long-lived local populations in diverse tissues [56,57,63] that are repurposed in the context of anticancer immunity. Delineating these two possibilities will be important, partly to fully understand the role of individual γδ TCR clonotypes in tumor recognition, especially given the clear co-upregulation of diverse stress receptors (e.g., NKRs) on adaptive-like $T_{effector}$ subsets [20], which may provide an entirely feasible route by which antitumor functionality could be initiated. The recent implication of $V\delta2^{neg}$ T cells in immune responses following checkpoint blockade in colorectal cancer [82] is one such setting where these arguments are highly pertinent to underlying molecular mechanisms.

ii. What effector function is best to employ against cancer cells?

Although γδ T cells demonstrate a range of different effector functions [84] (Fig. 3), which of these is/are likely to be most effective in the context of antitumor immunity and relevant indications is unclear. Cytotoxicity is among the most prevalent studied, alongside cytokine production, and can be elicited by both Vγ9Vδ2 and adaptive-like subsets when activated, with cytotoxicity demonstrated against diverse tumor targets. Moreover, the transcriptional profile of $V\delta2^{neg}$ $T_{effector}$ cells closely matches that of CD8 T cells [20], arguably the canonical tumor-reactive compartment, consistent with their recent implication in antitumor immunity following checkpoint blockade [82]. How best to elicit such responses in the tumor microenvironment, for example via TCR, NKR, or FasL/Fas stimulation, is also currently unclear. Alternatively, ADCC [104] remains an attractive route to elicit γδ T cell activation and cytotoxicity, which could potentially be exploited to synergize with antibody immunotherapy approaches.

iii. Can we exploit the orchestration functions of γδ T cells?

The majority of γδ T cell-based immunotherapy strategies are currently designed to exploit direct γδ T cell attack on the tumor target cell, as for many αβ and mAb-based approaches [6]. However, harnessing the ability of γδ T cells to orchestrate downstream immunity would be an attractive alternative approach justified by mouse models of infection, where γδ T cells can be critical to activation of neutrophils [94,95]; moreover, DETC interactions with Langerhans cells in the mouse skin are suggestive of modulation of DC functionality [97,100]. In humans, Vγ9Vδ2 T cells can potentiate activation of DCs in a cytokine and cell–cell-contact dependent manner [98], and reciprocally can become activated by immature DCs [99]. In a mouse model of *Mycobacterium tuberculosis* infection, activated Vγ9Vδ2 T cells have also been shown to relieve a pathogen-induced block in DC maturation [96]. Furthermore, it could be argued that the relatively low prevalence of γδ T cells in human tissues and tumors compared to other immune subsets may suggest a physiological role that in some scenarios at least, aligns more with both detection of cellular dysregulation/stress and downstream immunoregulation, as opposed to a primary role as antitarget effector. In the human setting, the ability of Vγ9Vδ2 T cells to upregulate their antigen presentation capabilities following antigen stimulation [101], a stimulus which appears to drive their in vivo distribution into tissues [60], may reflect such an exploitable orchestration capability, with potential impacts on downstream conventional αβ T cell responses, and efforts are underway to exploit this. Moreover, the observation that γδ T cell intratumoral prevalence correlates with strong favorable prognosis across several human tumors [156] could conceivably reflect such an orchestration role. Understanding the mechanistic basis of such effects is a priority and could lead to novel therapeutic approaches. In addition, recent studies on agonistic anti-BTN3A antibodies designed as potent activators of Vγ9Vδ2 T cells have highlighted their ability to drive a broader increase in intratumoral immunity [89], including elevated levels of CD4 and CD8 T cells, consistent with such an orchestration role. This emphasizes that translational efforts designed to elicit direct antitumor functionality may unexpectedly shed light on orchestration abilities. Determining the basis for such effects has significant implications for therapy development.

iv. How might γδ T cell immunogenicity impact therapeutic use?

Development of effective allogeneic "off-the-shelf" therapies is an area of intense interest, with potential to streamline delivery of CAR-T approaches that currently rely on αβ T cell-based engineering, and consequently on time-consuming autologous approaches to generate the cell therapy product that may fail for some patients [131]. As a non-MHC-restricted cell type, γδ T cells are strong candidates for development of such allogeneic off-the-shelf delivery approaches and would be expected to minimize dangerous alloreactivity of third-party cells against normal tissues. Moreover, an ongoing trial of a Vδ1 γδ T cell-based CD20-specific CAR-T approach [140] demonstrated promising results in B cell malignancies albeit with limited patient follow-up [141]. Nevertheless, in vivo persistence of such MHC-mismatched allogeneic cell therapy products is a major concern, as they would be expected to elicit potent host-versus-graft responses over time. Understanding the significance of this problem in the context of γδ T cell-based cellular therapies is important, and it is currently unclear how immunogenic γδ T cells are as targets for allo-mediated elimination. Of note, this is also a highly relevant issue for development of allogeneic NK cell therapy [160]. Conceivably, this could depend on the clinical setting and might be partially ameliorated in immunosuppressed scenarios, such as post-SCT, or in tolerising microenvironments such as the liver. Finally, developing improved γδ-based cell therapy products incorporating modifications to decrease immunogenicity is an area of interest.

v. Do γδ T cells present a reduced toxicity profile versus conventional T cells?

Immune-related toxicities are a cause of substantial morbidity in the context of cellular therapies such as CAR-T therapy [161] and stem cell transplantation (SCT) [162,163]. In allogeneic SCT, life-threatening graft-versus-host disease (GvHD) is dependent upon donor T cells in the graft, and results primarily from alloreactive responses elicited by donor-derived αβ T cells directed against recipient peptide–MHC targets. In contrast, γδ T cell reconstitution post-SCT is associated with decreased relapse and does not appear to drive GvHD [164–166], as would be expected from the absence of MHC-restriction within the γδ T cell compartment. γδ T cells therefore may provide a platform for development of reduced toxicity cellular therapy post-SCT, with potential for employment of engineering approaches to enhance tumor targeting, thereby selectively augmenting graft versus leukemia (GvL). In CAR-T cell therapy, cytokine release syndrome (CRS) is a serious potential complication associated with elevations in serum cytokine levels (most commonly including IFN-γ, IL-6, IL-8, IL-10, MCP-1, and MIP-1β), which can cause a range of symptoms from mild fever to life-threatening capillary leakage, tachycardia, hypotension, and hypoxia and can necessitate treatment with tocilizumab, an anti-IL-6 receptor antagonist [161]. In addition, CRS can be accompanied by neurotoxicity (immune effector cell–associated neurotoxicity syndrome, ICANS) [161]. Of relevance, an ongoing trial of CD20-directed allogeneic Vδ1-based CAR-T therapy approach [140] linked to Adicet Bio (ADI-001) for patients with heavily pretreated B-cell lymphoma is reported to elicit promising efficacy (complete responses observed in four of six patients treated), including in patients treated previously with autologous αβ T cell-based anti-CD19 CAR-T therapy, and a favorable safety profile with no severe (Grade 3) cases of CRS or ICANS [141]. Notwithstanding the requirement for additional follow-up to assess γδ T cell persistence and the durability of response, this raises the prospect that γδ T cell-based CAR-T approaches may demonstrate significantly lower immune-related toxicities than their αβ T cell counterparts, which is a priority for additional clinical trials in this area to resolve.

vi. Can we utilize γδ T cells in class I MHC-deficient and low mutational burden settings?

The highly distinct recognition modalities employed by the γδ TCR [11], enabling recognition of differences in intact, non-MHC, cell surface antigens on target cells, distinguishes the compartment from both αβ T cells and antibodies and suggests γδ T cells may represent "Nature's CAR-Ts" [5]. Underpinned by the defining feature of the compartment itself, the γδ

TCR, this aspect of γδ T cell function may explain evolutionary retention of the compartment over ~500 million years. In addition, distinct stress-recognition capabilities suggest γδ T cells may be able to contribute to antitumor immunity in settings where αβ T cells have limited impact.

One obvious setting is in immunosurveillance of MHC-negative niches. In the context of infectious stress, one relevant setting could be the spleen where γδ T cells are reported to localize to the splenic red pulp [62], rich in MHC-negative erythrocytes, a potential focus of infection for certain intracellular parasites such as *Plasmodium* species [103]. In the cancer setting, recent data linked to the NICHE trial suggest a role for $V\delta2^{neg}$ T cells in antitumor immunity to MHC-negative microsatellite unstable colorectal cancers following neoadjuvant checkpoint blockade therapy [82]. This intriguing study not only demonstrated enrichment of cytotoxic Vδ1 and Vδ3 T cells in MHC-deficient cancers but suggested such responses may contribute to clinical efficacy of checkpoint blockade, at least in MHC-negative settings. Although the exact mechanisms of γδ T cell stress sensing involved are unclear, particularly with respect to relative involvement of γδ TCR-mediated versus NKR-mediated pathways, understanding these in significant detail is an area of strong interest and may stimulate approaches aiming to potentiate $V\delta2^{neg}$ tumor targeting, including conceivably in low mutational burden settings poorly served by current checkpoint blockade approaches.

vii. Is γδ T cell immunosurveillance and homeostatic functionality harnessable to prevent cancer?

Studies on the mouse Vγ5Vδ1 DETC subset support an immunosurveillance role that has potential to limit cutaneous malignancy, and operates early in cancer development [106,107]. In addition, this subset, which is selected intrathymically by the presence of the BTN family member Skint-1 [167,168] and also requires Skint-2 for maturation [169], has recently been shown to play a role in normality sensing in the skin, triggering homeostatic responses that boost barrier protection and limit inflammation [91]. Although the DETC subset has no equivalent in humans, it is possible analogous functions may apply to other γδ subsets that do. One candidate is the intestinal Vγ4Vδ1 subset in humans, which is a component of intraepithelial lymphocytes, and is reported to display a dual phenotype incorporating both NKR expression and cytotoxic functionality, and also growth factor production, suggesting potential for both immunosurveillance and homeostatic roles [80,92]. Moreover, they have been shown to recognize the BTNL3/8 heterodimer selectively expressed in the intestine [55], which may provide a molecular signature of normality, the presence or absence of which enables them to pivot between these two functionalities. Moreover, skewing of intestinal γδ T cell functionality to enhanced AREG production has recently been suggested to suppress intratumoral immunity in colorectal cancer and correlate with poor prognosis [92]. Conversely, dysregulation in γδ IEL phenotype, TCR usage, and BTNL3/8 responsiveness has been linked to BTNL polymorphisms in humans, potentially reflecting defective γδ T cell-mediated normality sensing [170], and attenuated BTNL3/8-mediated TCR downregulation has been noted in γδ T cells from IBD patients versus normal controls [72]. Therefore, while exploitation of direct γδ T cell antitumor reactivity is highly desirable, manipulation of γδ T cell-mediated tissue immunosurveillance and homeostatic functions would be a novel therapeutic route forward worthy of exploration, in order to boost both ongoing cancer immunosurveillance, and also to limit inflammation, which itself can drive intestinal carcinogenesis.

Conclusions

The immunobiology of γδ T cells is highly distinct from that of conventional αβ T cells in a number of respects, some of which suggest advantages in the context of cancer immunotherapy approaches. Consequently, development of γδ T cell-based cancer immunotherapy approaches is an area of intense research interest, particularly given an expanding appetite for immunotherapy development in cancer in general, and the currently limited footprint for its successful application in standard of care therapy. Despite this, many aspects of this immunobiology are yet to be understood in granular detail, and realistically, answers to such questions may well differ in distinct tumor settings, given the intricacies of tumor microenvironmental regulation. Understanding the unique contributions of γδ T cells in nontumor settings such as immunosurveillance of infection may also shed light on fundamental principles and molecular axes that are exploitable in cancer.

Acknowledgments

This work was supported by the Wellcome Trust, UK (grant 221725/Z/20/Z to B.E.W.). We thank Dr. Youcef Mehellou for useful discussions.

References

[1] Bagchi S, Yuan R, Engleman EG. Immune checkpoint inhibitors for the treatment of cancer: clinical impact and mechanisms of response and resistance. Annu Rev Pathol 2021;16:223–49.
[2] Robert C. A decade of immune-checkpoint inhibitors in cancer therapy. Nat Commun 2020;11(1):3801.
[3] June CH, O'Connor RS, Kawalekar OU, Ghassemi S, Milone MC. CAR T cell immunotherapy for human cancer. Science 2018;359(6382):1361–5.
[4] Hayday AC. [gamma][delta] Cells: a right time and a right place for a conserved third way of protection. Annu Rev Immunol 2000;18:975–1026.
[5] Willcox CR, Mohammed F, Willcox BE. The distinct MHC-unrestricted immunobiology of innate-like and adaptive-like human gammadelta T cell subsets-nature's CAR-T cells. Immunol Rev 2020.
[6] Mensurado S, Blanco-Dominguez R, Silva-Santos B. The emerging roles of gammadelta T cells in cancer immunotherapy. Nat Rev Clin Oncol 2023;20(3):178–91.
[7] Hayday AC, Saito H, Gillies SD, Kranz DM, Tanigawa G, Eisen HN, et al. Structure, organization, and somatic rearrangement of T cell gamma genes. Cell 1985;40(2):259–69.
[8] Correa I, Bix M, Liao NS, Zijlstra M, Jaenisch R, Raulet D. Most gamma delta T cells develop normally in beta 2-microglobulin-deficient mice. Proc Natl Acad Sci USA 1992;89(2):653–7.
[9] Benveniste PM, Roy S, Nakatsugawa M, Chen ELY, Nguyen L, Millar DG, et al. Generation and molecular recognition of melanoma-associated antigen-specific human gammadelta T cells. Sci Immunol 2018;3(30): eaav4036. https://doi.org/10.1126/sciimmunol.aav4036 [PMID: 30552102].
[10] Deseke M, Rampoldi F, Sandrock I, Borst E, Boning H, Ssebyatika GL, et al. A CMV-induced adaptive human Vdelta1+ gammadelta T cell clone recognizes HLA-DR. J Exp Med 2022;219(9).
[11] Willcox BE, Willcox CR. Gammadelta TCR ligands: the quest to solve a 500-million-year-old mystery. Nat Immunol 2019;20(2):121–8.
[12] Asarnow DM, Kuziel WA, Bonyhadi M, Tigelaar RE, Tucker PW, Allison JP. Limited diversity of gamma delta antigen receptor genes of Thy-1+ dendritic epidermal cells. Cell 1988;55(5):837–47.
[13] Itohara S, Farr AG, Lafaille JJ, Bonneville M, Takagaki Y, Haas W, et al. Homing of a gamma delta thymocyte subset with homogeneous T-cell receptors to mucosal epithelia. Nature 1990;343(6260):754–7.
[14] Janeway Jr CA, Jones B, Hayday A. Specificity and function of T cells bearing gamma delta receptors. Immunol Today 1988;9(3):73–6.

[15] Kreslavsky T, Savage AK, Hobbs R, Gounari F, Bronson R, Pereira P, et al. TCR-inducible PLZF transcription factor required for innate phenotype of a subset of gammadelta T cells with restricted TCR diversity. Proc Natl Acad Sci USA 2009;106(30):12453–8.
[16] Nielsen MM, Witherden DA, Havran WL. Gammadelta T cells in homeostasis and host defence of epithelial barrier tissues. Nat Rev Immunol 2017;17(12):733–45.
[17] Ribot JC, deBarros A, Pang DJ, Neves JF, Peperzak V, Roberts SJ, et al. CD27 is a thymic determinant of the balance between interferon-gamma- and interleukin 17-producing gammadelta T cell subsets. Nat Immunol 2009;10(4):427–36.
[18] McVay LD, Carding SR. Extrathymic origin of human gamma delta T cells during fetal development. J Immunol 1996;157(7):2873–82.
[19] McVay LD, Jaswal SS, Kennedy C, Hayday A, Carding SR. The generation of human gammadelta T cell repertoires during fetal development. J Immunol 1998;160(12):5851–60.
[20] McMurray JL, von Borstel A, Taher TE, Syrimi E, Taylor GS, Sharif M, et al. Transcriptional profiling of human Vdelta1 T cells reveals a pathogen-driven adaptive differentiation program. Cell Rep 2022;39(8), 110858.
[21] Papadopoulou M, Dimova T, Shey M, Briel L, Veldtsman H, Khomba N, et al. Fetal public Vgamma9Vdelta2 T cells expand and gain potent cytotoxic functions early after birth. Proc Natl Acad Sci USA 2020;117(31):18638–48.
[22] Ravens S, Schultze-Florey C, Raha S, Sandrock I, Drenker M, Oberdorfer L, et al. Human gammadelta T cells are quickly reconstituted after stem-cell transplantation and show adaptive clonal expansion in response to viral infection. Nat Immunol 2017;18(4):393–401.
[23] Davey MS, Willcox CR, Hunter S, Kasatskaya SA, Remmerswaal EBM, Salim M, et al. The human Vdelta2(+) T-cell compartment comprises distinct innate-like Vgamma9(+) and adaptive Vgamma9(−) subsets. Nat Commun 2018;9(1):1760.
[24] Papadopoulou M, Tieppo P, McGovern N, Gosselin F, Chan JKY, Goetgeluk G, et al. TCR sequencing reveals the distinct development of fetal and adult human Vgamma9Vdelta2 T cells. J Immunol 2019;203(6):1468–79.
[25] Willcox CR, Davey MS, Willcox BE. Development and selection of the human Vgamma9Vdelta2(+) T-cell repertoire. Front Immunol 2018;9:1501.
[26] Perriman L, Tavakolinia N, Jalali S, Li S, Hickey PF, Amann-Zalcenstein D, et al. A three-stage developmental pathway for human Vgamma9Vdelta2 T cells within the postnatal thymus. Sci Immunol 2023;8(85):eabo4365.
[27] Gutierrez-Arcelus M, Teslovich N, Mola AR, Polidoro RB, Nathan A, Kim H, et al. Lymphocyte innateness defined by transcriptional states reflects a balance between proliferation and effector functions. Nat Commun 2019;10(1):687.
[28] Fergusson JR, Smith KE, Fleming VM, Rajoriya N, Newell EW, Simmons R, et al. CD161 defines a transcriptional and functional phenotype across distinct human T cell lineages. Cell Rep 2014;9(3):1075–88.
[29] Morita CT, Jin C, Sarikonda G, Wang H. Nonpeptide antigens, presentation mechanisms, and immunological memory of human Vgamma2Vdelta2 T cells: discriminating friend from foe through the recognition of prenyl pyrophosphate antigens. Immunol Rev 2007;215:59–76.
[30] Harly C, Guillaume Y, Nedellec S, Peigne CM, Monkkonen H, Monkkonen J, et al. Key implication of CD277/butyrophilin-3 (BTN3A) in cellular stress sensing by a major human gammadelta T-cell subset. Blood 2012;120(11):2269–79.
[31] Karunakaran MM, Willcox CR, Salim M, Paletta D, Fichtner AS, Noll A, et al. Butyrophilin-2A1 directly binds germline-encoded regions of the Vgamma9Vdelta2 TCR and is essential for phosphoantigen sensing. Immunity 2020;52(3):487–98 e6.
[32] Rigau M, Ostrouska S, Fulford TS, Johnson DN, Woods K, Ruan Z, et al. Butyrophilin 2A1 is essential for phosphoantigen reactivity by gammadelta T cells. Science 2020;367(6478).
[33] Vantourout P, Laing A, Woodward MJ, Zlatareva I, Apolonia L, Jones AW, et al. Heteromeric interactions regulate butyrophilin (BTN) and BTN-like molecules governing gammadelta T cell biology. Proc Natl Acad Sci USA 2018;115(5):1039–44.
[34] Salim M, Knowles TJ, Baker AT, Davey MS, Jeeves M, Sridhar P, et al. BTN3A1 discriminates gammadelta T cell phosphoantigens from nonantigenic small molecules via a conformational sensor in its B30.2 domain. ACS Chem Biol 2017;12(10):2631–43.
[35] Sandstrom A, Peigne CM, Leger A, Crooks JE, Konczak F, Gesnel MC, et al. The intracellular B30.2 domain of butyrophilin 3A1 binds phosphoantigens to mediate activation of human Vgamma9Vdelta2 T cells. Immunity 2014;40(4):490–500.
[36] Hsiao CH, Lin X, Barney RJ, Shippy RR, Li J, Vinogradova O, et al. Synthesis of a phosphoantigen prodrug that potently activates Vgamma9Vdelta2 T-lymphocytes. Chem Biol 2014;21(8):945–54.

[37] Hsiao CC, Nguyen K, Jin Y, Vinogradova O, Wiemer AJ. Ligand-induced interactions between butyrophilin 2A1 and 3A1 internal domains in the HMBPP receptor complex. Cell Chem Biol 2022;29(6):985–95 e5.

[38] Yuan L, Ma X, Yang Y, Li X, Ma W, Yang H, et al. Phosphoantigens are molecular glues that promote butyrophilin 3A1/2A1 association leading to Vγ9Vδ2 T cell activation. BioRxiv 2022;2923. https://doi.org/10.1101/2022.01.02.474068.

[39] Karunakaran MM, Subramanian H, Jin Y, Mohammed F, Kimmel B, Juraske C, et al. A distinct topology of BTN3A IgV and B30.2 domains controlled by juxtamembrane regions favors optimal human γδ T cell phosphoantigen sensing. Nat Commun 2023;14(1):7617. https://doi.org/10.1038/s41467-023-41938-8 [PMID: 37993425].

[40] Willcox CR, Salim M, Begley CR, Karunakaran MM, Easton EJ, von Klopotek C, et al. Phosphoantigen sensing combines TCR-dependent recognition of the BTN3A IgV domain and germline interaction with BTN2A1. Cell Rep 2023;42(4), 112321.

[41] Davey MS, Willcox CR, Joyce SP, Ladell K, Kasatskaya SA, McLaren JE, et al. Clonal selection in the human Vdelta1 T cell repertoire indicates gammadelta TCR-dependent adaptive immune surveillance. Nat Commun 2017;8:14760.

[42] Dechanet J, Merville P, Lim A, Retiere C, Pitard V, Lafarge X, et al. Implication of gammadelta T cells in the human immune response to cytomegalovirus. J Clin Invest 1999;103(10):1437–49.

[43] Kaminski H, Menard C, El Hayani B, Adjibabi AN, Marseres G, Courant M, et al. Characterization of a unique gammadelta T cell subset as a specific marker of CMV infection severity. J Infect Dis 2021;223(4):655–66. https://doi.org/10.1093/infdis/jiaa400 [PMID: 32622351].

[44] Farnault L, Gertner-Dardenne J, Gondois-Rey F, Michel G, Chambost H, Hirsch I, et al. Clinical evidence implicating gamma-delta T cells in EBV control following cord blood transplantation. Bone Marrow Transplant 2013;48(11):1478–9.

[45] Fujishima N, Hirokawa M, Fujishima M, Yamashita J, Saitoh H, Ichikawa Y, et al. Skewed T cell receptor repertoire of Vdelta1(+) gammadelta T lymphocytes after human allogeneic haematopoietic stem cell transplantation and the potential role for Epstein-Barr virus-infected B cells in clonal restriction. Clin Exp Immunol 2007;149(1):70–9.

[46] Boullier S, Dadaglio G, Lafeuillade A, Debord T, Gougeon ML. V delta 1 T cells expanded in the blood throughout HIV infection display a cytotoxic activity and are primed for TNF-alpha and IFN-gamma production but are not selected in lymph nodes. J Immunol 1997;159(7):3629–37.

[47] Hviid L, Kurtzhals JA, Adabayeri V, Loizon S, Kemp K, Goka BQ, et al. Perturbation and proinflammatory type activation of V delta 1(+) gamma delta T cells in African children with plasmodium falciparum malaria. Infect Immun 2001;69(5):3190–6.

[48] Ravens S, Fichtner AS, Willers M, Torkornoo D, Pirr S, Schoning J, et al. Microbial exposure drives polyclonal expansion of innate gammadelta T cells immediately after birth. Proc Natl Acad Sci USA 2020;117(31):18649–60.

[49] Rutishauser T, Lepore M, Di Blasi D, Dangy JP, Abdulla S, Jongo S, et al. Activation of TCR Vdelta1(+) and Vdelta1(−)Vdelta2(−) gammadelta T cells upon controlled infection with plasmodium falciparum in Tanzanian volunteers. J Immunol 2020;204(1):180–91.

[50] von Borstel A, Chevour P, Arsovski D, Krol JMM, Howson LJ, Berry AA, et al. Repeated *Plasmodium falciparum* infection in humans drives the clonal expansion of an adaptive gammadelta T cell repertoire. Sci Transl Med 2021;13(622):eabe7430.

[51] Holtmeier W, Rowell DL, Nyberg A, Kagnoff MF. Distinct delta T cell receptor repertoires in monozygotic twins concordant for coeliac disease. Clin Exp Immunol 1997;107(1):148–57.

[52] Davey MS, Willcox CR, Baker AT, Hunter S, Willcox BE. Recasting human Vdelta1 lymphocytes in an adaptive role. Trends Immunol 2018;39(6):446–59.

[53] Davey MS, Willcox CR, Hunter S, Oo YH, Willcox BE. Vdelta2(+) T cells-two subsets for the price of one. Front Immunol 2018;9:2106.

[54] Willcox CR, Pitard V, Netzer S, Couzi L, Salim M, Silberzahn T, et al. Cytomegalovirus and tumor stress surveillance by binding of a human gammadelta T cell antigen receptor to endothelial protein C receptor. Nat Immunol 2012;13(9):872–9.

[55] Willcox CR, Vantourout P, Salim M, Zlatareva I, Melandri D, Zanardo L, et al. Butyrophilin-like 3 directly binds a human Vgamma4(+) T cell receptor using a modality distinct from clonally-restricted antigen. Immunity 2019;51(5):813–25 e4.

[56] Khairallah C, Netzer S, Villacreces A, Juzan M, Rousseau B, Dulanto S, et al. Gammadelta T cells confer protection against murine cytomegalovirus (MCMV). PLoS Pathog 2015;11(3), e1004702.
[57] Sell S, Dietz M, Schneider A, Holtappels R, Mach M, Winkler TH. Control of murine cytomegalovirus infection by gammadelta T cells. PLoS Pathog 2015;11(2), e1004481.
[58] Bucht A, Soderstrom K, Esin S, Grunewald J, Hagelberg S, Magnusson I, et al. Analysis of gamma delta V region usage in normal and diseased human intestinal biopsies and peripheral blood by polymerase chain reaction (PCR) and flow cytometry. Clin Exp Immunol 1995;99(1):57–64.
[59] Glatzel A, Wesch D, Schiemann F, Brandt E, Janssen O, Kabelitz D. Patterns of chemokine receptor expression on peripheral blood gamma delta T lymphocytes: strong expression of CCR5 is a selective feature of V delta 2/V gamma 9 gamma delta T cells. J Immunol 2002;168(10):4920–9.
[60] Brandes M, Willimann K, Lang AB, Nam KH, Jin C, Brenner MB, et al. Flexible migration program regulates gamma delta T-cell involvement in humoral immunity. Blood 2003;102(10):3693–701.
[61] Zeglam AA. Exploring adaptive-like biology in human and murine γδ T cells., 2022, https://ethesesbhamacuk/id/eprint/12265/.
[62] Bordessoule D, Gaulard P, Mason DY. Preferential localisation of human lymphocytes bearing gamma delta T cell receptors to the red pulp of the spleen. J Clin Pathol 1990;43(6):461–4.
[63] Hunter S, Willcox CR, Davey MS, Kasatskaya SA, Jeffery HC, Chudakov DM, et al. Human liver infiltrating gammadelta T cells are composed of clonally expanded circulating and tissue-resident populations. J Hepatol 2018;69(3):654–65.
[64] Zakeri N, Hall A, Swadling L, Pallett LJ, Schmidt NM, Diniz MO, et al. Characterisation and induction of tissue-resident gamma delta T-cells to target hepatocellular carcinoma. Nat Commun 2022;13(1):1372.
[65] Bos JD, Teunissen MB, Cairo I, Krieg SR, Kapsenberg ML, Das PK, et al. T-cell receptor gamma delta bearing cells in normal human skin. J Invest Dermatol 1990;94(1):37–42.
[66] Holtmeier W, Pfander M, Hennemann A, Zollner TM, Kaufmann R, Caspary WF. The TCR-delta repertoire in normal human skin is restricted and distinct from the TCR-delta repertoire in the peripheral blood. J Invest Dermatol 2001;116(2):275–80.
[67] McCarthy NE, Bashir Z, Vossenkamper A, Hedin CR, Giles EM, Bhattacharjee S, et al. Proinflammatory Vdelta2 + T cells populate the human intestinal mucosa and enhance IFN-gamma production by colonic alphabeta T cells. J Immunol 2013;191(5):2752–63.
[68] Mikulak J, Oriolo F, Bruni E, Roberto A, Colombo FS, Villa A, et al. NKp46-expressing human gut-resident intraepithelial Vdelta1 T cell subpopulation exhibits high antitumor activity against colorectal cancer. JCI Insight 2019;4(24).
[69] Edelblum KL, Singh G, Odenwald MA, Lingaraju A, El Bissati K, McLeod R, et al. Gammadelta intraepithelial lymphocyte migration limits transepithelial pathogen invasion and systemic disease in mice. Gastroenterology 2015;148(7):1417–26.
[70] Di Marco BR, Roberts NA, Dart RJ, Vantourout P, Jandke A, Nussbaumer O, et al. Epithelia use butyrophilin-like molecules to shape organ-specific gammadelta T cell compartments. Cell 2016;167(1):203–18 e17.
[71] Melandri D, Zlatareva I, Chaleil RAG, Dart RJ, Chancellor A, Nussbaumer O, et al. The gammadeltaTCR combines innate immunity with adaptive immunity by utilizing spatially distinct regions for agonist selection and antigen responsiveness. Nat Immunol 2018;19(12):1352–65.
[72] Dart RJ, Vantourout P, Irving PM, Hayday AC. OP32 a novel mechanism of colonic epithelial-T-cell cross-talk is dysregulated in IBD. J Crohn's Colitis 2019;13:S023–4.
[73] Ryan PL, Sumaria N, Holland CJ, Bradford CM, Izotova N, Grandjean CL, et al. Heterogeneous yet stable Vdelta2(+) T-cell profiles define distinct cytotoxic effector potentials in healthy human individuals. Proc Natl Acad Sci USA 2016;113(50):14378–83.
[74] Lafont V, Liautard J, Liautard JP, Favero J. Production of TNF-alpha by human V gamma 9V delta 2 T cells via engagement of fc gamma RIIIA, the low affinity type 3 receptor for the fc portion of IgG, expressed upon TCR activation by nonpeptidic antigen. J Immunol 2001;166(12):7190–9.
[75] Das H, Wang L, Kamath A, Bukowski JF. Vgamma2Vdelta2 T-cell receptor-mediated recognition of aminobisphosphonates. Blood 2001;98(5):1616–8.
[76] Zuo J, Willcox CR, Mohammed F, Davey M, Hunter S, Khan K, et al. A disease-linked ULBP6 polymorphism inhibits NKG2D-mediated target cell killing by enhancing the stability of NKG2D ligand binding. Sci Signal 2017;10(481).
[77] Rincon-Orozco B, Kunzmann V, Wrobel P, Kabelitz D, Steinle A, Herrmann T. Activation of V gamma 9V delta 2 T cells by NKG2D. J Immunol 2005;175(4):2144–51.

[78] Couzi L, Pitard V, Sicard X, Garrigue I, Hawchar O, Merville P, et al. Antibody-dependent anti-cytomegalovirus activity of human gammadelta T cells expressing CD16 (FcgammaRIIIa). Blood 2012;119 (6):1418–27.

[79] Wu Y, Kyle-Cezar F, Woolf RT, Naceur-Lombardelli C, Owen J, Biswas D, et al. An innate-like $V\delta1^{+}\gamma\delta$ T cell compartment in the human breast is associated with remission in triple-negative breast cancer. Sci Transl Med 2019;11(513):eaax9364. https://doi.org/10.1126/scitranslmed.aax9364 [PMID: 31597756; PMCID: PMC6877350].

[80] Mayassi T, Ladell K, Gudjonson H, McLaren JE, Shaw DG, Tran MT, et al. Chronic inflammation permanently reshapes tissue-resident immunity in celiac disease. Cell 2019;176(5):967–81 e19.

[81] Sen Santara S, Lee DJ, Crespo A, Hu JJ, Walker C, Ma X, et al. The NK cell receptor NKp46 recognizes ecto-calreticulin on ER-stressed cells. Nature 2023;616(7956):348–56.

[82] de Vries NL, van de Haar J, Veninga V, Chalabi M, Ijsselsteijn ME, van der Ploeg M, et al. γδ T cells are effectors of immunotherapy in cancers with HLA class I defects. Nature 2023;613(7945):743–50. https://doi.org/10.1038/s41586-022-05593-1 [Epub 2023 Jan 11. PMID: 36631610; PMCID: PMC9876799].

[83] Rancan C, Arias-Badia M, Dogra P, Chen B, Aran D, Yang H, et al. Exhausted intratumoral Vdelta2(−) gammadelta T cells in human kidney cancer retain effector function. Nat Immunol 2023;24(4):612–24.

[84] Vantourout P, Hayday A. Six-of-the-best: unique contributions of gammadelta T cells to immunology. Nat Rev Immunol 2013;13(2):88–100.

[85] Silva-Santos B, Mensurado S, Coffelt SB. Gammadelta T cells: pleiotropic immune effectors with therapeutic potential in cancer. Nat Rev Cancer 2019;19(7):392–404.

[86] Kenna TJ, Davidson SI, Duan R, Bradbury LA, McFarlane J, Smith M, et al. Enrichment of circulating interleukin-17-secreting interleukin-23 receptor-positive gamma/delta T cells in patients with active ankylosing spondylitis. Arthritis Rheum 2012;64(5):1420–9.

[87] Maher CO, Dunne K, Comerford R, O'Dea S, Loy A, Woo J, et al. *Candida albicans* stimulates IL-23 release by human dendritic cells and downstream IL-17 secretion by Vdelta1 T cells. J Immunol 2015;194(12):5953–60.

[88] Cipriani B, Borsellino G, Poccia F, Placido R, Tramonti D, Bach S, et al. Activation of C-C beta-chemokines in human peripheral blood gammadelta T cells by isopentenyl pyrophosphate and regulation by cytokines. Blood 2000;95(1):39–47.

[89] De Gassart A, Le KS, Brune P, Agaugue S, Sims J, Goubard A, et al. Development of ICT01, a first-in-class, anti-BTN3A antibody for activating Vgamma9Vdelta2 T cell-mediated antitumor immune response. Sci Transl Med 2021;13(616):eabj0835.

[90] Jameson J, Ugarte K, Chen N, Yachi P, Fuchs E, Boismenu R, et al. A role for skin gammadelta T cells in wound repair. Science 2002;296(5568):747–9.

[91] McKenzie DR, Hart R, Bah N, Ushakov DS, Munoz-Ruiz M, Feederle R, et al. Normality sensing licenses local T cells for innate-like tissue surveillance. Nat Immunol 2022;23(3):411–22.

[92] Harmon C, Zaborowski A, Moore H, St Louis P, Slattery K, Duquette D, et al. Gammadelta T cell dichotomy with opposing cytotoxic and wound healing functions in human solid tumors. Nat Cancer 2023;4(8):1122–37.

[93] Hu MD, Golovchenko NB, Burns GL, Nair PM, TJT K, Agos J, et al. Gammadelta intraepithelial lymphocytes facilitate pathological epithelial cell shedding via CD103-mediated granzyme release. Gastroenterology 2022;162(3):877–89 e7.

[94] King DP, Hyde DM, Jackson KA, Novosad DM, Ellis TN, Putney L, et al. Cutting edge: protective response to pulmonary injury requires gamma delta T lymphocytes. J Immunol 1999;162(9):5033–6.

[95] Romagnoli PA, Sheridan BS, Pham QM, Lefrancois L, Khanna KM. IL-17A-producing resident memory gammadelta T cells orchestrate the innate immune response to secondary oral listeria monocytogenes infection. Proc Natl Acad Sci USA 2016;113(30):8502–7.

[96] Caccamo N, Sireci G, Meraviglia S, Dieli F, Ivanyi J, Salerno A. Gammadelta T cells condition dendritic cells in vivo for priming pulmonary CD8 T cell responses against *Mycobacterium tuberculosis*. Eur J Immunol 2006;36(10):2681–90.

[97] Chodaczek G, Papanna V, Zal MA, Zal T. Body-barrier surveillance by epidermal gammadelta TCRs. Nat Immunol 2012;13(3):272–82.

[98] Conti L, Casetti R, Cardone M, Varano B, Martino A, Belardelli F, et al. Reciprocal activating interaction between dendritic cells and pamidronate-stimulated gammadelta T cells: role of CD86 and inflammatory cytokines. J Immunol 2005;174(1):252–60.

[99] Devilder MC, Maillet S, Bouyge-Moreau I, Donnadieu E, Bonneville M, Scotet E. Potentiation of antigen-stimulated V gamma 9V delta 2 T cell cytokine production by immature dendritic cells (DC) and reciprocal effect on DC maturation. J Immunol 2006;176(3):1386–93.
[100] Strid J, Roberts SJ, Filler RB, Lewis JM, Kwong BY, Schpero W, et al. Acute upregulation of an NKG2D ligand promotes rapid reorganization of a local immune compartment with pleiotropic effects on carcinogenesis. Nat Immunol 2008;9(2):146–54.
[101] Brandes M, Willimann K, Moser B. Professional antigen-presentation function by human gammadelta T cells. Science 2005;309(5732):264–8.
[102] Strid J, Sobolev O, Zafirova B, Polic B, Hayday A. The intraepithelial T cell response to NKG2D-ligands links lymphoid stress surveillance to atopy. Science 2011;334(6060):1293–7.
[103] Junqueira C, Polidoro RB, Castro G, Absalon S, Liang Z, Sen Santara S, et al. Gammadelta T cells suppress plasmodium falciparum blood-stage infection by direct killing and phagocytosis. Nat Immunol 2021;22(3):347–57.
[104] Tokuyama H, Hagi T, Mattarollo SR, Morley J, Wang Q, So HF, et al. V gamma 9 V delta 2 T cell cytotoxicity against tumor cells is enhanced by monoclonal antibody drugs—rituximab and trastuzumab. Int J Cancer 2008;122(11):2526–34.
[105] Karunakaran MM, Gobel TW, Starick L, Walter L, Herrmann T. Vgamma9 and Vdelta2 T cell antigen receptor genes and butyrophilin 3 (BTN3) emerged with placental mammals and are concomitantly preserved in selected species like alpaca (*Vicugna pacos*). Immunogenetics 2014;66(4):243–54.
[106] Girardi M, Glusac E, Filler RB, Roberts SJ, Propperova I, Lewis J, et al. The distinct contributions of murine T cell receptor (TCR)gammadelta+ and TCRalphabeta+ T cells to different stages of chemically induced skin cancer. J Exp Med 2003;198(5):747–55.
[107] Girardi M, Oppenheim DE, Steele CR, Lewis JM, Glusac E, Filler R, et al. Regulation of cutaneous malignancy by gammadelta T cells. Science 2001;294(5542):605–9.
[108] Liu Z, Eltoum IE, Guo B, Beck BH, Cloud GA, Lopez RD. Protective immunosurveillance and therapeutic antitumor activity of gammadelta T cells demonstrated in a mouse model of prostate cancer. J Immunol 2008;180(9):6044–53.
[109] Street SE, Hayakawa Y, Zhan Y, Lew AM, MacGregor D, Jamieson AM, et al. Innate immune surveillance of spontaneous B cell lymphomas by natural killer cells and gammadelta T cells. J Exp Med 2004;199(6):879–84.
[110] Matsuda S, Kudoh S, Katayama S. Enhanced formation of azoxymethane-induced colorectal adenocarcinoma in gammadelta T lymphocyte-deficient mice. Jpn J Cancer Res 2001;92(8):880–5.
[111] Reis BS, Darcy PW, Khan IZ, Moon CS, Kornberg AE, Schneider VS, et al. TCR-Vgammadelta usage distinguishes protumor from antitumor intestinal gammadelta T cell subsets. Science 2022;377(6603):276–84.
[112] Coffelt SB, Kersten K, Doornebal CW, Weiden J, Vrijland K, Hau CS, et al. IL-17-producing gammadelta T cells and neutrophils conspire to promote breast cancer metastasis. Nature 2015;522(7556):345–8.
[113] Carmi Y, Rinott G, Dotan S, Elkabets M, Rider P, Voronov E, et al. Microenvironment-derived IL-1 and IL-17 interact in the control of lung metastasis. J Immunol 2011;186(6):3462–71.
[114] Kulig P, Burkhard S, Mikita-Geoffroy J, Croxford AL, Hovelmeyer N, Gyulveszi G, et al. IL17A-mediated endothelial breach promotes metastasis formation. Cancer Immunol Res 2016;4(1):26–32.
[115] Rei M, Goncalves-Sousa N, Lanca T, Thompson RG, Mensurado S, Balkwill FR, et al. Murine CD27(−) Vgamma6(+) gammadelta T cells producing IL-17A promote ovarian cancer growth via mobilization of protumor small peritoneal macrophages. Proc Natl Acad Sci USA 2014;111(34):E3562–70.
[116] Ma S, Cheng Q, Cai Y, Gong H, Wu Y, Yu X, et al. IL-17A produced by gammadelta T cells promotes tumor growth in hepatocellular carcinoma. Cancer Res 2014;74(7):1969–82.
[117] Hoeres T, Smetak M, Pretscher D, Wilhelm M. Improving the efficiency of Vgamma9Vdelta2 T-cell immunotherapy in cancer. Front Immunol 2018;9:800.
[118] Wilhelm M, Kunzmann V, Eckstein S, Reimer P, Weissinger F, Ruediger T, et al. Gammadelta T cells for immune therapy of patients with lymphoid malignancies. Blood 2003;102(1):200–6.
[119] Dieli F, Vermijlen D, Fulfaro F, Caccamo N, Meraviglia S, Cicero G, et al. Targeting human gammadelta T cells with zoledronate and interleukin-2 for immunotherapy of hormone-refractory prostate cancer. Cancer Res 2007;67(15):7450–7.
[120] Wiemer DF, Wiemer AJ. Opportunities and challenges in development of phosphoantigens as Vgamma9Vdelta2 T cell agonists. Biochem Pharmacol 2014;89(3):301–12.
[121] Davey MS, Malde R, Mykura RC, Baker AT, Taher TE, Le Duff CS, et al. Synthesis and biological evaluation of (E)-4-hydroxy-3-methylbut-2-enyl phosphate (HMBP) aryloxy triester phosphoramidate prodrugs as activators of Vgamma9/Vdelta2 T-cell immune responses. J Med Chem 2018;61(5):2111–7.

[122] Foust BJ, Poe MM, Lentini NA, Hsiao CC, Wiemer AJ, Wiemer DF. Mixed aryl phosphonate prodrugs of a butyrophilin ligand. ACS Med Chem Lett 2017;8(9):914–8.
[123] Kadri H, Taher TE, Xu Q, Sharif M, Ashby E, Bryan RT, et al. Aryloxy diester phosphonamidate prodrugs of phosphoantigens (ProPAgens) as potent activators of Vgamma9/Vdelta2 T-cell immune responses. J Med Chem 2020;63(19):11258–70.
[124] Kilcollins AM, Li J, Hsiao CH, Wiemer AJ. HMBPP analog prodrugs bypass energy-dependent uptake to promote efficient BTN3A1-mediated malignant cell lysis by Vgamma9Vdelta2 T lymphocyte effectors. J Immunol 2016;197(2):419–28.
[125] Shippy RR, Lin X, Agabiti SS, Li J, Zangari BM, Foust BJ, et al. Phosphinophosphonates and their Tris-pivaloyloxymethyl prodrugs reveal a negatively cooperative butyrophilin activation mechanism. J Med Chem 2017;60(6):2373–82.
[126] Matsumoto K, Hayashi K, Murata-Hirai K, Iwasaki M, Okamura H, Minato N, et al. Targeting cancer cells with a bisphosphonate prodrug. ChemMedChem 2016;11(24):2656–63.
[127] Tanaka Y, Iwasaki M, Murata-Hirai K, Matsumoto K, Hayashi K, Okamura H, et al. Anti-tumor activity and immunotherapeutic potential of a bisphosphonate prodrug. Sci Rep 2017;7(1):5987.
[128] Almeida AR, Correia DV, Fernandes-Platzgummer A, da Silva CL, da Silva MG, Anjos DR, et al. Delta one T cells for immunotherapy of chronic lymphocytic leukemia: clinical-grade expansion/differentiation and preclinical proof of concept. Clin Cancer Res 2016;22(23):5795–804.
[129] National Center for Biotechnology Information. PubChem Patent Summary for US-11655453-B2, Expansion of γδ T cells, compositions, and methods of use thereof. 2024. Retrieved July 28, 2024 from https://pubchem.ncbi.nlm.nih.gov/patent/US-11655453-B2.
[130] Di Lorenzo B, Simoes AE, Caiado F, Tieppo P, Correia DV, Carvalho T, et al. Broad cytotoxic targeting of acute myeloid leukemia by polyclonal delta one T cells. Cancer Immunol Res 2019;7(4):552–8.
[131] Aparicio C, Acebal C, Gonzalez-Vallinas M. Current approaches to develop "off-the-shelf" chimeric antigen receptor (CAR)-T cells for cancer treatment: a systematic review. Exp Hematol Oncol 2023;12(1):73.
[132] Deniger DC, Switzer K, Mi T, Maiti S, Hurton L, Singh H, et al. Bispecific T-cells expressing polyclonal repertoire of endogenous gammadelta T-cell receptors and introduced CD19-specific chimeric antigen receptor. Mol Ther 2013;21(3):638–47.
[133] Rischer M, Pscherer S, Duwe S, Vormoor J, Jurgens H, Rossig C. Human gammadelta T cells as mediators of chimeric-receptor redirected anti-tumour immunity. Br J Haematol 2004;126(4):583–92.
[134] Capsomidis A, Benthall G, Van Acker HH, Fisher J, Kramer AM, Abeln Z, et al. Chimeric antigen receptor-engineered human Gamma Delta T cells: enhanced cytotoxicity with retention of cross presentation. Mol Ther 2018;26(2):354–65.
[135] Zhai X, You F, Xiang S, Jiang L, Chen D, Li Y, et al. MUC1-Tn-targeting chimeric antigen receptor-modified Vgamma9Vdelta2 T cells with enhanced antigen-specific anti-tumor activity. Am J Cancer Res 2021;11(1):79–91.
[136] Frieling JS, Tordesillas L, Bustos XE, Ramello MC, Bishop RT, Cianne JE, et al. Gammadelta-enriched CAR-T cell therapy for bone metastatic castrate-resistant prostate cancer. Sci Adv 2023;9(18):eadf0108.
[137] Ang WX, Ng YY, Xiao L, Chen C, Li Z, Chi Z, et al. Electroporation of NKG2D RNA CAR improves Vgamma9Vdelta2 T cell responses against human solid tumor xenografts. Mol Ther Oncolytics 2020;17:421–30.
[138] Makkouk A, Yang XC, Barca T, Lucas A, Turkoz M, Wong JTS, et al. Off-the-shelf Vδ1 gamma delta T cells engineered with glypican-3 (GPC-3)-specific chimeric antigen receptor (CAR) and soluble IL-15 display robust antitumor efficacy against hepatocellular carcinoma. J Immunother Cancer 2021;9(12):e003441. https://doi.org/10.1136/jitc-2021-003441 [PMID: 34916256; PMCID: PMC8679077].
[139] Sánchez Martínez D, Tirado N, Mensurado S, Martínez-Moreno A, Romecín P, Gutiérrez Agüera F, et al. Generation and proof-of-concept for allogeneic CD123 CAR-Delta One T (DOT) cells in acute myeloid leukemia. J Immunother Cancer 2022;10(9):e005400. https://doi.org/10.1136/jitc-2022-005400 [PMID: 36162920; PMCID: PMC9516293].
[140] Nishimoto KP, Barca T, Azameera A, Makkouk A, Romero JM, Bai L, et al. Allogeneic CD20-targeted gammadelta T cells exhibit innate and adaptive antitumor activities in preclinical B-cell lymphoma models. Clin Transl Immunol 2022;11(2), e1373.
[141] Neelapu SS, Stevens DA, Hamadani M, Frank MJ, Holmes H, Jacobovits A, et al. A phase 1 study of ADI-001: anti-CD20 CAR-engineered allogeneic gamma delta (γδ) T cells in adults with B-cell malignancies [abstract]. J Clin Oncol 2022;40:7509.
[142] Fisher J, Abramowski P, Wisidagamage Don ND, Flutter B, Capsomidis A, Cheung GW, et al. Avoidance of on-target off-tumor activation using a co-stimulation-only chimeric antigen receptor. Mol Ther 2017;25(5):1234–47.

[143] Morris EC, Stauss HJ. Optimizing T-cell receptor gene therapy for hematologic malignancies. Blood 2016;127(26):3305–11.
[144] Marcu-Malina V, Heijhuurs S, van Buuren M, Hartkamp L, Strand S, Sebestyen Z, et al. Redirecting alphabeta T cells against cancer cells by transfer of a broadly tumor-reactive gammadeltaT-cell receptor. Blood 2011;118(1):50–9.
[145] Johanna I, Straetemans T, Heijhuurs S, Aarts-Riemens T, Norell H, Bongiovanni L, et al. Evaluating in vivo efficacy—toxicity profile of TEG001 in humanized mice xenografts against primary human AML disease and healthy hematopoietic cells. J Immunother Cancer 2019;7(1):69.
[146] Braham MVJ, Minnema MC, Aarts T, Sebestyen Z, Straetemans T, Vyborova A, et al. Cellular immunotherapy on primary multiple myeloma expanded in a 3D bone marrow niche model. Onco Targets Ther 2018;7(6), e1434465.
[147] Harly C, Joyce SP, Domblides C, Bachelet T, Pitard V, Mannat C, et al. Human γδ T cell sensing of AMPK-dependent metabolic tumor reprogramming through TCR recognition of EphA2. Sci Immunol 2021;6(61): eaba9010. https://doi.org/10.1126/sciimmunol.aba9010 [PMID: 34330813].
[148] Marlin R, Pappalardo A, Kaminski H, Willcox CR, Pitard V, Netzer S, et al. Sensing of cell stress by human gammadelta TCR-dependent recognition of annexin A2. Proc Natl Acad Sci USA 2017;114(12):3163–8.
[149] Tapia-Galisteo A, Alvarez-Vallina L, Sanz L. Bi- and trispecific immune cell engagers for immunotherapy of hematological malignancies. J Hematol Oncol 2023;16(1):83.
[150] Ganesan R, Chennupati V, Ramachandran B, Hansen MR, Singh S, Grewal IS. Selective recruitment of gammadelta T cells by a bispecific antibody for the treatment of acute myeloid leukemia. Leukemia 2021;35(8):2274–84.
[151] Oberg HH, Peipp M, Kellner C, Sebens S, Krause S, Petrick D, et al. Novel bispecific antibodies increase gammadelta T-cell cytotoxicity against pancreatic cancer cells. Cancer Res 2014;74(5):1349–60.
[152] de Bruin RCG, Veluchamy JP, Lougheed SM, Schneiders FL, Lopez-Lastra S, Lameris R, et al. A bispecific nanobody approach to leverage the potent and widely applicable tumor cytolytic capacity of Vgamma9Vdelta2-T cells. Onco Targets Ther 2017;7(1), e1375641.
[153] de Weerdt I, Lameris R, Ruben JM, de Boer R, Kloosterman J, King LA, et al. A bispecific single-domain antibody boosts autologous Vgamma9Vdelta2-T cell responses toward CD1d in chronic lymphocytic leukemia. Clin Cancer Res 2021;27(6):1744–55.
[154] de Weerdt I, Lameris R, Scheffer GL, Vree J, de Boer R, Stam AG, et al. A bispecific antibody antagonizes prosurvival CD40 signaling and promotes Vgamma9Vdelta2 T cell-mediated antitumor responses in human B-cell malignancies. Cancer Immunol Res 2021;9(1):50–61.
[155] van Diest E, Hernández López P, Meringa AD, Vyborova A, Karaiskaki F, Heijhuurs S, et al. Gamma delta TCR anti-CD3 bispecific molecules (GABs) as novel immunotherapeutic compounds. J Immunother Cancer 2021;9(11): e003850. https://doi.org/10.1136/jitc-2021-003850 [Erratum in: J Immunother Cancer 2021;9(12):e003850corr1. https://doi.org/10.1136/jitc-2021-003850corr1. PMID: 34815357; PMCID: PMC8611453].
[156] Gentles AJ, Newman AM, Liu CL, Bratman SV, Feng W, Kim D, et al. The prognostic landscape of genes and infiltrating immune cells across human cancers. Nat Med 2015;21(8):938–45.
[157] Halary F, Pitard V, Dlubek D, Krzysiek R, de la Salle H, Merville P, et al. Shared reactivity of V{delta}2(neg) {gamma}{delta} T cells against cytomegalovirus-infected cells and tumor intestinal epithelial cells. J Exp Med 2005;201(10):1567–78.
[158] Lal N, Willcox CR, Beggs A, Taniere P, Shikotra A, Bradding P, et al. Endothelial protein C receptor is overexpressed in colorectal cancer as a result of amplification and hypomethylation of chromosome 20q. J Pathol Clin Res 2017;3(3):155–70.
[159] Wu Y, Biswas D, Usaite I, Angelova M, Boeing S, Karasaki T, et al. A local human Vdelta1 T cell population is associated with survival in nonsmall-cell lung cancer. Nat Cancer 2022;3(6):696–709.
[160] Kennedy PR, Felices M, Miller JS. Challenges to the broad application of allogeneic natural killer cell immunotherapy of cancer. Stem Cell Res Ther 2022;13(1):165.
[161] Freyer CW, Porter DL. Cytokine release syndrome and neurotoxicity following CAR T-cell therapy for hematologic malignancies. J Allergy Clin Immunol 2020;146(5):940–8.
[162] Malard F, Holler E, Sandmaier BM, Huang H, Mohty M. Acute graft-versus-host disease. Nat Rev Dis Primers 2023;9(1):27.
[163] Malard F, Mohty M. Updates in chronic graft-versus-host disease management. Am J Hematol 2023;98 (10):1637–44. https://doi.org/10.1002/ajh.27040 [Epub 2023 Jul 22. PMID: 37483142].

[164] Gaballa A, Arruda LCM, Uhlin M. Gamma delta T-cell reconstitution after allogeneic HCT: a platform for cell therapy. Front Immunol 2022;13, 971709.

[165] Godder KT, Henslee-Downey PJ, Mehta J, Park BS, Chiang KY, Abhyankar S, et al. Long term disease-free survival in acute leukemia patients recovering with increased gammadelta T cells after partially mismatched related donor bone marrow transplantation. Bone Marrow Transplant 2007;39(12):751–7.

[166] Lamb Jr LS, Henslee-Downey PJ, Parrish RS, Godder K, Thompson J, Lee C, et al. Increased frequency of TCR gamma delta+ T cells in disease-free survivors following T cell-depleted, partially mismatched, related donor bone marrow transplantation for leukemia. J Hematother 1996;5(5):503–9.

[167] Barbee SD, Woodward MJ, Turchinovich G, Mention JJ, Lewis JM, Boyden LM, et al. Skint-1 is a highly specific, unique selecting component for epidermal T cells. Proc Natl Acad Sci USA 2011;108(8):3330–5.

[168] Boyden LM, Lewis JM, Barbee SD, Bas A, Girardi M, Hayday AC, et al. Skint1, the prototype of a newly identified immunoglobulin superfamily gene cluster, positively selects epidermal gammadelta T cells. Nat Genet 2008;40(5):656–62.

[169] Jandke A, Melandri D, Monin L, Ushakov DS, Laing AG, Vantourout P, et al. Butyrophilin-like proteins display combinatorial diversity in selecting and maintaining signature intraepithelial gammadelta T cell compartments. Nat Commun 2020;11(1):3769.

[170] Dart RJ. The human colonic gamma delta T cell compartment and its regulation by butyrophilin-like molecules in health and inflammatory bowel disease. Ethosbluk; 2019. https://ethos.bl.uk/OrderDetails.do?uin=uk.bl.ethos.784529.

CHAPTER

2

Examining γδ T cell receptor (γδ-TCR) structure and signaling in the context of cellular immunotherapy design

John Anderson[a], *Gaya Nair*[b], *and Marta Barisa*[a]

[a]UCL Great Ormond Street Institute of Child Health, University College London, London, United Kingdom [b]Faculty of Mathematical and Physical Sciences, University College London, London, United Kingdom

Abstract

The ways in which human γδ T cell receptor (TCR) structure, signaling, and function are different from αβ T cell counterparts is poorly understood. An appreciation for these parameters of γδ T cell biology is crucial to the optimal design of γδ T cell immunotherapy. The herein illustrated review summarizes the latest evidence on what is known about the human γδ-TCR and discusses it in the context of chimeric antigen receptor design for use as cellular therapy against cancer.

Abbreviations

Ag antigen
B-ALL B cell acute lymphoblastic leukemia
BCMA B cell maturation antigen
BRS basic amino acid-rich sequence
BTN butyrophilin
BTNL butyrophilin ligand
CAR chimeric antigen receptor
CD cluster of differentiation
CDR complementarity-determining region
Csk C-terminal Src kinase
CTL cytotoxic lymphocyte
DLBCL diffuse large B cell lymphoma

https://doi.org/10.1016/B978-0-443-21766-1.00008-4

EBV Epstein-Barr virus
FcR-γ Fc-γ receptor
FDA U.S. Food and Drug Administration
FL follicular lymphoma
HLA human leukocyte antigen
HV4 hypervariable region 4
ITAM immunoreceptor tyrosine-based activation motif
ITK inducible T cell kinase
LAT linker for activation of T cells
LBCL large B cell lymphoma
Lck lymphocyte-specific protein tyrosine kinase
MCL mantle cell lymphoma
MHC major histocompatibility complex
MM multiple myeloma
Nck noncatalytic region of tyrosine kinase
NCR natural cytotoxicity receptor TLR—toll-like receptor
NK natural killer
PBMC peripheral blood mononuclear cells
PI3K phosphoinositide 3-kinase
PLC-γ phospholipase C gamma
pMHC peptide-loaded major histocompatibility complex
PRS proline-rich sequence
PTLD posttransplant lymphoproliferative disorder
RKM receptor kinase motif
scFv short-chain variable fragment
SLP76 SH2-domain-containing leukocyte protein of 76 kDa
TCR T cell receptor
TIL tumor-infiltrating lymphocyte

Conflict of interest

G.N. declares no conflict of interest. M.B. and J.A. hold patents in CAR-T technology development. J.A. holds founder's shares in Autolus Therapeutics.

Introduction

The interest in γδ T cells as a vehicle for cellular immunotherapy in the oncology context derives from three features of their biology. The first is γδ T cell innate reactivity against transformed cells via a range of innate-like receptors on their cell surface. The second is their capacity to home to and reside within tissues. The third is their unique γδ T cell receptor (γδ-TCR), which, in contrast to the canonical αβ-TCR, is major histocompatibility complex (MHC)-unrestricted and nonalloreactive. These features render γδ T cells an attractive lymphocyte candidate for allogeneic, and especially solid tumor, immunotherapy.

The treatment paradigm-shifting success of αβ T cell chimeric antigen receptor (αβ-CAR) therapy for B cell malignancies was largely the result of our increasingly sophisticated understanding of αβ T cell signaling and mode of antigen recognition. A firm grasp of how αβ T cells are activated via their TCR and co-receptors enabled synthetic intervention or "hacking" of these signaling axes via first- and then second-generation CAR designs. The resulting successes of both Kymriah and Yescarta anti-CD19 αβ-CAR therapies [1–4] precipitated an explosion of interest in TCR and co-stimulatory receptor signaling networks within canonical αβ T cells as a means to interpret emerging therapeutic successes and failures, enabling educated improvements to CAR-T synthetic construct design [5]. This has, in turn, driven a vast range of empirically evaluated αβ T cell synthetic engineering approaches, which aim to regulate all aspects of cell behavior—from proliferation and persistence, to cytotoxicity, cytokine production, stemness and metabolic fitness [6]. Our understanding of γδ T cell signaling in this same context is vastly inferior. Even basic features of γδ-TCR signaling remain elusive, especially so in the human context. This is a major bottleneck in the intelligent design and translational evaluation of γδ T cell immunotherapeutic interventions. The herein contained review aims to summarize the knowns and unknowns of the γδ-TCR, its signaling and putative co-receptors. It also explores what is known about the ligands that human γδ-TCRs are thought to engage, with a particular emphasis on the central mediator of these interactions in the human Vγ9Vδ2 subset—butyrophilin (BTN) molecules on the surface of target cells. Where possible, human data were used and, if alternative model organisms are discussed, this will be made explicit.

Global cancer incidence is on the rise [7], despite improvements in therapies and increased awareness leading to earlier detection. Typical treatment for solid cancer combines surgical excision of the tumor with or without chemotherapy and radiotherapy [8]. These interventions typically elicit significant injury to healthy tissue. There is an unmet need to develop less toxic treatments, and for these treatments to be affordable, given the rapid rise in cancer rates across developing countries [9]. Immunotherapy offers a more targeted treatment approach with fewer putative toxicities, which can be more durable than traditional cancer treatments in the cellular format, as the therapeutic cells can persist in patients for many years [10]. A downside of cellular immunotherapies for cancer is, however, that they can be substantially more expensive and logistically complex than traditional cancer treatments. This is especially true for autologous gene-modified cell therapies like αβ-CAR T cells, the manufacturing of which is on a per-patient basis [11].

One of the paths to decreasing the logistical complexity and expense of modified cellular immunotherapies is opting for off-the-shelf, allogeneic solutions over autologous, patient-tailored approaches (Fig. 1). In this context, the majority of preclinical development and translational efforts have focused on rendering canonical αβ T cells nonalloreactive (often by knockout of the TCR) [12,13], or utilizing naturally nonalloreactive NK cells [14,15] or γδ T cells [16,17]. Here, γδ T cells have theoretical advantages over NK cells in targeting tumors that reside in the tissues, as, in contrast to γδ T cells, NK cells are primarily blood-resident [18]. γδ T-cells constitute 0.5%–5% of human peripheral T cells but are enriched at sites exposed to the external environment such as the skin and gut [19]. There is further speculation that γδ T cells are capable of greater therapeutic persistence than NK cells due to well-established yearslong persistence of virally reactive γδ T cell clones in patient circulation [20,21], though this is yet to be proven in the clinical adoptive cell transfer context.

The anatomy of a chimeric antigen receptor

The basic structure of a CAR aims to bridge TCR signaling to the antigen recognition ability of an ectodomain, most commonly an antibody-derived short-chain variable fragment (scFv), with specificity for a cell surface antigen. This enables TCR-based immune activation in the absence of MHC-dependent antigen presentation, overcoming a common tumor evasion strategy—the downregulation or loss of cell surface molecules [22]. It also foregoes the

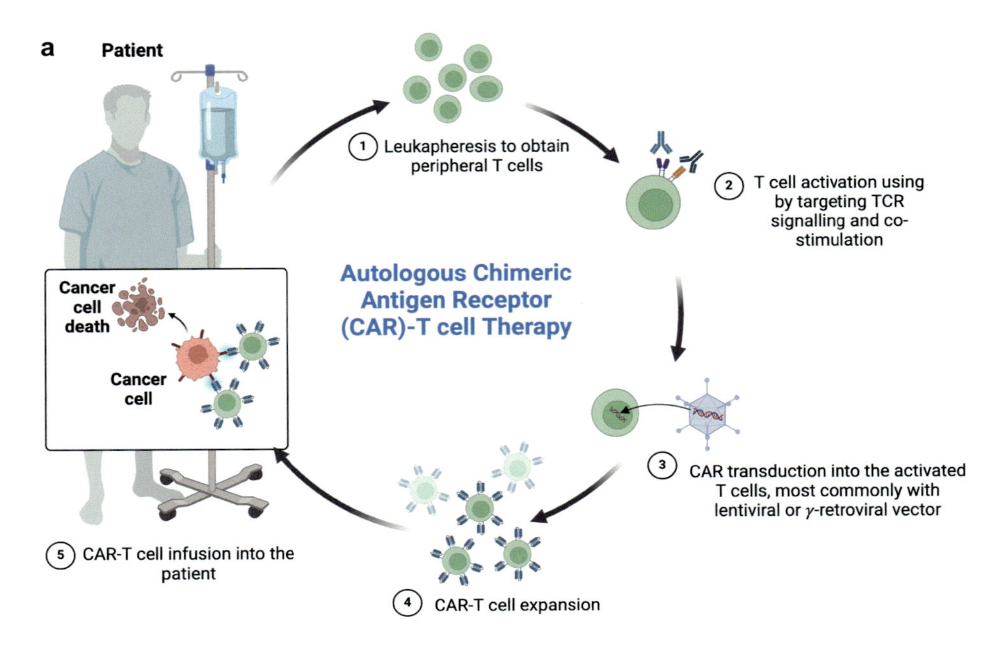

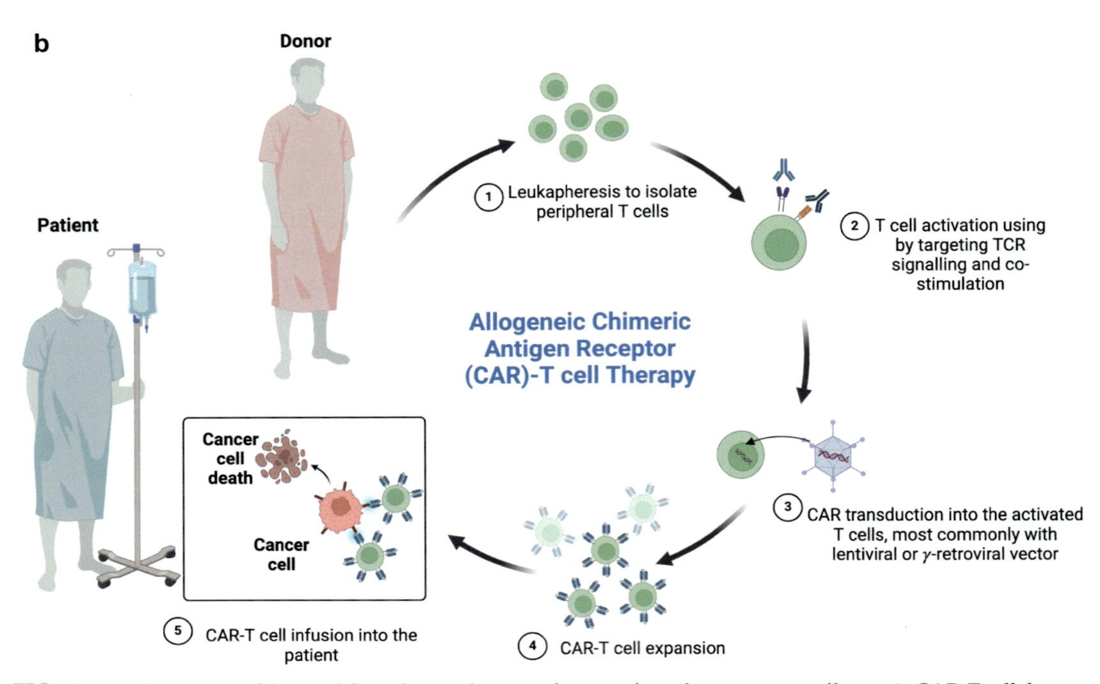

FIG. 1 An illustration of the workflow during the manufacture of autologous versus allogeneic CAR-T cell therapy for cancer. (A) Autologous manufacture involves the manufacture of therapeutic products from T cells that are derived from the cancer patient. (B) Allogeneic manufacture utilizes T cells from healthy donors which are then infused into a cancer patient.

requirement for TCR—human leukocyte antigen (HLA; human MHC)-matching, which places a substantial limit on the broad applicability of TCR-engineered adoptive cell therapies. One of the best established tumor-associated antigen (TAA)-targeting synthetic TCRs (clone ID: MART-1) can only recognize Melan-A peptide in the context of an HLA-A2 subtype [23], which, while being one of the most common HLA types, is present in only ~25% of the human population [24].

There are four key regions that define a CAR: an antigen-recognition domain, a hinge domain that connects the recognition domain to the cell membrane, a transmembrane domain, and a signaling domain. The antigen-recognition domain of a CAR is classically derived from combining a chosen clone IgG heavy and light variable chains with a synthetic linker to create an scFv (illustrated in Fig. 2). The scFv binds to a complementary antigen found on the surface of target cells. Connecting this extracellular region to the transmembrane region is a hinge region, typically from naturally occurring homodimers, in which cysteine residues required for covalent dimerization are retained; these are required for effective CAR-T functionality [25–27]. The length and flexibility of the hinge region may be altered to optimize the binding to each antigen. The transmembrane domain is derived from a naturally occurring transmembrane protein and is often derived from the same protein as the hinge region [28,29]. The signaling domain has been the target for many modifications over the years, resulting in

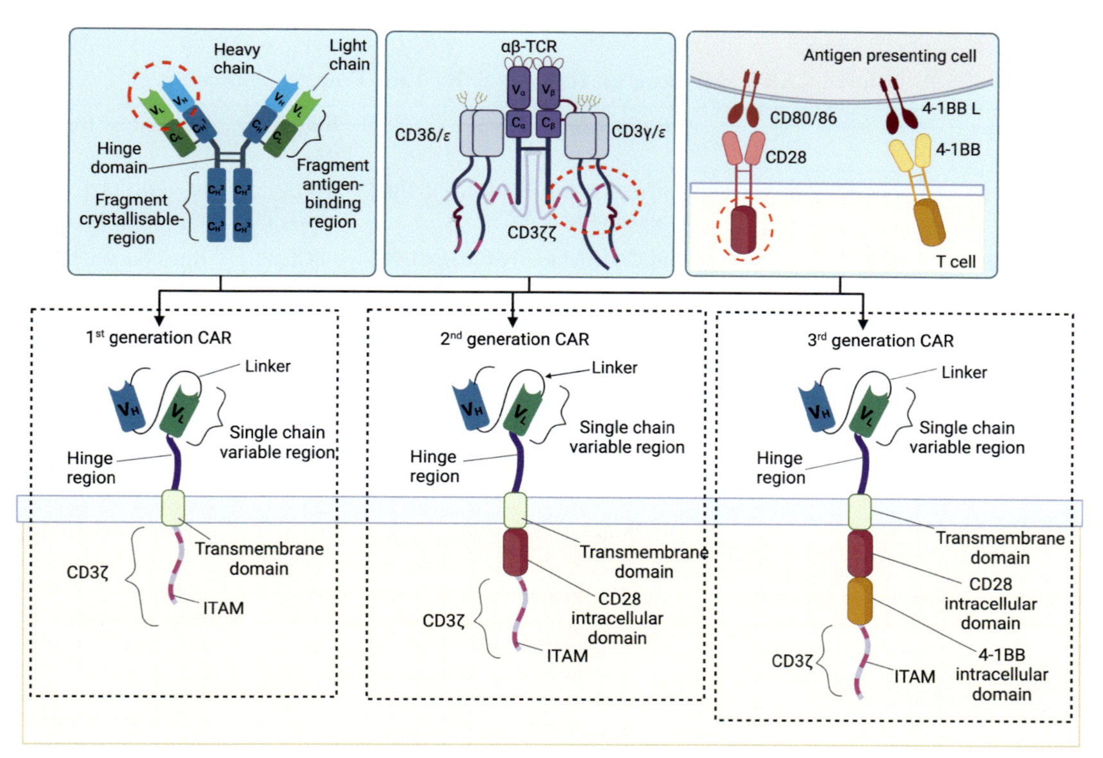

FIG. 2 The basic principles of chimeric antigen receptor (CAR) design. CARs are synthetic receptors that combine into a single, continuous molecule the antigen specificity of antibodies, the signaling capacity of T cell receptors (TCRs) and the co-stimulatory signaling of TCR co-receptors, such as CD28 and 4-1BB.

numerous "generations" of CARs [30]. If optimal conditions are met and the CAR-T cell binds to its target antigen, it will proliferate, release cytokines, and engage in cytotoxicity against target cells.

The first-generation of CARs emerged in 1993, when Eshhar and colleagues successfully combined a TCR complex chain with the specific targeting of an antibody in a single-gene transfer [31]. This very earliest iteration of CAR technology employed the ζ-chain of the CD3 complex, since the ζ-chain is the complex component with the highest number (three) of immunoreceptor tyrosine-based activation motifs (ITAMs) required to confer signal 1 of T cell activation. They also evaluated an alternative source of ITAM signal that is known to mediate cytotoxic responses—the γ-chain of an Fc-γ receptor (FcR-γ). Both receptors were examined in the αβ T cells and NK cells [31]. Remarkably, the team even postulated that co-stimulation is likely an important missing component from the construct and that its inclusion would boost important effector functions including IL-2 production [31]. Such a structure is now referred to as second-generation CAR-T, and it is interesting to reflect, then, what 30 following years of research in the oncology adoptive cell transfer setting have yielded: six autologous second-generation CD3ζ αβ-CAR-T products for hematological malignancies, and, as of 2024, one allogeneic unmodified αβ T cell therapy for lymphoproliferative disease and the first solid tumor-targeting cell therapy with autologous tumor-infiltrating lymphocytes (TILs) for melanoma (summarized in Table 1). Notable by their

TABLE 1 FDA-approved CAR-T cell therapies. FDA – U.S. Food and Drug Administration.

Cell therapy	Construct and cell type	Indication (s)	Offered by (ref)
Tisagenlecleucel (Kymriah)—approved by FDA in 2017	Anti-CD19 (clone: FMC63), 4-1BBζ autologous αβ-CAR	B-ALL, LBCL, FL	Novartis [32]
Axicabtagene ciloleucel (Yescarta)—approved by FDA in 2017	Anti-CD19 (clone: FMC63), CD28ζ autologous αβ-CAR	LBCL, FL	Kite Pharma, Inc [33]
Brexucabtagene autoleucel (Tecartus)—approved by FDA in 2020	Anti-CD19 (clone: FMC63), CD28ζ autologous αβ-CAR	B-ALL, MCL	Kite Pharma, Inc [34]
Lisocabtagene maraleucel (Breyanzi)—approved by FDA in 2021	Anti-CD19 (clone: FMC63), 4-1BBζ autologous αβ-CAR	LBCL, DLBCL	Juno Therapeutics, Inc. (BMS) [35]
Idecabtagene vicleucel (Abecma)—approved by FDA in 2021	Anti-BCMA (clone: C11D5.3), autologous 4-1BBζ αβ-CAR	MM	Celgene Corporation (BMS) [36]
Ciltacabtagene autoleucel (Carvykti)—approved by FDA in 2022	Anti-BCMA (VHH clone), 4-1BBζ autologous αβ-CAR	MM	Janssen Biotech, Inc [37]
Tabelecleucel (Ebvallo)—approved by FDA in 2023	Allogeneic EBV CTLs	PTLD	Atara Biotherapeutics [38,39]
Lifileucel (Amtagvi)—approved by FDA in 2024	Expanded, activated TILs	Melanoma	Iovance Biotherapeutics, Inc [40,41]

B-ALL: acute lymphoblastic leukemia; LBCL: large B cell lymphoma; FL: follicular lymphoma; MCL: mantle cell lymphoma; DLBCL: diffuse large B cell leukemia; MM: multiple myeloma; PTLD: posttransplant lymphoproliferative disorder; EBV: Epstein-Barr virus; CTL: cytotoxic lymphocytes; BCMA: B cell maturation antigen.

absence are constructs containing anything other than the basic scFv-CD3ζ design, and cell therapies based on cell types that are not predominantly made up of peripheral blood mononuclear cell (PBMC)-derived αβ T cells.

We postulate that one of the driving factors for the vast discrepancy in the type of FDA-approved cancer therapies for αβ T cells versus other cytotoxic lymphocytes (CTLs) like γδ T cells or NK cells is the far greater scientific insight we possess for the factors that govern αβ T cell activation, persistence, proliferation, exhaustion, and effector function. An important unresolved issue is whether innate lymphocytes equipped with appropriate CAR designs are fundamentally capable of the clinical benefits that have so far been demonstrated with αβ T cells, and indeed whether additional innate properties may ultimately make them more favorable. Continued evaluation of γδ T cell (and NK cell)-based immunotherapy products in clinical trials is the most direct way to address the latter possibility. The remainder of this review will focus on summarizing the knowns and unknowns of γδ-TCR signaling and respective co-stimulation, with a view to inform the design of efficacy-boosting synthetic constructs for γδ T cell immunotherapeutic interventions.

TCR ectodomains and transmembrane regions

αβ T and γδ T cells differ in the structure and composition of their TCRs, their downstream receptor signaling, homing, and physiological function [42]. Both TCRs are composed of two TCR chain heterodimers that consist of immunoglobulin-like extracellular constant (C) and variable (V) domains on each chain. The V region is membrane-distal and confers the TCR with its binding specificity, while the membrane-proximal C region—encoded by a single gene segment—contains a connecting peptide, transmembrane region (TM), and a cytoplasmic domain. Both TCRs hydrophobically associate with heteromers of cell membrane–embedded CD3 chains that confer the TCR ectodomains with signal transduction capacity [43–45].

CD3γ, CD3δ, and CD3ε chains are structurally similar to one another, with an acidic residue that fixes them in the cellular membrane, and a short cytoplasmic tail that contains a single ITAM. CD3ζ chains are composed of a larger intracellular signaling domain that contain three ITAMs each. The canonical αβ-TCR consists of a TCRα/β heterodimer, associated with CD3δε and CD3γε chains symmetrically positioned on either side, and CD3ζζ chains that reside at the center of the complex [46] (Fig. 3A). The γδ-TCR, meanwhile, is asymmetrical; only the CD3ζζ chains are central to the TCR heterodimer, with remaining CD3 chains sequestered to one side [47] (Fig. 3B). This difference is thought to derive from distinct amino acid sequences in the transmembrane region of the αβ- versus γδ-TCR altering the pattern of CD3 association with the complex [48].

While the V regions of both types of TCRs are largely structurally alike, the C regions differ significantly [44]. The spatial separation of the V and C immunoglobulin-like domains as well as the connecting peptide are longer in the TCRγ/δ than the TCRα/β [44,49,50]. At the same time, the TCR-γ chain possesses a substantially shorter and less flexible FG loop than the TCR-β chain (Fig. 3B). TCRγ/δ and α/β C regions further possess a different shape and distribution of polarity [44], which may also contribute to the different ways by which each TCR associates with CD3 chains in the cell membrane. Based on the reduced size of the TCRδ chain FG loop,

a

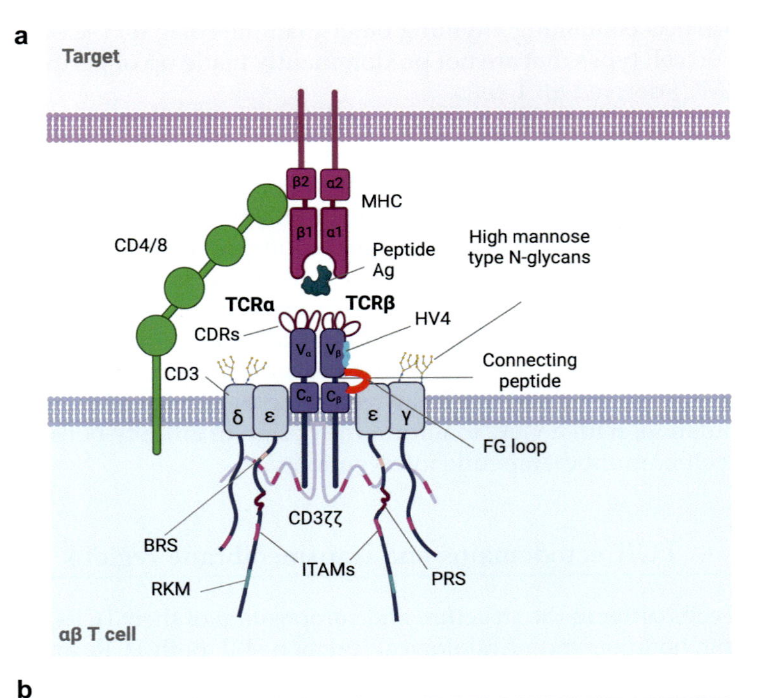

b

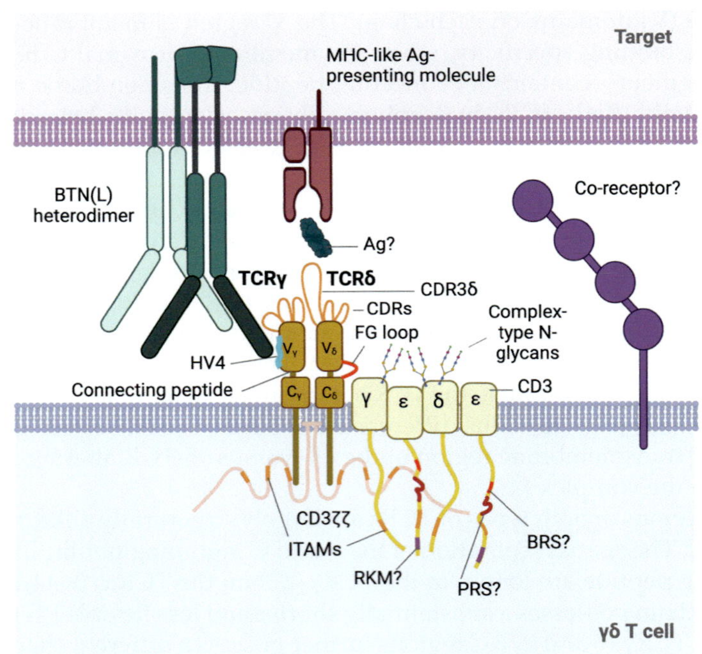

FIG. 3 The basic structure of γδ- and αβ-heterodimer T cell receptors (TCRs). The basic TCR structure is illustrated within the context of associated respective CD3 complexes, putative co-receptors and TCR targets: (A) shows a canonical, peripheral blood-resident pMHC-reactive αβ-TCR and (B) illustrates the proposed structure and model of engagement for the γδ-TCR. CDRs: complementarity-determining regions; BRS: basic amino acid-rich sequence; RKM: receptor kinase motif; ITAMs: immunoreceptor tyrosine-based activation motif; PRS: proline-rich sequence; HV4: hypervariable region 4; Ag: antigen; MHC: major histocompatibility complex; BTN: butyrophilin; BTNL: butyrophilin-like.

Mallis and colleagues propose that mechanotransduction may not be involved in γδ-TCR signal transduction or, alternatively, that the Cδ chain within the γδ-TCR complex may provide mechanosensing itself directly [51].

Differences between the TCRs extend to their associated CD3 complexes, with CD3γ and CD3δ chains in each type of T cell displaying different glycosylation patterns. CD3δ in γδ T cells display N-linked oligosaccharides, whereas CD3δ in αβ T cells contains mannose carbohydrates. This is thought to result from the different spatial relationships of the CD3δ chain with the respective αβ and γδ TCRs, leading to differing accessibility to the respective glycosyltransferase enzymes in the Golgi apparatus. The implications of this are unclear, although it is speculated to correlate with differences in γδ T cell compared to αβ T cell signaling [52,53]. Indeed, in murine γδ T cells, once activated and proliferating, the TCR glycosylation pattern changes, allowing for TCR internalization [48,54]. Dietrich et al. performed site-directed mutagenesis of glycosylated CD3δ residues and observed that this had no effect on TCR assembly, subsequently hypothesizing instead that the glycosylation may be involved in peptide recognition [55]. This was supported by work from Rossi and colleagues, who showed that downregulation of the glycosylation pattern in an αβ T cell CD3δ chain altered the antigen-binding ability of the cell [56]. It is, therefore, tempting to speculate that the alternative glycosylation pattern found in γδ-TCR/CD3 complexes contributes to the different ligands recognized by γδ T cells compared to αβ T cells. Much further work is required to elucidate the effects of glycosylation patterns on TCR complex signaling, particularly in human TCRs.

Mode of TCR/antigen engagement

In contrast to αβ T cells, the majority of γδ T cell subtypes are MHC-unrestricted and rarely express CD4 and CD8 peptide-loaded major histocompatibility complex (pMHC)-docking co-receptors [54]. αβ-TCRs are encoded by V region sequences of relatively uniform length and size but possess an enormous diversity of sequence identities in the peptide-specificity-determining CDR3 regions. γδ-TCRs, on the other hand, possess a high diversity of V region sequence lengths and frameworks, suggesting that, in contrast to αβ-TCRs that mediate pMHC recognition, different γδ-TCRs mediate recognition of different types of ligands presented in the context of different types of antigen-presenting modalities [57]. Indeed, studies show that different γδ-TCRs can be activated by a range of antigen types presented in a range of different contexts, including phosphorylated molecules, alkyl amines, lipids, or stress-induced MHC-like molecules, the upregulation of which are often by-products of DNA damage or intracellular infection [48,58].

A prominent difference in the otherwise similar V regions of γδ- and αβ-TCRs resides in their respective membrane-distal complementarity-determining regions (CDRs). Specifically, compared to αβ-TCRs, γδ-TCRs possess a protruding and extended CDR3δ loop [59] (Fig. 3). Adams et al. showed that this loop—albeit in murine TCRs—is engaged in antigen recognition in a manner that is largely autonomous from the other CDR3 loops, in sharp contrast to αβ-TCR ligand recognition where all three uniform-size CDR3 loops appear to contribute equally. The specific antigen in question was MHC class Ib molecule T22, and the murine TCRs—G8 and KN6 [60–62]. The degree to which this is true for human CDR3δ loops remains to be established [63].

The groups of Hayday, Hermann, Uldrich, Behren, Godfrey, and Willcox have transformed our understanding of the γδ-TCR as a receptor that is primarily engaged via its CDR3 loops. They have shown, first in murine and then human TCR clones, that the germline-encoded hypervariable 4 (HV4) region of the TCR-γ chain can be activated in the absence of CDR1–3 engagement [64–67]. Indeed, they have shown that it is the germline-encoded nonvariable framework regions out with the CDRs within V segments that engaged the best-characterized of γδ-TCR targets—the invariant B7-like butyrophilins and butyrophilin-like molecules (BTN and BTNL) on transformed and "stressed" self-target cells. The most striking feature of this mode of antigen engagement is that—while highly activatory to the T cell—it does not mediate clonal enrichment of the population, expanding the T cells bearing a particular and invariant framework region instead [65]. This mechanism of activation is the putative mediator of phosphoantigen (pAg)-driven Vγ9Vδ2 proliferation. Exogenously added or induced pAg's drive BTN conformational changes on "stressed" target cell surface, and the altered BTN (2A and 3A heterodimer) molecules engage an invariant region on the Vγ9-TCR chain to produce potent Vγ9Vδ2 cell proliferation in a manner that does not clonally enrich for particular CDR3s [66,68,69] (illustrated in Fig. 4A). This yields an oligoclonal, CDR3-unfocused but Vγ9Vδ2-TCR-enriched population, which remains the most commonly utilized means of γδ T cell activation for immunotherapeutic evaluation.

This BTN(L) interaction bears resemblance to how invariant NK-T (iNK-T) cells engage CD1d molecules loaded with glycolipids [70], or how αβ T cells can be polyclonally activated by bacterial superantigens [71]. In the case of αβ T cells, this interaction is mediated by the HV4 framework region located on the TCR-β chain. Similar to iNK-T and αβ T cells, the HV4-bearing γδ-TCR are also able to engage antigens clonally via their re-arranged CDRs. Melandri et al. showed that the human Vγ4Vδ1-TCR can engage both BTNL-3 and BTNL-8 via its germline Vγ4 HV4 domain and can also be activated via its CDRs 1–3 by sulfatide-loaded CD1d antigen-presenting molecules [65] (illustrated in Fig. 4B). The same paper identified that a number of mouse TCRs in intraepithelial lymphocytes also employ germline encoded HV4 regions to interact with mouse btnl family members. This dual mode of TCR engagement may explain how γδ-TCRs are capable of oligoclonal activation against stimuli of cellular "stress" and transformation while maintaining rearranged CDR-dependent reactivity to unique antigens. Examples of human γδ-TCR target antigens with confirmed CDR3 involvement include endothelial protein C receptor (CDR3γ and CDR3δ of a Vγ4Vδ5-TCR involved) [72], ephrin type-A receptor 2 (CDR3δ of a Vγ9Vδ1-TCR involved-published in abstract form) [73], and tRNA-synthetases (CDR3γ and CDR3δ of a Vγ3Vδ2-TCR involved) [74].

TCR signaling machinery

As αβ- and γδ-TCRs have an equal number of CD3 chains, which—despite differences in other areas—possess the same number of ITAMs (Fig. 3), both types of TCRs associate with 10 ITAMs in total. These can be phosphorylated to change the conformation of the complex and activate downstream signaling. In αβ T cells, following CD4 or CD8 docking onto a pMHC, the complementarity determining regions (CDRs) within the variable domains of the TCRα/β heterodimer bind to the presented peptide/MHC complex. The interaction, if of sufficient affinity, is then thought to mechanically extend the length of the TCR complex by about

a

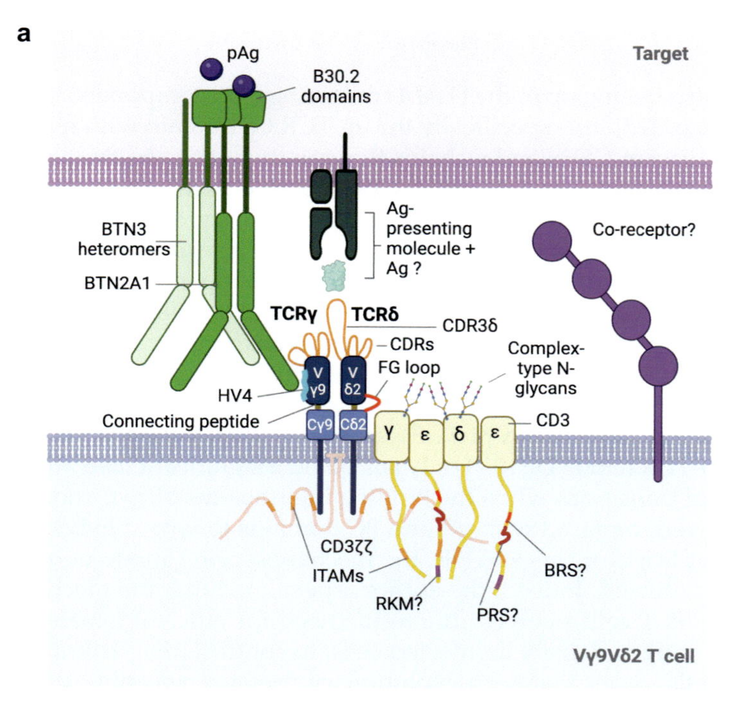

b

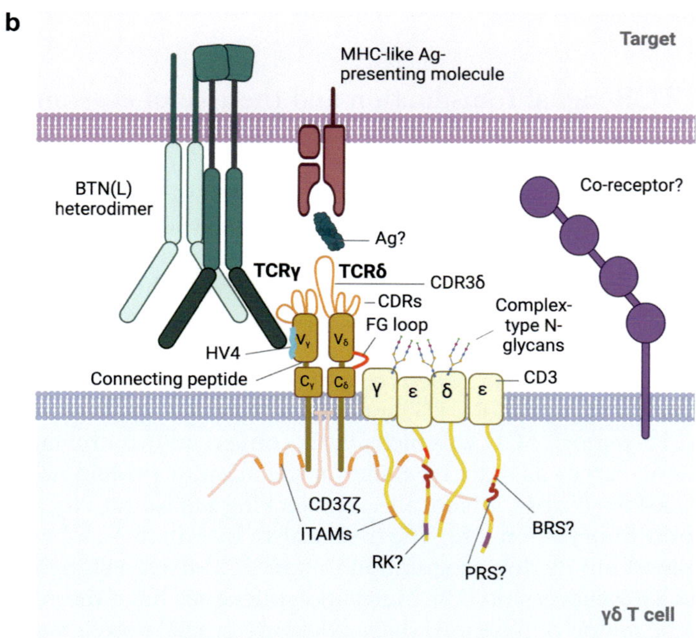

FIG. 4 Examples of γδ-TCR/antigen engagement models. Panel (A) illustrates the putative mode of antigen engagement by the peripherally enriched Vγ9Vδ2-TCR γδ T cell subset. While BTN2A1/BTN3 heteromer engagement by the TCR γ-chain has been described extensively, specific ligands or antigen-presenting moieties that engage the Vγ9Vδ2-TCR CDR3 regions have remained elusive. Panel (B) illustrates Vγ4Vδ1-TCR target engagement, for which both butyrophilin-like/HV4 interactions and CDR3/sulfatide antigen-loaded CD1d interactions have been described. CDRs: complementarity-determining regions; BRS: basic amino acid-rich sequence; RKM: receptor kinase motif; ITAMs: immunoreceptor tyrosine-based activation motif; PRS: proline-rich sequence; HV4: hypervariable region 4; Ag: antigen; MHC: major histocompatibility complex; BTN: butyrophilin; BTNL: butyrophilin-like.

19 nM, resulting in a freeing up of the ITAM containing CD3 components as downstream conformational changes [75]. It is specifically the αβ-TCR Cβ domain with its associated FG loop in concert with adjacent CD3γ and ε chains that are thought to be the primary mediators of this mechanosensing-driven release of the CD3ζ-chain ITAMs to a conformation permissive for their phosphorylation by lymphocyte-specific protein tyrosine kinase (Lck).

Mallis and colleagues evaluated models of signal transduction mediated by αβ- and γδ-TCR. Molecular domain simulations showed that antigen binding to the CDRs on the Vβ chain of αβ T cells leads to a force on the TCR, thereby unfolding the FG loop, which connects the Vβ and Cβ regions [51]. The force was then transferred to the Cβ chain-adjacent CD3γ and CD3ε chains, pushing their cytoplasmic tails downward and exposing them to downstream kinases which effect phosphorylation. Remarkably, the team showed that the same was not true for a concrete human γδ-TCR. Specifically, this sulfoglycolipid sulfatide-loaded CD1d-reactive γδ-TCR (clone DP10.7) only sustained a significant load and underwent force-induced structural transitions when the binding interface-distal γδ C module comprising γC and δ_C domains were replaced with αC and βC domains to form a hybrid TCR. The authors concluded that, at least for this γδ-TCR, the mechanosensing mode of αβ T cells for ligand recognition is not utilized. Indeed, the authors speculated that the mechanosensing mechanism in adaptive αβ T cells evolved due to the need for efficient expansion against targets presented at low antigen density in infected cells. In contrast, the plethora of other innate receptors in γδ T cells meant that such evolutionary pressure was not exerted on the γδ-TCR [51]. An alternative mechanism for γδ-TCR signal transduction has not yet been established.

Early TCR signal transduction and the role of co-stimulation

In the αβ-TCR, once a pMHC has bound to the TCR with sufficient avidity, the cytoplasmic tails of the CD3 chains become accessible for phosphorylation. The CD3ε chain is key to initial interaction with the Src family kinase Lck which is the most commonly employed Src family member to initiate phosphorylation of ITAMs. CD3ε contains a number of unique interaction motifs that play a central role in early αβ-TCR signaling. From the N- to the C-terminus, these include a basic amino acid-rich sequence (BRS), a proline-rich sequence (PRS), and a receptor kinase (RK) motif. Along with the BRS and single ITAM of the CD3ε chain, the RK motif is only exposed in the conformationally active state following αβ-TCR engagement. Following antigen binding, the BRS [76] and RK [77] interact with early signaling mediator, Lck, via its SH3 domain. The PRS region of CD3ε, which also becomes conformationally available following TCR engagement, serves as a docking site for the adaptor protein noncatalytic region of adaptor protein Nck-beta (Grb4), which, although lacking kinase activity, is a requirement for full CD3ε ITAM phosphorylation [78]. The mechanism by which Nck-beta augments signaling poorly understood but its downregulation impair TCR-driven signaling and subsequent reduced IL-2 and CD69 expression [79]. Hartl and colleagues have recently shown that both Lck and noncatalytic region of tyrosine kinase (Nck) are crucial for early αβ-TCR signal propagation [80], with the role of Nck-beta being to stabilize Lck recruitment to the exposed CD3ε chain.

Nck is further involved in the downstream phosphorylation of protein kinases such as ZAP70 [80], acting in concert with the Lck protein tyrosine kinase to initiate phosphorylation cascades during proximal TCR signaling. Gil et al. have showed additional means by which the recruitment of Nck is required for the full activation of αβ T cells by acting on downstream effectors such as WASP, which is important for actin reorganization [78].

This same exposure of a PRS in the CD3ε cytoplasmic tail may play a different role in γδ-TCR activation. For example, in murine γδ T cells, exposed to known stimulatory antigens, a PRS mutation negatively impacted signal transduction [81]. In contrast, Dopfer and co-workers demonstrated that in both mouse G8 and human Vγ9Vδ2-TCR, ligand engagement did not mediate typical CD3 conformational change to expose PRS, BRS, or RK motifs [82]. Given the multimodal way in which γδ-TCRs can engage antigens (via their HV4 and CDR1–3 domains), and the broad range of ligands that they can respond to, it is nonetheless impossible to exclude CD3ε chain conformational changes, leading to exposure of the ITAMs and regulatory motifs as key mediators of signal propagation in all circumstances.

Full αβ T cell activation is achieved via productive TCR engagement, as well as of co-stimulatory molecules such as CD28, which act to alter to quantitative and qualitative aspects of TCR downstream signaling, through engagement of downstream effectors such as phospholipase C-γ (PLC-γ) [42]. In αβ T cells, naïve T cell activation via the TCR in the absence of co-stimulation is indicative of peptide recognition in the absence of professional antigen presentation and implies pathological autoreactivity. TCR activation in the absence of co-stimulation can, therefore, cause the activated T cell to undergo induction of anergy or apoptosis. This mechanism of self-censorship illustrates the importance of the qualitative aspects of αβ-TCR signaling and forms a pillar of maintaining peripheral tolerance. The provision of co-stimulation to αβ-TCR signaling is known as "licensing" and is a prerequisite for canonical T cell activation, effector function, and proliferation.

Given that the preponderance of known γδ-TCR clones is not autoreactive [58], it is tempting to speculate that a similar "licensing" safety switch does not operate in γδ T cells. In γδ T cells, there is less evidence for the role of co-stimulation to avoid anergy during priming, but the importance of dendritic cells for the initiation of γδ T cell responses has been documented [83,84]. Nonetheless, significant safety constraints on human γδ T cell responses have been placed, albeit via a different route. αβ T cells that have been activated via their TCR in the presence of licensing co-stimulation undergo significant clonal expansion and differentiate into potent long-term effectors. The clonal proliferation is self-sustained by αβ T cell production of common γ-chain cytokines like IL-2, which supports T cell proliferation in an autocrine and paracrine manner [85]. The same is not true for human γδ T cells. While able to become activated in the absence of co-stimulatory licensing, data suggest that γδ T cells are incapable of manufacturing proliferation-supporting common γ-chain cytokines, whereby their proliferation collapses upon withdrawal of exogenous cytokine support [86]. In situ, this likely serves to limit their functionality to the specific circumstance where external sources of common γ-chain cytokines are available. While IL-2 is primarily derived from delayed-onset adaptive lymphocyte responses, proliferation-supporting common γ-chain cytokines such as IL-7 and IL-15 are produced by stromal cells of various lymphoid organs for the former and tissue-resident macrophages and DCs for the latter [87,88].

Although priming and co-stimulatory requirements for expansion of "naïve" γδ T cell are incompletely understood, the differences with the biology of αβ-TCR signal transduction are supportive of the hypothesis of Mallis and colleagues [51] that posits the γδ-TCR as a structurally distinct receptor from the αβ-TCR, evolved thus to fulfill a distinct purpose.

Specifically, that, in contrast to the αβ-TCR, the γδ-TCR serves a role akin to a pattern-recognition receptor, evolved to become activated in the presence of frequently encountered evolutionarily conserved molecular patterns that are associated with cellular stress, transformation, or intracellular infection. Given that these patterns are so conserved, restrictive licensing by professional antigen-presenting cells may not be required, with γδ T cell responses being checked only at subsequent stages of the response by provision or deprivation of proliferation-sustaining cytokines. Consistent with this, human and murine γδ T cell effector function in the absence of TCR engagement has been described in various settings, for example, by cross-linking the activatory IgG opsonin receptor, CD16, cross-linking natural killer cell (NK)-associated receptor NKG2D, or exposure to IL-12 and IL-18 cytokines [89–94]. This suggests that γδ-TCR activation is not tightly regulated by other immune cells but also that it is not uniquely essential for γδ T cell activation, acting instead as one of a plethora of ways in which a γδ T cell can engage effector function. This paints γδ T cell activation as far more akin to that of NK cells, the engagement of which depends on a wide variety and balance of stimulatory and inhibitory inputs [14,95], than to αβ T cells for which productive TCR signaling is an obligate node for full effector function. Consistent with these concepts, although γδ T cells can play a contributory role in autoimmunity, there is no evidence that they can be drivers of autoimmune pathology, again keeping with the different processes of T cell priming and selection compared with αβ T cells [96–98].

On the one hand, the optionality of TCR engagement for γδ T cell function is a potential advantage for their use as an immunotherapeutic chassis, allowing for their activation in response to a broad range of evolutionarily conserved "danger signals" that are likely to be present in a majority of malignancies. On the other hand, this comes at the expense of their ability to sustain an ongoing immune response in the absence of exogenous cytokine support. In the context of immunotherapy design for cancer, therapeutic persistence and proliferation by adoptively transferred γδ T cells, therefore, may necessitate a reliance on the patient's own body to produce the required proliferation-supporting cytokines, the infusion of off-the-shelf cytokines into the patient, or synthetic engineering of the γδ T cells to force their own production of said cytokines.

TCR signal propagation

Following productive TCR engagement, an activatory signal is communicated from the membrane-embedded CD3 chains to secondary and tertiary membrane-proximal signaling molecules downstream of the TCR. The canonical αβ-TCR signaling pathway is far better understood than that of the γδ-TCR pathway. Membrane-proximal Lck and Nck downstream signaling in αβ T cells is propagated by the recruitment of Src-family protein kinases (SFKs), which include Src, Lck, and Fyn as well as engagement of the adapter protein Nck [99,100]. There is a degree of functional redundancy between the SFKs for phosphorylation of CD3 ITAMs. For example, $Lck^{-/-}$ mice are incapable of fully fledged T cell activation, whereas, $Fyn^{-/-}$ animals do have T cells capable of activation by way of compensatory signal mediation from alternative SFKs [57].

Lck is brought into proximity with the αβ-TCR/CD3 complex by CD4 or CD8 co-receptor to which it is noncovalently bound docking onto the same pMHC. Lck associates with CD4 and CD8 cytoplasmic tails via its N-terminal region. The enhanced accessibility of CD3 ITAMs following TCR ligation, as envisioned through the mechanosensing hypothesis, results in their accessibility to Lck [101]. Lck is then able to phosphorylate all 10 of the CD3-associated ITAMs. This then allows the Syk family protein tyrosine kinase ZAP-70 to bind to the ITAMs via its SH2 domain and is itself phosphorylated at the tyrosine residues by Lck [42]. Thereafter, ZAP-70 phosphorylates two key scaffold proteins: linker for T cell activation (LAT) and Src-homology-containing 76-kDa leukocyte protein (SLP76) [42]. A simplified map of the main membrane-proximal signaling events in αβ-TCR activation is shown in Fig. 5.

These scaffold proteins recruit further downstream effector molecules in a specific order and location to co-ordinate several more-or-less overlapping downstream signaling pathways. The key pathways are [1] activation of PLC-γ which induces calcium influx, [2] activation of phosphoinositide 3-kinase (PI3K) with its subsequent signaling cascade, [3] activation of VAV to enhance motility through F-actin recruitment, and [4] activation of ADAP leading to greater integrin-mediated T cell adhesion. PLC-γ binds to phosphorylated LAT in order to hydrolyze membrane-bound phosphatidylinositol 4,5-bisphosphate (PIP2) into the secondary messengers diacylglycerol (DAG) and inositol-3-phosphate (IP_3) [42].

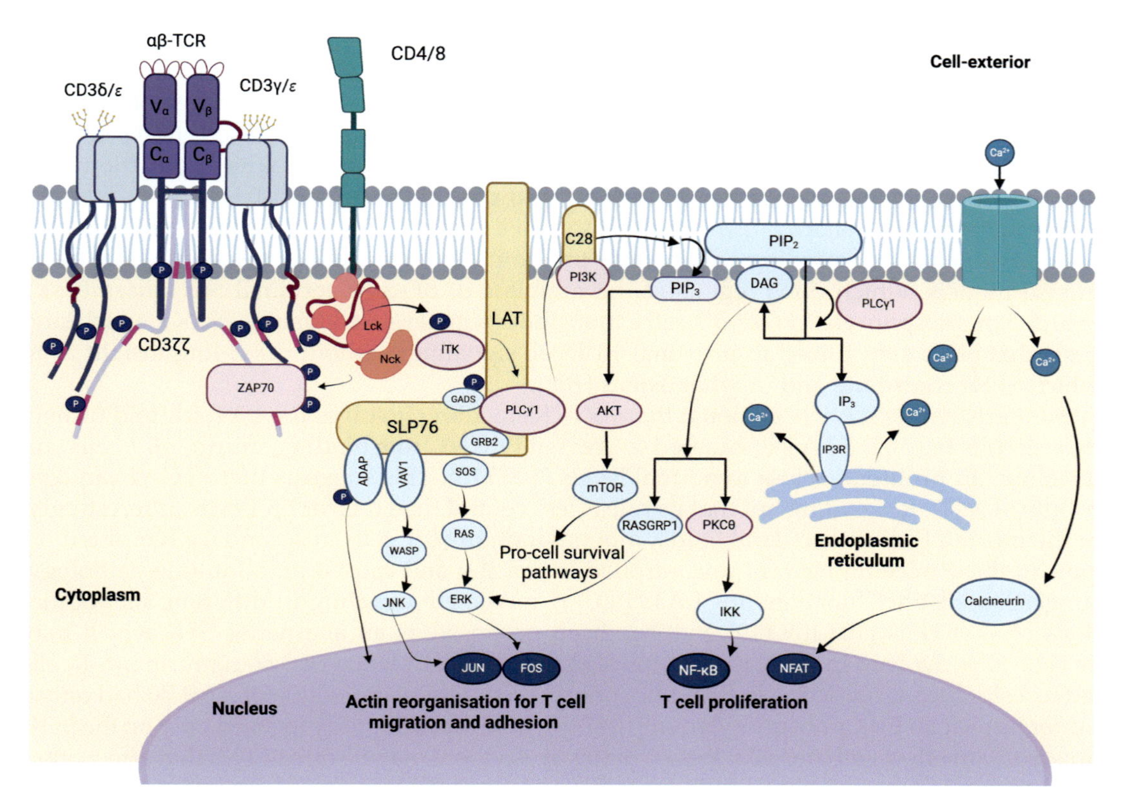

FIG. 5 The early, membrane-proximal signaling events of αβ-TCR engagement.

These have distinct downstream effects (as shown in Fig. 5) but ultimately serve to induce calcium influx with consequent NFAT transcription factor activation via the calmodulin pathway, which activates genes implicit in αβ T cell activation, effector function, and proliferation. One of the earliest downstream functions of this signaling acts via VAV on the actin cytoskeletal network, which undergoes a dramatic reorganization to polarize the cell prior to migration [102].

Many of these elements in the γδ-TCR activation cascade remain ambiguous, with key effectors unknown [103]. Muro and colleagues propose that the γδ-TCR can signal through two different pathways, depending on whether the TCR is engaged with an activatory ligand with sufficient avidity, though much of the data underpinning this hypothesis derives from mouse models. At rest (when not in contact with an activating ligand), weak signals from the murine TCR direct γδ T cells to produce interleukin-17 (IL-17), which is replaced with IFN-γ production upon full TCR engagement with complementary ligand [104]. In contrast to αβ-TCR signaling, where ZAP70 is the main signaling mediator following phosphorylation of ITAM tyrosines, Syk tyrosine kinases were found to be necessary and irreplaceable for γδ TCR signal transduction. The group therefore proposed that Syk, rather than ZAP-70, is of primary importance for early γδ T cell TCR signaling. The degree to which this is true for human γδ T cells remains unclear, especially given that a similar IL-17/IFN-γ phenotypic dichotomy does not appear to exist in humans.

It appears that the majority of the differences in αβ- and γδ-TCR signaling lie in the membrane-proximal events, with relative convergence at the downstream membrane-distal events [105]. This is often true for immune cell signaling at large, with a range of different receptor inputs at the surface of a range of cell types converging onto transcriptional effector pathways that mediate the basic immune functions of cell motility, cytokine production, proliferation, and, in specialist cases, cytotoxicity and phagocytosis [106,107].

For instance, the majority of γδ T cell subsets lack CD4 or CD8 co-receptors, which in αβ T cells are crucial for providing Lck in close proximity to engaged TCRs. Experiments using mouse models, which downregulated the inhibitor of SFKs, C-terminal Src kinase (Csk), resulted in phosphorylation of γδ T cell extracellular signal related kinases (ERK), which suggests that SFKs are utilized in proximal γδ T cell signaling, without providing insight as to which SFKs contribute and to what extent [108].

Similarly, while it has been noted that γδ-TCR signaling involves the LAT scaffold protein that recruits PLC-γ1, if LAT is mutated to prevent PLC-γ1 from binding, murine γδ T cells still develop and proliferate close to normal levels [109,110]. This suggests that PLC-γ1 can contribute to γδ-TCR signaling but that its role is less central than it is for αβ T cells. Interestingly, in the murine PLC-γ1 knockout study, a population of γδ T cells in secondary lymphoid organs in these mice underwent uncontrolled expansion and caused autoimmune pathology, suggesting distinct functions for LAT/PLC-γ1-mediated signaling in different subpopulations of γδ T cells [103,109,110]. Syk has been proposed as an alternative SFK which may be centrally involved in γδ TCR proximal signaling [104]. Deletion of Syk in murine γδ T cells led to a significant reduction in ERK phosphorylation, whereas deletion of ZAP-70 had only a minor impact on ERK phosphorylation [104]. We note that this study focused only on the SFK-downstream effects on the MAP kinase pathway, so there may be other SFKs that affect other pathways. We note also that similar data for human γδ-TCR membrane-proximal signaling are yet to be generated. A map of the putative early signaling events in γδ-TCR activation proximal to the cell membrane is shown in Fig. 6.

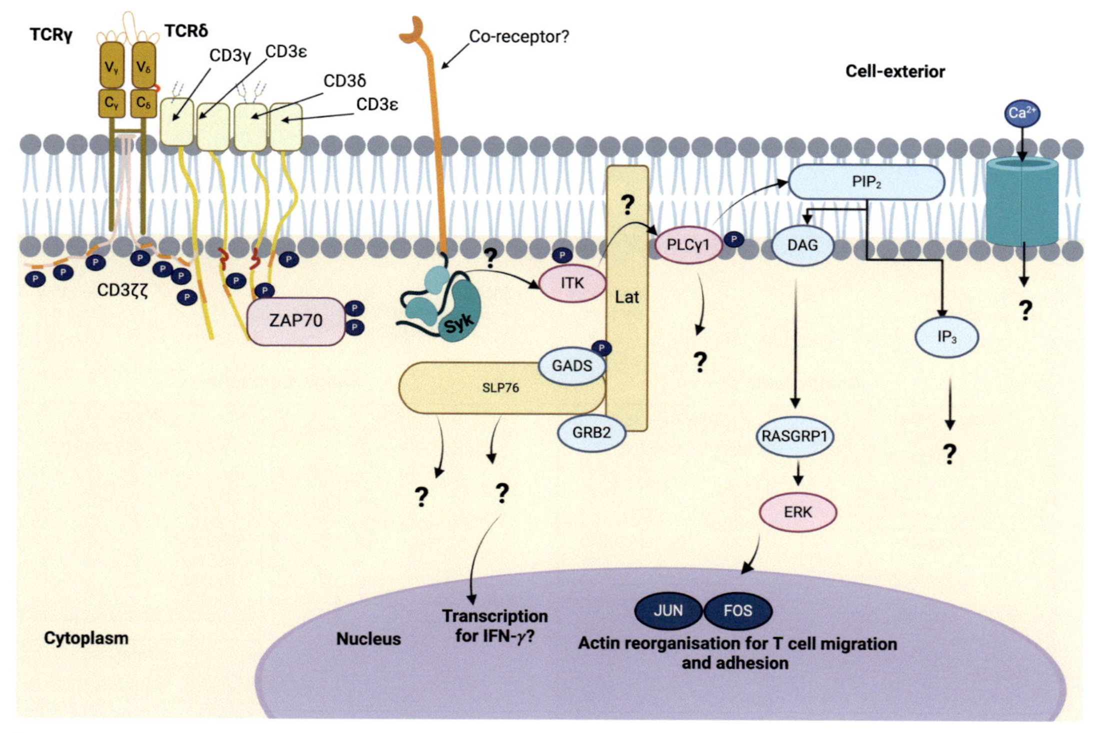

FIG. 6 The knowns and unknowns of the early, membrane-proximal signaling events of γδ-TCR engagement.

The TCR-mediated immune synapse

If a TCRα/β-pMHC interaction is of sufficient avidity, the resulting conformational changes and early signaling events described earlier will result in "inside-out" signaling that will strengthen and stabilize the forming immune synapse. This is primarily mediated by adhesion molecules on the T cell, such as integrins, which bind to receptors on the target cell and pull it closer, eventually establishing the classic "bullseye" shape of a stabilized immune synapse.

The synapse "bullseye" center consists of concentric circles of αβ-TCR/CD3 complexes, CD4 or CD8, and co-stimulatory receptors (such as CD28) on the T cell and corresponds to clusters of pMHC on the target cell side (Fig. 7A). This region of the synapse on the T cell is known as a central supramolecular activation complex (cSMAC) [111]. The cSMAC is surrounded by the secretory domain into which CTLs exocytose cytokines, perforins, and granzymes, in a manner that is directed by Lck signal-dependent centrosome docking onto the T cell membrane [112,113]. Surrounding the cSMAC is the peripheral SMAC (pSMAC), which contains proteins involved in cell adhesion, such as integrin LFA-1, talin, and ICAM1. Membrane-proximal intracellular signaling mediators and domains associated with the cSMAC are Lck, PKC-θ, SHP-2, and co-stimulatory receptor (such as CD28 or ICOS) endodomains. In addition to activatory co-receptors, the αβ T cell cSMAC can also contain co-inhibitory receptors such as CTLA-4 and PD-1, which dampen T cell activation [114].

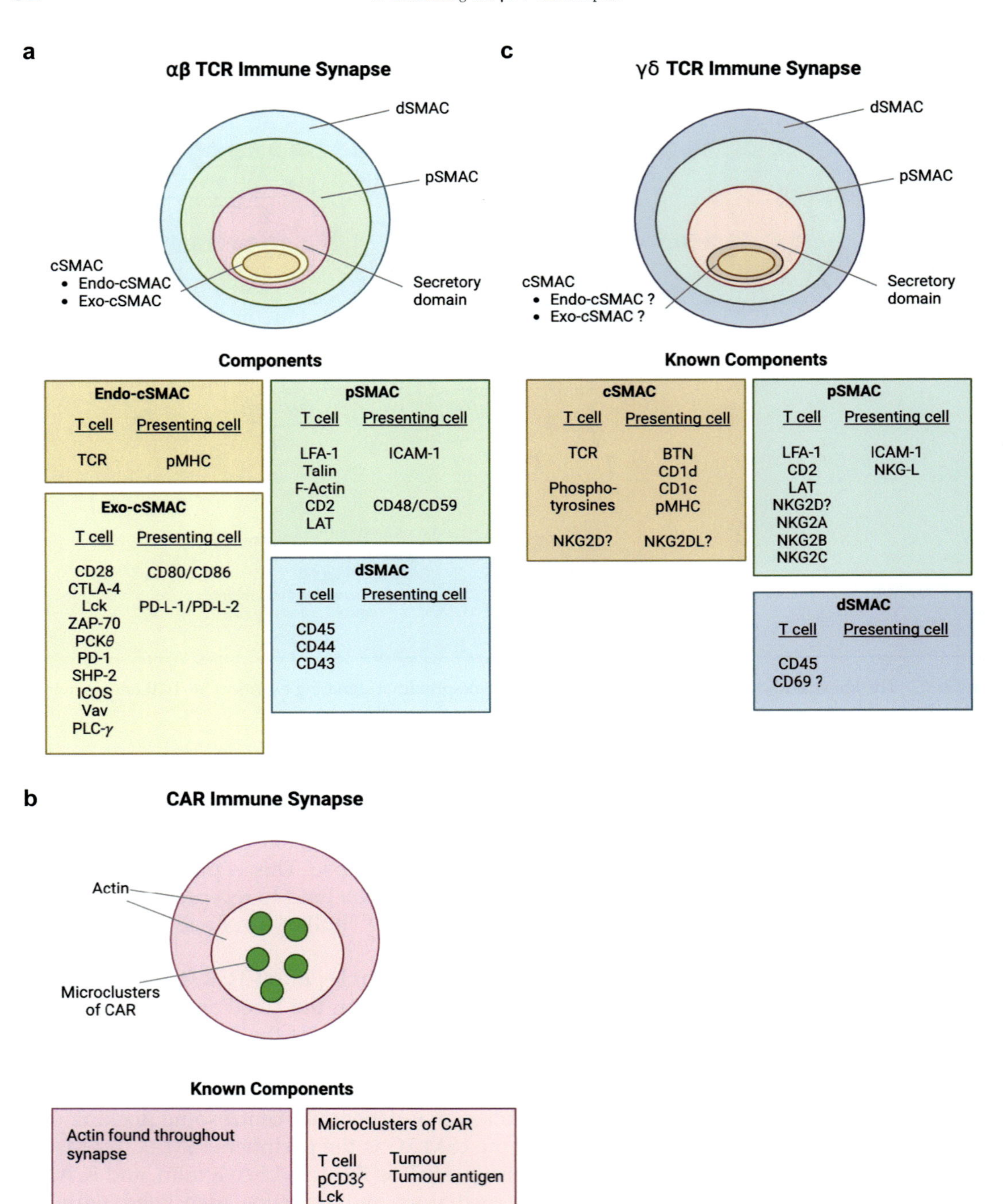

FIG. 7 The structure of αβ T cell, γδ T cell, and chimeric antigen receptor (CAR) immune synapses. (A) The αβ T cell synapse is by far the best understood T cell synapse. (B) The CAR synapse is poorly understood compared to the natural T cell receptor (TCR)-governed synapse. Data indicate that its organization differs substantially from TCR synapses. (C) Relative understanding of the γδ-TCR cell synapse is poorer still, with little human data available. The illustrated existence of the γδ T cell *endo-* and *exo-*cSMACs in this diagram is speculative.

Physically larger cell surface molecules, such as CD43 and inhibitory phosphatase CD45, are excluded from the pSMAC and make up the distal SMAC (dSMAC), thereby removing inhibitory effects that normally limit T cell tonic signaling [115].

Interestingly, while data are sparse, the αβ-CAR/target cell synapse is thought not to resemble the αβ-TCR/pMHC immune synapse much. Compared to the highly organized natural immune TCR synapse, CAR/target immune synapses appear to consist largely of microclusters of CARs and Lck and correspond to respective antigen microclusters on target cells, all surrounded by a disorganized and diffuse ring of actin [115] (Fig. 7B). While the CAR immune synapse is likely to vary between different CAR designs, nature of target cells, and types of antigens targeted, the fact that CAR synapses do exist and can mediate potent, sustained, and therapeutically efficacious T cell functionality suggests that different types of T cell/target synapses can be productive.

Published literature on the γδ T cell immune synapse is sparse and often examines mouse rather than human γδ T cells. For example, Mantri et al. described that murine cells form immune synapses not only with virally infected target cells but, as importantly, with other immune cells (mast cells, in this case), to orchestrate the local immune response [116]. One of the few studies that has examined human γδ T cell immune synapses with targets examined the interaction between Vγ9Vδ2 cells and THP-1 myelomonocytic cancer cell line. While fine resolution of the synapse was lacking, the group was able to establish that the synapse was characterized by the enrichment of phosphotyrosines, CD2, and the γδ-TCR together, with exclusion of CD45. The CD94 and NKG2D receptors were also recruited to the synapse, while CD69 was excluded [117]. Of note, this synapse formed even when THP-1 cells were untreated with exogenous pAg, while Vγ9Vδ2 cell intracellular signaling and effector function against THP-1 cells was observed only when the targets were treated with pAg. The authors concluded, "[h]ence the molecular interactions at the γδ T cell/THP-1 target cell interface are sufficient to induce the formation of an [immune synapse], but cytokine production requires the full engagement of γδ-TCR by a strong agonist. Thus in γδ T cells, formation of the [immune synapse] is uncoupled from its functional outcome." [117]

Another, more recent, study demonstrated that lung cancer cell line treatment with DNA methyltransferase inhibitor drugs upregulated tumor cell expression of various human γδ T cell-activating ligands (including MIC-A, MIC-B, ULB-P2, and ULB-P3), which have been described as both γδ T cell NKG2D receptor and γδ-TCR targets [58]. Upregulation of these ligands enhanced Vδ1 γδ T cell synapse formation with the lung cancer cell lines in a manner that involved ICAM-1, but failed to explicitly implicate the TCR/CD3 complex as the primary ITAM signal provider. Nonetheless, the synapse was associated with enrichment for LAT, phosphotyrosinases and LFA-1 integrins [118]. This finding once again points to the multimodal way in which γδ T cells can be activated, in a manner where it is impossible to argue that the TCR is more important to activatory signaling than a "co-stimulatory" receptor like NKG2D. The γδ T cell surface receptor that was found to be uniquely crucial to the formation of the synapse in this interaction was not the TCR nor NKG2D, but the adhesion molecule, ICAM-1 [118]. A similar enhancement of the number and strength of Vδ1 γδ T cell/tumor target cell synapses was observed upon mantle cell lymphoma target pretreatment with immunomodulatory drug, lenalidomide, putatively by the upregulation of CD1c and adhesion molecule expression on the tumor cell surface, though detailed synapse characterization was lacking [119]. We have summarized the literature on the human γδ T cell/tumor target immune synapse in Fig. 7C.

Discussion and conclusion

The herein review summarizes what is known about γδ-TCR structure and signaling, relative to its better-known αβ-TCR counterpart. Review of the relevant literature leads us to draw two main conclusions. The first is that our understanding of the γδ-TCR structure and function remains poor, despite decades of research and a surge of recent interest in γδ T cell-based adoptive immunotherapy for cancer. Second, data about γδ-TCR signaling that have emerged to date point to significantly different properties and, possibly, a different evolutionary purpose than the canonical αβ-TCR.

In broad terms, pMHC-engaging αβ-TCRs found on peripheral T cells are receptors designed to recognize rare peptide antigens with exquisite CDR3-mediated precision. Once these antigens are engaged, productive αβ-TCR signaling leads to an explosive and self-sustaining (via cytokine production) primary acute response against these antigens, and the formation of immunological memory to enable a potential secondary acute response, as appropriate. Because of the sensitivity and potency of these responses, αβ-TCR signaling and αβ T cell activation in general are tightly regulated via a number of routes, including central and peripheral tolerance mechanisms, a requirement for professional antigen presentation and co-stimulation. These layers of regulation act as safety checks on αβ T cell activation and reduce the likelihood of αβ-TCR activation in inappropriate settings or against inappropriate (e.g., "self") antigens.

While much remains unknown as to how this paradigm maps onto human γδ T cell signaling, current mechanistic understanding of the γδ-TCR suggests a different role and regulation of activity. Similar to NK cells, γδ T cells express a range of pattern recognition receptors that engage conserved patterns of cellular stress that are consistent with, among other things, infection and malignant transformation. These receptors include invariant receptors such as NKGD2, DNAM-1, natural cytotoxicity receptors (NCRs), and toll-like receptors (TLRs), as well as FcRγ that enable responses against opsonized targets. We postulate that, in terms of function and regulation, what is known about the γδ-TCR aligns it more closely to these pattern recognition receptors than the αβ-TCR. Since γδ-TCRs can be activated by commonly encountered, evolutionarily conserved markers of "stress," there is less stringent requirement for tight regulation of TCR signaling. This may be one of the reasons why obligate professional antigen presentation and a requirement for co-stimulatory receptor signaling have not been described for the γδ-TCR in the way that they have been for the αβ-TCR.

The possible implications of this for γδ T cell engineering in the context of immunotherapy for cancer are manifold. For one, the role of co-stimulatory domains in CAR construct design for use in γδ T cells requires empirical evaluation; it is conceivable that αβ-CAR-derived 28ζ or 4-1BBζ designs are not optimal for use in γδ T cells. It is, moreover, important to determine whether a classical ζ-chain ITAM design is best suited for use in γδ T cells. Given the mechanistic differences in γδ-TCR versus αβ-TCR activation, it is possible that the synthetic CAR ζ-chain-mediated activation provides a signal that is either too strong or too weak for naturalistic γδ T cell activation. Indeed, if the γδ-TCR is designed to recognize commonly occurring and densely encountered ligands of cellular stress, it may be the case that γδ T cell immunotherapy would benefit from synthetic constructs that mimic this type of interaction. For example, instead of expressing an αβ T cell-optimized high-avidity CAR that mediates explosive activation against a range of antigen densities via its synthetic ζ and co-stimulatory domains, perhaps a γδ T cell CAR would function more naturalistically if pitted against densely expressed antigens with lower avidity interactions, possibly in a manner that does not require synthetically derived co-stimulatory signals. Perhaps CAR-γδ cells do not require synthetic ζ endodomains at all and can function optimally relying on other stress-recognition receptor domains, including alternative sources of ITAM signaling such as the adaptor domains of DAP12 and FcRγ.

The inability of most human γδ T cells to self-sustain proliferation via common γ-chain cytokine secretion limits the response duration of any adoptively transferred γδ T cell product. This, therefore, urges further consideration of re-infusion, cytokine armoring of the product using gene-engineering, or therapeutic co-infusion of exogenous cytokines as strategies to prolong therapeutic efficacy. The requirement for this therapeutic augmentation might be a key difference between engineered γδ and αβ T cell products.

Taken together, the differences between αβ-TCR and γδ-TCR signaling, combined with the paucity of knowledge surrounding γδ-TCR activation, form a substantial barrier to the informed design of engineered γδ T cell products for use in cancer immunotherapy. Nonetheless, this divergence also offers an opportunity to develop γδ T cell immunotherapies with properties distinct from those constructed using αβ T cells, using new and subset-specific engineering approaches.

References

[1] Novartis presents results from first global registration trial of CTL019 in pediatric and young adult patients with r/r B-ALL. Novartis; 2016. https://www.novartis.com/news/media-releases/novartis-presents-results-from-first-global-registration-trial-ctl019-pediatric-and-young-adult-patients-rr-b-all.

[2] Schuster SJ, et al. Long-term clinical outcomes of tisagenlecleucel in patients with relapsed or refractory aggressive B-cell lymphomas (JULIET): a multicentre, open-label, single-arm, phase 2 study. Lancet Oncol 2021;22:1403–15.

[3] Neelapu SS, et al. Axicabtagene Ciloleucel CAR T-cell therapy in refractory large B-cell lymphoma. N Engl J Med 2017;377:2531–44.

[4] Neelapu SS, et al. Five-year follow-up of ZUMA-1 supports the curative potential of axicabtagene ciloleucel in refractory large B-cell lymphoma. Blood 2023;141:2307–15.

[5] Cappell KM, Kochenderfer JN. Long-term outcomes following CAR T cell therapy: what we know so far. Nat Rev Clin Oncol 2023;20:359–71.

[6] Labanieh L, Mackall CL. CAR immune cells: design principles, resistance and the next generation. Nature 2023;614:635–48.

[7] Torre LA, Siegel RL, Ward EM, Jemal A. Global Cancer incidence and mortality rates and trends—an update. Cancer Epidemiol Biomarkers Prev 2016;25:16–27.

[8] Miller KD, et al. Cancer treatment and survivorship statistics, 2022. CA Cancer J Clin 2022;72:409–36.

[9] Jemal A, Center MM, DeSantis C, Ward EM. Global patterns of cancer incidence and mortality rates and trends. Cancer Epidemiol Biomarkers Prev 2010;19:1893–907.

[10] Melenhorst JJ, et al. Decade-long leukaemia remissions with persistence of CD4+ CAR T cells. Nature 2022;602:503–9.

[11] Cliff ERS, et al. High cost of chimeric antigen receptor T-cells: challenges and solutions. Am Soc Clin Oncol Educ Book 2023;e397912. https://doi.org/10.1200/EDBK_397912.

[12] Qasim W. Allogeneic CAR T cell therapies for leukemia. Am J Hematol 2019;94:S50–4.

[13] Qasim W. Genome-edited allogeneic donor "universal" chimeric antigen receptor T cells. Blood 2023;141:835–45.

[14] Laskowski TJ, Biederstädt A, Rezvani K. Natural killer cells in antitumour adoptive cell immunotherapy. Nat Rev Cancer 2022;22:557–75.

[15] Marin D, et al. Safety, efficacy and determinants of response of allogeneic CD19-specific CAR-NK cells in CD19+ B cell tumors: a phase 1/2 trial. Nat Med 2024;1–13. https://doi.org/10.1038/s41591-023-02785-8.
[16] Dolgin E. Unconventional γδ T cells 'the new black' in cancer therapy. Nat Biotechnol 2022;40:805–8.
[17] Mensurado S, Blanco-Domínguez R, Silva-Santos B. The emerging roles of γδ T cells in cancer immunotherapy. Nat Rev Clin Oncol 2023;1–14. https://doi.org/10.1038/s41571-022-00722-1.
[18] Morvan MG, Lanier LL. NK cells and cancer: you can teach innate cells new tricks. Nat Rev Cancer 2016;16:7–19.
[19] Wo J, et al. The role of Gamma-Delta T cells in diseases of the central nervous system. Front Immunol 2020;11. https://doi.org/10.3389/fimmu.2020.580304.
[20] Davey MS, et al. Clonal selection in the human Vδ1 T cell repertoire indicates γδ TCR-dependent adaptive immune surveillance. Nat Commun 2017;8.
[21] Ho M, et al. Polyclonal expansion of peripheral gamma delta T cells in human plasmodium falciparum malaria. Infect Immun 1994;62:855–62.
[22] Gavriil A, Barisa M, Halliwell E, Anderson J. Engineering solutions for mitigation of chimeric antigen receptor T-cell dysfunction. Cancer 2020;12:2326.
[23] Romero P, Speiser DE, Rufer N. Deciphering the unusual HLA-A2/Melan-a/MART-1-specific TCR repertoire in humans. Eur J Immunol 2014;44:2567–70.
[24] Olivier T, Haslam A, Tuia J, Prasad V. Eligibility for human leukocyte antigen–based therapeutics by race and ethnicity. JAMA Netw Open 2023;6, e2338612.
[25] Sterner RC, Sterner RM. CAR-T cell therapy: current limitations and potential strategies. Blood Cancer J 2021;11:69.
[26] Salzer B, et al. Engineering AvidCARs for combinatorial antigen recognition and reversible control of CAR function. Nat Commun 2020;11:4166.
[27] Hirabayashi K, et al. Dual-targeting CAR-T cells with optimal co-stimulation and metabolic fitness enhance antitumor activity and prevent escape in solid tumors. Nat Cancer 2021;2:904–18.
[28] Asmamaw Dejenie T, et al. Current updates on generations, approvals, and clinical trials of CAR T-cell therapy. Hum Vaccin Immunother 2022;18:2114254.
[29] Zhang C, Liu J, Zhong JF, Zhang X. Engineering CAR-T cells. Biomark Res 2017;5.
[30] Tomasik J, Jasiński M, Basak GW. Next generations of CAR-T cells—new therapeutic opportunities in hematology? Front Immunol 2022;13:1034707.
[31] Eshhar Z, Waks T, Gross G, Schindler DG. Specific activation and targeting of cytotoxic lymphocytes through chimeric single chains consisting of antibody-binding domains and the gamma or zeta subunits of the immunoglobulin and T-cell receptors. Proc Natl Acad Sci USA 1993;90:720–4.
[32] Research, C. for B. E. KYMRIAH (tisagenlecleucel). FDA; 2022.
[33] Research, C. for B. E. YESCARTA (axicabtagene ciloleucel). FDA; 2022.
[34] Research, C. for B. E. TECARTUS (brexucabtagene autoleucel). FDA; 2023.
[35] Research, C. for B. E. BREYANZI (lisocabtagene maraleucel). FDA; 2022.
[36] Research, C. for B. E. ABECMA (idecabtagene vicleucel). FDA; 2021.
[37] Research, C. for B. E. CARVYKTI. FDA; 2023.
[38] Mahadeo KM, et al. Tabelecleucel for allogeneic haematopoietic stem-cell or solid organ transplant recipients with Epstein–Barr virus-positive post-transplant lymphoproliferative disease after failure of rituximab or rituximab and chemotherapy (ALLELE): a phase 3, multicentre, open-label trial. Lancet Oncol 2024;25 (3):376–87. https://doi.org/10.1016/S1470-2045(23)00649-6.
[39] Keam SJ. Tabelecleucel: first approval. Mol Diagn Ther 2023;27:425–31.
[40] Research, C. for B. E. FDA grants accelerated approval to lifileucel for unresectable or metastatic melanoma. FDA; 2024.
[41] Chesney J, et al. Efficacy and safety of lifileucel, a one-time autologous tumor-infiltrating lymphocyte (TIL) cell therapy, in patients with advanced melanoma after progression on immune checkpoint inhibitors and targeted therapies: pooled analysis of consecutive cohorts of the C-144-01 study. J Immunother Cancer 2022;10, e005755.
[42] Shah K, Al-Haidari A, Sun J, Kazi JU. T cell receptor (TCR) signaling in health and disease. Sig Transduct Target Ther 2021;6:1–26.
[43] Allison JP, Lanier LL. Identification of antigen receptor-associated structures on murine T cells. Nature 1985;314:107–9.
[44] Allison TJ, Winter CC, Fournié J-J, Bonneville M, Garboczi DN. Structure of a human γδ T-cell antigen receptor. Nature 2001;411:820–4.

[45] Sušac L, et al. Structure of a fully assembled tumor-specific T cell receptor ligated by pMHC. Cell 2022;185:3201–3213.e19.

[46] Mariuzza RA, Agnihotri P, Orban J. The structural basis of T-cell receptor (TCR) activation: an enduring enigma. J Biol Chem 2020;295:914–25.

[47] Wucherpfennig KW, Gagnon E, Call MJ, Huseby ES, Call ME. Structural biology of the T-cell receptor: insights into receptor assembly, ligand recognition, and initiation of signaling. Cold Spring Harb Perspect Biol 2010;2, a005140.

[48] Morath A, Schamel WW. αβ and γδ T cell receptors: similar but different. J Leukoc Biol 2020;107:1045–55.

[49] Touma M, et al. The TCR C beta FG loop regulates alpha beta T cell development. J Immunol 2006;176:6812–23.

[50] Sasada T, et al. Involvement of the TCR Cbeta FG loop in thymic selection and T cell function. J Exp Med 2002;195:1419–31.

[51] Mallis RJ, et al. Molecular design of the γδT cell receptor ectodomain encodes biologically fit ligand recognition in the absence of mechanosensing. Proc Natl Acad Sci USA 2021;118, e2023050118.

[52] Krangel MS, et al. T3 glycoprotein is functional although structurally distinct on human T-cell receptor gamma T lymphocytes. Proc Natl Acad Sci USA 1987;84:3817–21.

[53] Alarcon B, et al. The T-cell receptor gamma chain-CD3 complex: implication in the cytotoxic activity of a CD3+ CD4- CD8- human natural killer clone. Proc Natl Acad Sci USA 1987;84:3861–5.

[54] Deseke M, Prinz I. Ligand recognition by the γδ TCR and discrimination between homeostasis and stress conditions. Cell Mol Immunol 2020;17:914–24.

[55] Dietrich J, et al. Role of CD3 gamma in T cell receptor assembly. J Cell Biol 1996;132:299–310.

[56] Rossi NE, et al. Differential antibody binding to the surface alphabetaTCR.CD3 complex of CD4+ and CD8+ T lymphocytes is conserved in mammals and associated with differential glycosylation. Int Immunol 2008;20:1247–58.

[57] Hwang J-R, Byeon Y, Kim D, Park S-G. Recent insights of T cell receptor-mediated signaling pathways for T cell activation and development. Exp Mol Med 2020;52:750–61.

[58] Willcox BE, Willcox CR. γδ TCR ligands: the quest to solve a 500-million-year-old mystery. Nat Immunol 2019;20:121–8.

[59] Adams EJ, Strop P, Shin S, Chien Y-H, Garcia KC. An autonomous CDR3delta is sufficient for recognition of the nonclassical MHC class I molecules T10 and T22 by gammadelta T cells. Nat Immunol 2008;9:777–84.

[60] Bluestone JA, Cron RQ, Cotterman M, Houlden BA, Matis LA. Structure and specificity of T cell receptor gamma/delta on major histocompatibility complex antigen-specific CD3+, CD4-, CD8- T lymphocytes. J Exp Med 1988;168:1899–916.

[61] Bonneville M, et al. Recognition of a self major histocompatibility complex TL region product by gamma delta T-cell receptors. Proc Natl Acad Sci USA 1989;86:5928–32.

[62] Ito K, et al. Recognition of the product of a novel MHC TL region gene (27b) by a mouse γδ T cell receptor. Cell 1990;62:549–61.

[63] Wong WK, Leem J, Deane CM. Comparative analysis of the CDR loops of antigen receptors. Front Immunol 2019;10.

[64] Willcox CR, et al. Butyrophilin-like 3 directly binds a human Vγ4+ T cell receptor using a modality distinct from clonally-restricted antigen. Immunity 2019;51:813–825.e4.

[65] Melandri D, et al. The γδTCR combines innate immunity with adaptive immunity by utilizing spatially distinct regions for agonist selection and antigen responsiveness. Nat Immunol 2018;19:1352–65.

[66] Karunakaran MM, et al. Butyrophilin-2A1 directly binds germline-encoded regions of the Vγ9Vδ2 TCR and is essential for phosphoantigen sensing. Immunity 2020;52:487–498.e6.

[67] Rigau M, et al. Butyrophilin 2A1 is essential for phosphoantigen reactivity by γδ T cells. Science 2020;367: eaay5516.

[68] Barisa M, et al. *E. coli* promotes human Vγ9Vδ2 T cell transition from cytokine-producing bactericidal effectors to professional phagocytic killers in a TCR-dependent manner. Nat Sci Rep 2017;7. https://doi.org/10.1038/s41598-017-02886-8.

[69] Fichtner AS, et al. TCR repertoire analysis reveals phosphoantigen-induced polyclonal proliferation of Vγ9Vδ2 T cells in neonates and adults. J Leukoc Biol 2020;107:1023–32.

[70] Paletta D, et al. The hypervariable region 4 (HV4) and position 93 of the α chain modulate CD1d-glycolipid binding of iNKT TCRs. Eur J Immunol 2015;45:2122–33.

[71] Fields BA, et al. Crystal structure of a T-cell receptor β-chain complexed with a superantigen. Nature 1996;384:188–92.
[72] Willcox CR, et al. Cytomegalovirus and tumor stress surveillance by binding of a human γδ T cell antigen receptor to endothelial protein C receptor. Nat Immunol 2012;13:872–9.
[73] Silva-Santos B, Schamel WWA, Fisch P, Eberl M. γδ T-cell conference 2012: close encounters for the fifth time. Eur J Immunol 2012;42:3101–5.
[74] Bruder J, et al. Target specificity of an autoreactive pathogenic human γδ-T cell receptor in myositis. J Biol Chem 2012;287:20986–95.
[75] Chen Y, Ju L, Rushdi M, Ge C, Zhu C. Receptor-mediated cell mechanosensing. Mol Biol Cell 2017;28:3134–55.
[76] Li L, et al. Ionic CD3−Lck interaction regulates the initiation of T-cell receptor signaling. Proc Natl Acad Sci USA 2017;114:E5891–9.
[77] Hartl FA, et al. Noncanonical binding of Lck to CD3ε promotes TCR signaling and CAR function. Nat Immunol 2020;21:902–13.
[78] Gil D, Schamel WWA, Montoya M, Sánchez-Madrid F, Alarcón B. Recruitment of Nck by CD3ϵ reveals a ligand-induced conformational change essential for T cell receptor signaling and synapse formation. Cell 2002;109:901–12.
[79] Non-overlapping functions of Nck1 and Nck2 adaptor proteins in T cell activation. Cell Commun Signal 2014. https://doi.org/10.1186/1478-811X-12-21. Full Text.
[80] Hartl FA, et al. Cooperative interaction of Nck and Lck orchestrates optimal TCR signaling. Cells 2021;10:834.
[81] Blanco R, Borroto A, Schamel W, Pereira P, Alarcon B. Conformational changes in the T cell receptor differentially determine T cell subset development in mice. Sci Signal 2014;7:ra115.
[82] Dopfer EP, et al. The CD3 conformational change in the γδ T cell receptor is not triggered by antigens but can be enforced to enhance tumor killing. Cell Rep 2014;7:1704–15.
[83] Takamizawa M, Fagnoni F, Mehta-Damani A, Rivas A, Engleman EG. Cellular and molecular basis of human gamma delta T cell activation. Role of accessory molecules in alloactivation. J Clin Invest 1995;95:296–303.
[84] Ye Z, Haley S, Gee AP, Henslee-Downey PJ, Lamb LS. In vitro interactions between gamma deltaT cells, DC, and CD4+ T cells; implications for the immunotherapy of leukemia. Cytotherapy 2002;4:293–304.
[85] Toumi R, et al. Autocrine and paracrine IL-2 signals collaborate to regulate distinct phases of CD8 T cell memory. Cell Rep 2022;39, 110632.
[86] Fowler D, et al. Payload-delivering engineered γδ T cells display enhanced cytotoxicity, persistence, and efficacy in preclinical models of osteosarcoma. Sci Transl Med 2024;16:eadg9814. https://doi.org/10.1126/scitranslmed.adg9814.
[87] Chen D, Tang T-X, Deng H, Yang X-P, Tang Z-H. Interleukin-7 biology and its effects on immune cells: mediator of generation, differentiation, survival, and homeostasis. Front Immunol 2021;12, 747324.
[88] Perera P-Y, Lichy JH, Waldmann TA, Perera LP. The role of Interleukin-15 in inflammation and immune responses to infection: implications for its therapeutic use. Microbes Infect 2012;14:247–61.
[89] Tokuyama H, et al. V gamma 9 V delta 2 T cell cytotoxicity against tumor cells is enhanced by monoclonal antibody drugs—rituximab and trastuzumab. Int J Cancer 2008;122:2526–34.
[90] Capietto A-H, Martinet L, Fournié J-J. Stimulated γδ T cells increase the in vivo efficacy of trastuzumab in HER-2+ breast cancer. J Immunol 2011;187:1031–8.
[91] Gertner-Dardenne J, et al. Bromohydrin pyrophosphate enhances antibody-dependent cell-mediated cytotoxicity induced by therapeutic antibodies. Blood 2009;113:4875–84.
[92] Schilbach K, et al. In the absence of a TCR signal IL-2/IL-12/18-stimulated γδ T cells demonstrate potent antitumoral function through direct killing and senescence induction in cancer cells. Cancers (Basel) 2020;12:130.
[93] Domae E, Hirai Y, Ikeo T, Goda S, Shimizu Y. Cytokine-mediated activation of human ex vivo-expanded Vγ9Vδ2 T cells. Oncotarget 2017;8:45928–42.
[94] Xu B, et al. Crystal structure of a γδ T-cell receptor specific for the human MHC class I homolog MICA. Proc Natl Acad Sci USA 2011;108:2414–9.
[95] Chan CJ, Smyth MJ, Martinet L. Molecular mechanisms of natural killer cell activation in response to cellular stress. Cell Death Differ 2014;21:5–14.
[96] Bank I. The role of Gamma Delta T cells in autoimmune rheumatic diseases. Cells 2020;9:462.
[97] Malik S, Want MY, Awasthi A. The emerging roles of gamma–delta T cells in tissue inflammation in experimental autoimmune encephalomyelitis. Front Immunol 2016;7:257–69.
[98] Shiromizu CM, Jancic CC. γδ T lymphocytes: an effector cell in autoimmunity and infection. Front Immunol 2018;9:2389.

[99] Brownlie RJ, Zamoyska R. T cell receptor signalling networks: branched, diversified and bounded. Nat Rev Immunol 2013;13:257–69.
[100] Src Family Kinase—an overview. ScienceDirect Topics; 2011. https://www.sciencedirect.com/topics/biochemistry-genetics-and-molecular-biology/src-family-kinase.
[101] Wang J-H. T cell receptors, mechanosensors, catch bonds and immunotherapy. Prog Biophys Mol Biol 2020;153:23–7.
[102] Kumari S, Curado S, Mayya V, Dustin ML. T cell antigen receptor activation and actin cytoskeleton remodeling. Biochim Biophys Acta 2014;1838:546–56.
[103] Ribeiro ST, Ribot JC, Silva-Santos B. Five layers of receptor signaling in γδ T-cell differentiation and activation. Front Immunol 2015;6:15.
[104] Muro R, et al. γδTCR recruits the Syk/PI3K axis to drive proinflammatory differentiation program. J Clin Invest 2018;128:415–26.
[105] Muro R, Takayanagi H, Nitta T. T cell receptor signaling for γδT cell development. Inflamm Regener 2019;39:6.
[106] Fung NH, et al. Understanding and exploiting cell signalling convergence nodes and pathway cross-talk in malignant brain cancer. Cell Signal 2019;57:2–9.
[107] Pereira BI, Akbar AN. Convergence of innate and adaptive immunity during human aging. Front Immunol 2016;7. https://doi.org/10.3389/fimmu.2016.00445.
[108] Tan YX, et al. Inhibition of the kinase Csk in thymocytes reveals a requirement for actin remodeling in the initiation of full TCR signaling. Nat Immunol 2014;15:186–94.
[109] Hayes SM, Shores EW, Love PE. An architectural perspective on signaling by the pre-, αβ and γδ T cell receptors. Immunol Rev 2003;191:28–37.
[110] Sullivan SA, et al. The role of LAT–PLCγ1 interaction in γδ T cell development and homeostasis. J Immunol 2014;192:2865–74.
[111] Dustin ML, Chakraborty AK, Shaw AS. Understanding the structure and function of the immunological synapse. Cold Spring Harb Perspect Biol 2010;2, a002311.
[112] Stinchcombe JC, Majorovits E, Bossi G, Fuller S, Griffiths GM. Centrosome polarization delivers secretory granules to the immunological synapse. Nature 2006;443:462–5.
[113] Tsun A, et al. Centrosome docking at the immunological synapse is controlled by Lck signaling. J Cell Biol 2011;192:663–74.
[114] Yokosuka T, et al. Spatiotemporal basis of CTLA-4 costimulatory molecule-mediated negative regulation of T cell activation. Immunity 2010;33:326–39.
[115] Watanabe K, Kuramitsu S, Posey AD, June CH. Expanding the therapeutic window for CAR T cell therapy in solid tumors: the knowns and unknowns of CAR T cell biology. Front Immunol 2018;9.
[116] Mantri CK, John ALS. Immune synapses between mast cells and γδ T cells limit viral infection. J Clin Invest 2019;129:1094–108.
[117] Favier B, et al. Uncoupling between immunological synapse formation and functional outcome in human γδ T lymphocytes. J Immunol 2003;171:5027–33.
[118] Weng RR, et al. Epigenetic modulation of immune synaptic-cytoskeletal networks potentiates γδ T cell-mediated cytotoxicity in lung cancer. Nat Commun 2021;12:2163.
[119] Gaidarova S, et al. Lenalidomide enhances anti-tumor effect of γδ T cells against mantle cell lymphoma. Blood 2008;112:2616.

CHAPTER

3

γδ T cell immunotherapy: Requirement for combinations?

Anna Maria Corsale[a,b,*], Marta Di Simone[a,c,*], Francesco Dieli[a,c], and Serena Meraviglia[a,c]

[a]Central Laboratory of Advanced Diagnosis and Biomedical Research (CLADIBIOR), University of Palermo, Palermo, Italy [b]Department of Health Promotion, Mother and Child Care, Internal Medicine and Medical Specialties, University of Palermo, Palermo, Italy [c]Department of Biomedicine, Neurosciences and Advanced Diagnosis, University of Palermo, Palermo, Italy

Abstract

γδ T cells are often compromised in cancer patients due to various resistance mechanisms induced by tumors and the presence of immunosuppressive molecules in the tumor microenvironment (TME). This leads to a decrease in the percentage and absolute number of tumor-infiltrating and circulating γδ T cells, accompanied by higher rates of apoptosis and an exhausted immunophenotype. The TME, composed of immunoregulatory cell populations such as myeloid-derived suppressor cells, regulatory T cells (Tregs), neutrophils, and tumor-associated macrophages, further suppresses the function of γδ T cells through the secretion of immunosuppressive cytokines and the expression of inhibitory molecules. Metabolic and hypoxia-mediated mechanisms also impact γδ T cells, as tumor cells compete for essential nutrients, leading to metabolic exhaustion and limited antitumor responses. To restore the antitumoral properties of γδ T cells, several approaches have been explored. Targeting immunosuppressive pathways, including immune checkpoint inhibitors such as PD-1, BTLA, and TIM-3, shows promise in enhancing γδ T cell proliferation and activation and improving their cytotoxicity against tumor cells. Combinatorial approaches with immunotherapy and chemotherapy have successfully enhanced the therapeutic benefits and γδ T cell killing. Collectively, these findings highlight the potential of immunomodulatory strategies to restore the antitumoral capabilities of γδ T cells in cancer. Combining checkpoint inhibitors with γδ T cell-based therapies and chemotherapy offers a promising avenue to enhance antitumor immunity and improve patient outcomes.

*AMC and MDS share first authorship.

https://doi.org/10.1016/B978-0-443-21766-1.00010-2

Abbreviations

ADCC	antibody-dependent cellular cytotoxicity
AML	acute myeloid leukemia
AMPK	adenosine monophosphate-activated protein kinase
CSC	colon cancer stem cells
EM	effector memory
GBM	glioblastoma
HCC	hepatocellular carcinoma
HIF-1α	hypoxia-inducible factor-1α
HMBPP	(E)-4-hydroxy-3-methyl-but-2-enyl pyrophosphate
HVEM	herpesvirus entry mediator
ICI	immunological checkpoint inhibitor
IDO	indoleamine-2,3-dioxgenase
IPP	isopentenyl pyrophosphate
LDL	low-density lipoprotein
mAb	monoclonal antibody
MDSC	myeloid-derived suppressor cell
MM	multiple myeloma
MMR-d	mismatch repair deficiency
MVA	mevalonate
NSCLC	nonsmall cell lung cancer
ORR	objective response rate
OS	overall survival
OSCC	oral squamous cell carcinoma
PAgs	phosphoantigens
PDAC	pancreatic ductal adenocarcinoma
PFS	progression-free survival
RCC	renal cell carcinoma
scRNAseq	single-cell RNA sequencing
TAMs	tumor-associated macrophages
TCR	T cell receptor
TEMRA	terminally differentiated
TEXs	tumor-derived exosomes
TILs	tumor-infiltrate lymphocytes
TME	tumor microenvironment
Tregs	T regulatory
TRM	tissue-resident memory
ZA	zoledronate

Conflict of interest

No potential conflicts of interest were disclosed.

Introduction

Cancer patients often experience a decrease in the percentage and absolute number of tumor-infiltrating and circulating γδ T cells, accompanied by a higher rate of cellular apoptosis and an exhausted immunophenotype. Different intrinsic resistance mechanisms induced by tumors and the presence of an immunosuppressive tumor microenvironment (TME) have been shown to have adverse effects on γδ T cells, switching their functional antitumoral properties toward a protumoral state. Consequently, one major challenge lies in overcoming the dysfunction of γδ T cells and restoring their antitumor potential [1].

Tumor microenvironment suppresses γδ T cells

TME is composed of different populations of immunoregulatory cells, including myeloid-derived suppressor cells (MDSCs), regulatory T cells (Tregs), neutrophils, and tumor-associated macrophages (TAMs). These cell populations can directly or indirectly suppress the function of γδ T cells by secreting immunosuppressive cytokines, such as TGF-β, IL-4, and IL-10 [2], and expressing inhibitory molecules like indoleamine 2,3-dioxygenase (IDO) and arginase. Specifically, granulocytic MDSCs (PMN-MDSCs) impact the IFN-γ production and cytotoxicity (degranulation) by antigen-activated γδ T cells independently of the modulation of primary inhibitory and activating NK receptors. However, they do not affect Vδ2 T cell proliferation. The diminished cytotoxic activity of γδ T cells against Burkitt lymphoma Daudi and Jurkat cell lines caused by PMN-MDSCs can be restored by using a specific inhibitor of Arginase I (nor-NOHA). Interestingly, restoration is also observed in the absence of PMN-MDSCs, indicating that Daudi cells can produce Arginase I [3]. In hepatocellular carcinoma (HCC), the cytotoxicity of tumor-infiltrating γδ T cells is reduced as a consequence of decreased degranulation and downregulation of IFN-γ production, which is directly influenced by $CD4^{+}CD25^{+}$ Tregs in a TGF-β and IL-10-dependent manner. Partial restoration of cytotoxicity can be achieved by anti-TGFβ or anti-IL-10 neutralizing antibodies. Moreover, recombinant soluble TGF-β and IL-10 mimic the inhibitory effect of Tregs on γδ T cells, suggesting a mechanism that relies on soluble factors [4]. Furthermore, the presence of TGF-β and IL-10 leads to the differentiation of γδ T cells into suppressive cells that promote tumor growth, such as γδ17 T cells and γδ Tregs [5–7]. Additionally, there is an overexpression of ectonucleoside triphosphate diphosphohydrolase-1 (CD39) and ecto-5′-nucleotidase (CD73) in these cells, which indicates their acquired state of exhaustion [8]. Despite its well-known inhibitory effect, TGF-β has been found to enhance the cytotoxic activity of expanded γδ T cells against solid tumor cells expressing E-cadherin. This enhancement is achieved by upregulating CD54, CD103, IFN-γ, IL-9, and granzyme B expression [9,10].

Metabolic-mediated suppression of γδ T cells

Tumor cells exhibit a heightened metabolic demand, leading to the depletion of essential nutrients within the TME. The limited availability of crucial nutrients like glucose and amino acids can harm the function and survival of γδ T cells. Tumor cells possess the ability to outcompete γδ T cells for these resources, resulting in metabolic exhaustion and compromised antitumor responses of γδ T cells [11].

During tumorigenesis, tumor cells undergo metabolic reprogramming, including upregulation of the mevalonate (MVA) pathway, which, in turn, increases phosphoantigen (PAg) production. PAgs are generally derived from either the microbial nonmevalonate Rohmer cycle (e.g., (E)-4-hydroxy-3-methyl-but-2-enyl pyrophosphate or HMBPP) or the endogenous mevalonate pathway (e.g., isopentenyl pyrophosphate or IPP) [12]. The dysregulation of the MVA pathway in cancer can be attributed to several mechanisms: (1) abnormal regulation of the enzyme hydroxy-methyl-glutaryl-CoA reductase, influenced by various transcription factors such as hypoxia-inducible factor 1α (HIF-1α); (2) mutations or abnormal activation of sterol regulatory element-binding proteins, which interact with mutated tumor suppressor proteins like p53 [13]; (3) decreased activation of AMP-activated protein kinase (AMPK); (4) increased activation of signaling pathways, including PI3K/AKT/mTORC1, JAK/STAT3 [9], and the Hippo signaling pathway (YAP-TAZ) [14].

In patients with type 2 diabetes mellitus, hyperglycemia and the consequent accumulation of lactate inhibit AMPK activation, thereby reducing the antitumor activity of the major subset of circulating γδ T cells that express the Vγ9Vδ2 T cell receptor (TCR) [15].

Evidence suggests that γδ T cells engage in bidirectional interactions with lipid metabolism, influencing and being influenced by various lipid molecules and pathways. These interactions hold profound implications for both immune responses and metabolic homeostasis [16]. For instance, the Vγ9Vδ2 TCR recognizes lipid-related ligands expressed on tumor cells, such as apolipoprotein A1 (apoA-1), abundant in high-density lipoproteins, and ATP synthase/F1-ATPase (a high-affinity apoA-1 receptor), promoting tumor recognition [17]. Furthermore, during influenza virus infection, $CD1d^+$ B1a cells present host-derived lipids, inducing IL-17A production by lung γδ T cells through γδ TCR-mediated IRF4-dependent transcription [18].

In pancreatic ductal adenocarcinomas (PDAC), the overexpression of IDO and its downstream metabolite, kynurenine, suppresses T cell activation and proliferation [19]. Recombinant kynurenine significantly reduces the cytotoxicity of γδ T cells, while IDO inhibitors (1-methyl-levotryptophan and 1-methyl-dextrotryptophan) enhance their cytotoxic activity against PDAC cell lines (Panc89 and PancTu-I), particularly when stimulated with bromohydrin diphosphate or in the presence of a tribody $[(HER2)_2 \times V\gamma9]$ [20]. Another intrinsic mechanism tumors utilize to escape the activity of tumor-infiltrating γδ T cells involves the production of the galactoside-binding protein galectin-3. Galectin-3 is expressed by PDAC cells as well as αβ and γδ T cells, albeit with weak secretion by each cell type [21]. Through cell-to-cell interaction, galectin-3 inhibits the initial proliferation of resting peripheral blood and tumor-infiltrating Vγ9Vδ2 T lymphocytes. Interestingly, the physiological addition of recombinant galectin-3 or the release of galectin-3 by tumor cells did not further impair T cell cytotoxicity against PDAC cells or induce T cell death in vitro. The interaction between galectin-3 and the α3β1 integrin (CD49c/CD29) expressed by Vγ9Vδ2 T cells has been implicated in the inhibition of γδ T cell proliferation. A bispecific antibody [HER2xCD3] designed to bind HER-2 present on PDAC cells and CD3-expressing T cells augmented the cytotoxic activity of γδ T cells against PDAC cells without releasing galectin-3. Additionally, Rodrigues et al. described the influence of low-density lipoprotein (LDL) uptake on immune responses mediated by γδ T cells [22]. They found that activated Vγ9Vδ2 T cells expressed the LDL receptor, and the binding of LDL to their receptor resulted in alterations of Vγ9Vδ2 T cell activation and functions. Treatment of PAg-expanded Vγ9Vδ2 T cells with LDL-cholesterol inhibited IFN-γ, NKG2D, and DNAM-1 expression. Furthermore, in an in vivo xenograft mouse model study, it was observed that LDL-cholesterol treatment did not affect the levels of γδ T cells in blood or tumor tissue. Still, these γδ T cells displayed reduced efficacy against tumor cells and failed to control tumor growth. Hence, the uptake of LDL-cholesterol appears to act as an inhibitor of the antitumor functions of Vγ9Vδ2 T cells.

Hypoxia-mediated suppression of γδ T cells

Hypoxia, characterized by reduced oxygen availability, is a prominent feature of cancer and arises from an imbalance between oxygen demand and supply. Abnormal angiogenesis and heightened oxygen consumption contribute to hypoxia, transforming the TME into a pro-tumorigenic and pro-metastatic milieu [23,24]. Hypoxia also induces changes in the transcriptome profiles of immune cells, impairing immune responses and limiting the effectiveness of immunotherapy [25]. Additionally, the poor oxygenation of tumors is associated with increased resistance to radiation and chemotherapy and an elevated risk of metastatic spread, ultimately leading to treatment failure [24,26].

γδ T cells that infiltrate solid tumors are associated with the hypoxia gene signature (SRPX, IL6, and HOXB9) and exhibit higher levels in the high-risk group, which correlates with a poorer survival rate [27]. Recent studies have examined the impact of hypoxia on the antitumor effector functions of γδ T cells in both solid and hematologic tumors, revealing a controversial role [28–33].

The cytotoxicity of γδ T cells against multiple myeloma stem-like cells (MM-HA) cultured under hypoxic conditions (1% O_2, 5% CO_2, and 94% N_2) was suppressed due to a reduction in the accumulation of IPP and a decrease in the expression of mevalonate decarboxylase and farnesyl diphosphate synthase proteins in MM-HA cells. Consequently, the mevalonate pathway, crucial for recognizing and activating γδ T cells, was negatively regulated. However, there were no differences in the expression levels of several adhesion molecules (such as MICA/B, ICAM-1, and PVR in MM cells; NKG2D, LFA-1, and DNAM-1 in γδ T cells), as well as perforin, granzyme B, IFN-γ, and CD107a by γδ T cells [31]. In patients with advanced oral cancer (stages III and IV) and gliomas, infiltrating γδ T cells exhibited elevated levels of HIF-1α [28,29]. Compared to normoxia, exposure to hypoxic conditions significantly impaired the ability of expanded γδ T cells from oral cancer patients to kill PAgs-treated oral cancer cell lines [28]. This reduction was attributed to decreased calcium efflux and CD107a expression. Similarly, hypoxia-exposed oral cancer and breast tumor cell lines developed resistance to γδ T cell-mediated killing [30]. However, this resistance caused by MICA/B shedding from the cancer cells was nullified when γδ T cells were precultured under hypoxic conditions for 48 h and the breast cancer cell lines were maintained in normoxia [30]. In oral cancer, infiltrating γδ T cells predominantly consisted of IL17A-producing γδ T cells (γδ17 T cells), whose differentiation was induced in the presence of hypoxia [28]. Hypoxia-mediated activation of the cyclic AMP-PKA signaling pathway in γδ T cells reduced NKG2A receptor expression and inhibited cytotoxic activity against brain tumors [29]. However, metformin treatment reduced tumor cell oxygen uptake, increased oxygen availability for tumor-infiltrating lymphocytes (TILs), and restored γδ T cell antitumor functions. This was accompanied by upregulation of NKG2A, CD107a, granzyme B, and IFN-γ production [32]. Treating ex vivo expanded γδ T cells with metformin or a HIF-1α inhibitor in a glioma xenograft model resulted in approximately 75% tumor-free survival [29].

Solid tumors frequently experience hypoxia, which is known to impact tumor-derived exosomes (TEXs) and contribute to the development of hypoxic TME. In normoxic conditions, exosomes derived from oral squamous cell carcinoma (OSCC) tumors can promote the growth and functions of γδ T cells in a dendritic cell-independent manner. However, under hypoxic conditions, TEXs have diminished stimulating effects on γδ T cell activity while enhancing the suppressive function of MDSCs through a miR-21/PTEN/PD-L1-axis-dependent mechanism. Consequently, in immunocompetent mice bearing OSCC, the treatment strategy can benefit from the combined targeting of miR-21 and PD-L1 to address these dysregulated pathways [33].

Hypoxia can influence the composition of the gut microbiota by amplifying the secretion of angiogenin-4, an antimicrobial peptide produced by intestinal epithelial cells. Upon exposure to phospholipid antigens originating from the hypoxia-induced intestinal microbiota, particularly *Desulfovibrio*, intraepithelial γδ T cells exhibit a propensity for selective release of IL-17A. This cytokine, in turn, exacerbates hypoxia-induced intestinal damage. The mediation of γδ T cell expansion and activation within the small intestine epithelial layer is primarily orchestrated by *Desulfovibrio* through CD1d signaling. This suggests that preventing and treating acute mountain sickness may involve interventions to inhibit the activation of γδ T cells by the microbiota during hypoxia [34].

It is important to note that the described immunosuppressive mechanisms are not limited to γδ T cells but can also impact other immune cell populations within the TME. Targeting these immunosuppressive pathways offers a potential strategy to restore γδ T cell function and enhance antitumor immunity.

Immune checkpoint-mediated suppression of γδ T cells

Combination therapies involving immunological checkpoint inhibitors (ICIs) and strategies to activate γδ T cells hold great promise for enhancing the effectiveness of cancer immunotherapy. γδ T cells have been found to express various ICs, including BTLA, TIGIT, PD-1, TIM-3, CD39, LAG-3, and NKG2A, in both solid (such as melanoma, neuroblastoma, colorectal cancer, breast cancer, ovarian cancer, head and neck squamous cell carcinoma, and kidney cancer) and hematological tumors (such as multiple myeloma, acute myeloid leukemia, and lymphomas) [8,35–53]. Tumor cells also express high levels of immunological checkpoint ligands, which, upon engagement by their receptors, transmit inhibitory signals that dampen γδ T cell activation and lead to anergy or exhaustion. Tumors exploit this interaction as a mechanism to evade immune surveillance. For example, in MM patients, bone marrow Vγ9Vδ2 T cells coexpress PD-1, TIM-3, and LAG-3, resulting in a state of super anergy [35].

A significant proportion of tumor-infiltrating Vγ9$^-$ γδ T cells were identified in both lymphomas and solid tumors, showing a quantitative correlation with tissue residency, exhaustion, and responsiveness to immune checkpoint therapy [53]. Specifically, the frequency of basal cell carcinoma and melanoma infiltrating Vγ9$^-$ γδ T cells notably increased following anti-PD1 therapy. Furthermore, bone marrow-infiltrating exhausted γδ and CD8 T cells shared a signature of 14 up-regulated genes involved in T cell exhaustion, including LAG3, TIGIT, and PDCD1 [53].

In both MM and acute myeloid leukemia (AML), terminally differentiated (TEMRA) Vδ1 T cells derived from the bone marrow displayed an elevated frequency of PD-1, TIGIT, TIM-3, and CD39 expression [42].

The expression of PD-1 on γδ T cells in AML was significantly increased between day 2 and day 4 after being stimulated with Zoledronate (ZA) and IL-2 [50], a combination leading to

PAgs production and $V\gamma9V\delta2$ T cell activation. Moreover, $PD\text{-}1^+Foxp3^+$ and $TIGIT^+CD266^-$ $\gamma\delta$ T cells were found in de novo AML patients, and those patients with a higher presence of these specific subsets exhibited a poor clinical outcome [43,44]. However, PD-1 expression decreased in patients who achieved complete remission after chemotherapy [54]. TIM-3-expressing $V\gamma9V\delta2$ T cells exhibit impaired proliferative capacity in AML upon IL-21 stimulation, which can be restored by inhibiting TIM-3 [52]. Coexpression of TIM-3 and PD-1 on $V\gamma9V\delta2$ T cells is accompanied by lower production of proinflammatory cytokines such as IFN-γ and TNF-α.

Elevated levels of 4-1BBL, TIM3, and LAG3 expression on infiltrating $\gamma\delta$ T cells, along with increased PD1 expression on circulating counterparts, were found to be correlated with earlier relapse and shorter overall survival (OS) in patients with melanoma [51]. Notably, BTLA was found to be highly expressed by $V\gamma9V\delta2$ T cells in the lymph nodes of lymphoma patients, and its engagement with herpesvirus entry mediator (HVEM) on primary tumors suppresses their proliferation [48].

In ovarian cancer, infiltrating effector memory (EM) and TEMRA $V\delta1$ T cells showed increased coexpression of PD-1, TIM-3, CD39, and TIGIT and produced modest levels of IFN-γ, TNF-α, and MIP-1β in response to stimulation, indicating a lack of antitumor effects [8]. Notably, patients whose $\gamma\delta$ T cells produced higher TNF-α levels had a greater OS rate than those with lower TNF-α production [40]. In breast cancer patients, peripheral $V\gamma9V\delta2$ T cells with a TEMRA phenotype increased the expression of PD-1, which was significantly associated with the invasion of tumor-draining lymph nodes [55]. Furthermore, approximately 20% of $V\delta1^+$ T cells infiltrating breast tumors expressed PD-1 [39].

TIM-3-expressing $V\gamma9V\delta2$ T cells exhibit reduced cytotoxicity against colon cancer cell lines [47]. In an experimental mouse model of colon cancer, it was observed that $\gamma\delta$ T cells expressing PD-1 and lacking CD8$\alpha\alpha$ exhibited a reduced expression of genes related to cytotoxic functions. Conversely, this cell subset demonstrated an upregulation of genes associated with IL-17 production and activities promoting tumor growth [56]. Activation of TIM-3 with galectin-9 has been observed to diminish the cytotoxicity of $V\gamma9V\delta2$ T cells against colon cancer cell lines. This effect is achieved by inhibiting the production of perforin and granzyme B through the ERK1/2 pathway [47].

In HCC, glioblastoma (GBM), and nonsmall-cell lung cancer (NSCLC), tumor-infiltrating $V\gamma9V\delta2$ T cells consistently expressed high levels of NKG2A, and their activation was tightly regulated by the interaction between cancer cells and its ligand HLA-E, which had a significant impact on the OS of patients. Specifically, an increased frequency of infiltrating $NKG2A^+$ $V\gamma9V\delta2$ T cells correlated with improved OS in patients with NSCLC and HCC. This association was observed even though HLA-E levels were either reduced or comparable to those found in normal tissue. On the contrary, the presence of $NKG2A^+$ $V\gamma9V\delta2$ T cells did not exhibit any significant clinical influence on patient OS in tumors characterized by high expression of HLA-E, as seen in GBM [37]. However, the functional capabilities of $V\gamma9V\delta2$ T cells can be restored even in HLA-E^+ malignancies by masking or knocking out NKG2A through CRISPR/Cas9 gene editing, allowing them to exert their effector activities [37,57].

These findings have provided evidence that the dysfunction of $\gamma\delta$ T cells within the TME can be reversed by targeting ICs. This targeted approach improves $\gamma\delta$ T cell proliferation and activation, enhances cytotoxicity, and ultimately exerts antitumor effects in solid tumors and hematological malignancies. Not all $\gamma\delta$ T cells are dysfunctional despite exhibiting a gradual increase in the expression of multiple ICs; instead, they can maintain their functional capacity, similar to CD8 T cells [58,59]. In HCC, both Vδ1 and Vγ9Vδ2 tissue-resident memory T cells (TRM) expressed more PD-1 than their non-TRM counterparts. However, despite their higher PD-1 expression, Vδ1 and Vγ9Vδ2 T cells retained their ability to produce IFN-γ and IL-2 [58]. Similar findings have been reported in kidney cancer, where infiltrating $V\delta2^-$ T cells coexpressing PD-1, TIGIT, and TIM-3 maintained the expression of effector cytokines and perforin at comparable levels to nonexhausted cells. Moreover, these cells maintained the capability to effectively eliminate autologous tumor cells in ex vivo coculture experiments [41]. In mismatch repair-deficient (MMR-d) cancers, γδ T cells, predominantly Vδ1 and Vδ3 subsets, displayed elevated levels of PD-1 expression along with other activation markers and cytotoxic molecules. Studies in vitro revealed that $PD\text{-}1^+$ γδ T cells exhibited heightened reactivity toward HLA-class-I-negative MMR-d colon cancer cell lines and B2M-knockout patient-derived tumor organoids compared to antigen-presentation-proficient cells [38]. The efficacy of γδ T cell-based immunotherapy for cancer can be heightened by interfering with IC signaling within these cells. Consequently, this strategy amplified the cytotoxic potential of γδ T cells, rendering them more pivotal agents in achieving successful antitumor immune responses following ICI therapy. Some examples of the efficacy of checkpoint-inhibiting treatments in restoring γδ T lymphocyte antitumoral functions are listed as follows:

PD1

The PD-1/PD-L1 signaling pathway might be pivotal in blocking the expansion and activation of Vγ9Vδ2 T cells in cancer patients. However, there is potential for restoring the compromised proliferation of $PD\text{-}1^+$ γδ T cells and triggering their activation by obstructing the PD-1 signaling pathway. For instance, BM Vγ9Vδ2 T cells derived from MM patients exhibited an impaired response to PAgs stimulation. Nevertheless, their proliferation rate was significantly augmented when treated with ZA and anti-PD-1 therapy [35].

Through a 3D coculture assay, it has been shown that inhibiting PD-1 can enhance the antibody-dependent cellular cytotoxicity (ADCC) of follicular lymphoma cells by $CD16^+$ Vγ9Vδ2 lymphocytes. This indicates that blocking PD-1 can boost immune cell cytotoxicity against follicular lymphoma, potentially leading to increased efficacy in targeting the disease [60].

When γδ T cells were directly treated with ZA or exposed to ZA-treated primary AML cells, the anti-PD-1 monoclonal antibody (mAb) pembrolizumab demonstrated the ability to increase the production of IFN-γ. Nevertheless, pembrolizumab neither impacted the proliferation or expression of activation markers nor altered the cytotoxic activity of γδ T cells against leukemia cell lines [50].

A metaanalysis of single-cell RNA sequencing (scRNAseq) data was conducted on melanoma patients who received anti-PD-1 therapy to investigate the differences between responders and nonresponders. The analysis revealed the presence of a specific population of γδ T cells and $TREM2^{hi}$ macrophages that correlated with nonresponder patients and could potentially contribute to resistance to ICI therapy [61].

In the presence of PD-L1$^+$ tumor cells, Y111, a bispecific antibody that targets both PD-L1 and CD3, induced a dose-dependent activation of the expanded Vγ9Vδ2 T cells and enhanced their cytotoxicity against various NSCLC-derived tumor cell lines, resulting in the release of granzyme B, IFN-γ, and TNF-α. Additionally, when Vγ9Vδ2 T cells were adoptively transferred along with Y111, they effectively inhibited the growth of established xenografts in immunodeficient mice [62]. Moreover, administering anti-PD-L1 mAb assigned ADCC capability to γδ T cells, enhancing their cytotoxic effect against cancer cells expressing high levels of PD-L1 [63]. When combining PD-1 blockade with ZA, MM-derived Vγ9Vδ2 T cells exhibited a fivefold rise in cytotoxicity. This effect was coupled with an elevation in the expression of the degranulation marker CD107, along with an increase in the percentage of CD107$^+$ Vγ9Vδ2 T cells [35].

BTLA

BTLA signaling is crucial in regulating human peripheral γδ T cell proliferation and cytotoxicity, operating independently from the PD-1 signaling pathway [64]. Inhibiting BTLA signaling using mAb caused an enhancement in the signal transduction of the PAg/TCR pathway, resulting in the upregulation of γδ T cell proliferation without affecting their cytotoxicity [48].

TIM3

Blocking TIM-3 signaling led to elevated cytokine production by Vγ9Vδ2 T cells without impacting their proliferation in breast cancer [45]. By inhibiting the interaction between TIM-3 and its ligands (e.g., galectin-9), using a TIM-3 inhibitor or the fusion protein TIM-3-Fc, dysfunctional γδ T cells can be revitalized. Additionally, in combination with the bispecific antibody MT110 (anti-CD3 × anti-EpCAM), γδ T cells demonstrate increased cytotoxicity against breast cancer cells in vitro and enhanced accumulation at tumor sites in vivo. When the TIM-3 inhibitor and MT110 are combined with γδ T-cell adoptive transfer, the antitumor efficacy of the infused γδ T cells is further enhanced [45].

Conversely, TIM-3 blocking improved Vγ9Vδ2 T cell proliferation and STAT1 phosphorylation in response to IL-21 in AML [52]. Treatment with anti-TIM-3 mAb or a combination of anti-TIM-3 and anti-PD-1 mAbs increases the levels of proinflammatory cytokines, unlike treatment with anti-PD-1 alone [46].

The transcriptional signature of tumor-infiltrating Vδ2$^-$ cells, which exhibits differentially elevated coexpression of PDCD1, TIGIT, HAVCR2, KLRD1, MATK, AOAH, NKG7, CRTAM, CCL5, XCL1, XCL2, CD7, and ZNF683, has been associated with a significantly higher 5-year survival rate in patients with renal cell carcinoma (RCC). Additionally, this Vδ2$^-$ gene signature strongly correlated with clinical responses in a cohort of 263 patients with metastatic RCC who received treatment with the combination of atezolizumab (anti-PD-L1 mAb) + bevacizumab (anti-VEGF mAb) in the IMmotion150 kidney cancer study. Furthermore, in the IMvigor210 bladder cancer trial, which included 168 patients with locally advanced and metastatic urothelial carcinoma treated with atezolizumab, the Vδ2$^-$ gene signature was also found to be significantly associated with improved clinical responses [41].

ICI therapy or chemotherapy in combination with other therapy

γδ T cell-based therapy has proven to be an effective adjuvant to conventional cancer treatments. Notably, a combined approach utilizing autologous γδ T cells and chemotherapy has demonstrated both safety and efficacy in treating patients with gastric carcinoma. This comprehensive treatment enhanced patients' quality of life and displayed promising potential in preventing recurrence, especially for those patients with advanced-stage gastric carcinoma, poorly differentiated carcinoma, or lymph node metastasis [65]. Low-dose carboplatin with γδ T cell therapy has significantly increased cytotoxicity against urothelial bladder cancer cells [66].

It has recently been demonstrated that ZA may enhance the therapeutic benefits of ICIs. In particular, ZA and anti-PD-1 inhibitors increased the objective response rate (ORR), OS, and progression-free survival (PFS) in patients with advanced NSCLC. Moreover, the combination of ZA and anti-PD-1 inhibitors exhibited more potent antitumor effects in a model of lung cancer [67]. During the administration of ICI therapy, there is a potential upregulation of compensatory and alternative IC receptors on γδ T cells [46]. Moreover, the lack of IC receptor signaling can trigger γδ T cells to secrete tumor-promoting cytokines, including IL-17, and transition into a pro-tumoral phenotype [68,69]. These findings strongly indicate the potential involvement of γδ T cells in various mechanisms contributing to ICI resistance in cancer patients.

Data from clinical studies have suggested that combining chemotherapy with immunotherapy has more survival benefits than chemotherapy alone [70]. Our previous study [71] showed that many colon cancer stem cell (CSC) lines were resistant to the cytotoxic activity of Vγ9Vδ2 T cells, but their pretreatment with low, sublethal concentrations of chemotherapeutic drugs 5-fluorouracil and doxorubicin sensitized CSC targets to Vγ9Vδ2 T cell killing, resulting in additive cytotoxicity activity. Previous data have demonstrated that pretreatment of tumors in the clinical setting with both chemotherapy and ZA would significantly improve the probability of a therapeutic benefit following the activation of Vγ9Vδ2 T cells as part of anticancer therapy. This combination of therapies can be broadly applicable across a range of malignancies, with minimal treatment-related toxicity [72].

Conclusion

In conclusion, the dysfunction of γδ T cells within the TME poses a significant challenge in cancer immunotherapy. Various immunosuppressive mechanisms, including inhibitory molecules and cytokines, metabolic alterations, hypoxia, and immune checkpoint signaling, contribute to the compromised antitumor capabilities of γδ T cells. A detailed understanding of these mechanisms is crucial in designing strategies to restore the efficacy of γδ T cells and enhance their antitumoral properties. Targeting the immunosuppressive pathways and inhibitory molecules in the TME offers a potential approach to overcome the dysfunction of γδ T cells (Fig. 1). Combination therapies involving ICIs and strategies to activate γδ T cells have shown promise in enhancing the effectiveness of cancer immunotherapy. For example, blocking PD-1, BTLA, TIM-3, and other ICI receptors can improve γδ T cell proliferation and

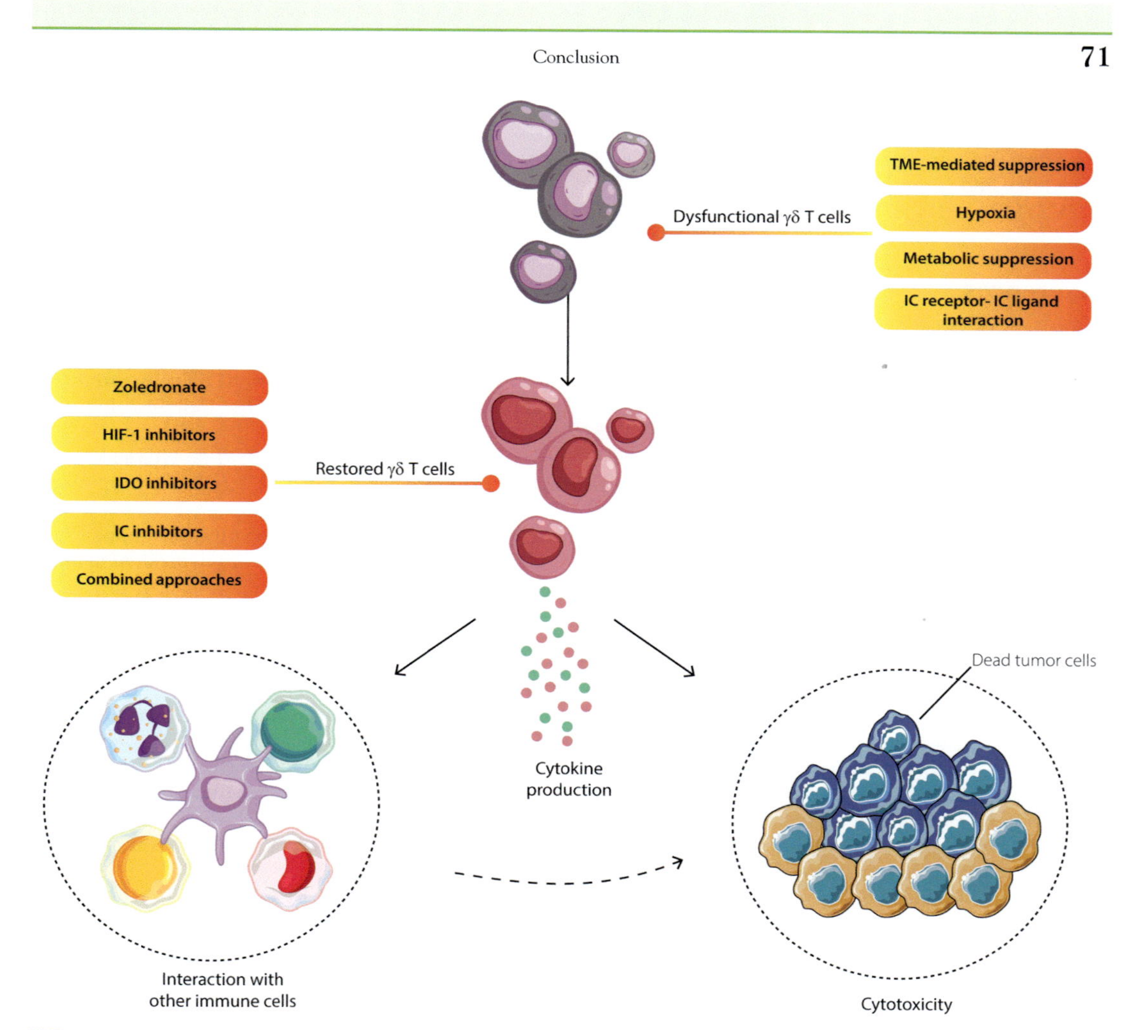

FIG. 1 Approaches to counteract the immunosuppressive state of γδ T cells'. In TME, tumor cells suppress the function of γδ T cells in the presence of immunosuppressive molecules, metabolic changes, hypoxia, and immune checkpoint signaling. Targeting these inhibitory pathways provides a potential strategy to reverse the dysfunction of γδ T cells and enhance antitumor immunity. Abbreviations: *TME*, tumor microenvironment; *IC*, immune checkpoint; *HIF*, hypoxia-inducible factor; *IDO*, indoleamine 2,3-dioxygenase. *Created by the authors with smart.servier.com and freepik.com.*

cytotoxicity. Furthermore, combining γδ T cell therapy with chemotherapy or other targeted therapies may lead to synergistic effects and improved therapeutic outcomes. By exploiting the unique features of γδ T cells and employing targeted interventions, it is possible to harness their antitumor potential and improve the prognosis for cancer patients. Further research and clinical trials will be crucial to fully unleashing the therapeutic potential of γδ T cell-based immunotherapies.

Acknowledgments

This work was supported by grants from the Italian Ministry of Health (Grant No. GR 2016-02364931) to Serena Meraviglia and from the Ministry of Education and Research to Francesco Dieli (Grant No. PRIN 2017-2017M8YMR8_001) and Serena Meraviglia (Grant No. PRIN 2017-2017ALPCM).

References

[1] Wesch D, Kabelitz D, Oberg H. Tumor resistance mechanisms and their consequences on Γδ T cell activation. Immunol Rev 2020;298:84–98. https://doi.org/10.1111/imr.12925.

[2] Mao Y, Yin S, Zhang J, Hu Y, Huang B, Cui L, Kang N, He W. A new effect of IL-4 on human Γδ T cells: promoting regulatory Vδ1 T cells via IL-10 production and inhibiting function of Vδ2 T cells. Cell Mol Immunol 2016;13:217–28. https://doi.org/10.1038/cmi.2015.07.

[3] Sacchi A, Tumino N, Sabatini A, Cimini E, Casetti R, Bordoni V, Grassi G, Agrati C. Myeloid-derived suppressor cells specifically suppress IFN-γ production and antitumor cytotoxic activity of Vδ2 T cells. Front Immunol 2018;9:1271. https://doi.org/10.3389/fimmu.2018.01271.

[4] Yi Y, He H-W, Wang J-X, Cai X-Y, Li Y-W, Zhou J, Cheng Y-F, Jin J-J, Fan J, Qiu S-J. The functional impairment of HCC-infiltrating Γδ T cells, partially mediated by regulatory T cells in a TGFβ- and IL-10-dependent manner. J Hepatol 2013;58:977–83. https://doi.org/10.1016/j.jhep.2012.12.015.

[5] Casetti R, Agrati C, Wallace M, Sacchi A, Martini F, Martino A, Rinaldi A, Malkovsky M. Cutting edge: TGF-B1 and IL-15 induce FOXP3+ Γδ regulatory T cells in the presence of antigen stimulation. J Immunol 2009;183:3574–7. https://doi.org/10.4049/jimmunol.0901334.

[6] Li X, Kang N, Zhang X, Dong X, Wei W, Cui L, Ba D, He W. Generation of human regulatory Γδ T cells by TCRγδ stimulation in the presence of TGF-β and their involvement in the pathogenesis of systemic lupus erythematosus. J Immunol 2011;186:6693–700. https://doi.org/10.4049/jimmunol.1002776.

[7] Caccamo N, Todaro M, Sireci G, Meraviglia S, Stassi G, Dieli F. Mechanisms underlying lineage commitment and plasticity of human Γδ T cells. Cell Mol Immunol 2013;10:30–4. https://doi.org/10.1038/cmi.2012.42.

[8] Weimer P, Wellbrock J, Sturmheit T, Oliveira-Ferrer L, Ding Y, Menzel S, Witt M, Hell L, Schmalfeldt B, Bokemeyer C, et al. Tissue-specific expression of TIGIT, PD-1, TIM-3, and CD39 by Γδ T cells in ovarian cancer. Cells 2022;11:964. https://doi.org/10.3390/cells11060964.

[9] Peters C, Meyer A, Kouakanou L, Feder J, Schricker T, Lettau M, Janssen O, Wesch D, Kabelitz D. TGF-β enhances the cytotoxic activity of Vδ2 T cells. Onco Targets Ther 2019;8, e1522471. https://doi.org/10.1080/2162402X.2018.1522471.

[10] Beatson RE, Parente-Pereira AC, Halim L, Cozzetto D, Hull C, Whilding LM, Martinez O, Taylor CA, Obajdin J, Luu Hoang KN, et al. TGF-B1 potentiates Vγ9Vδ2 T cell adoptive immunotherapy of cancer. Cell Rep Med 2021;2, 100473. https://doi.org/10.1016/j.xcrm.2021.100473.

[11] Corsale AM, Di Simone M, Lo Presti E, Picone C, Dieli F, Meraviglia S. Metabolic changes in tumor microenvironment: how could they affect Γδ T cells functions? Cells 2021;10:2896. https://doi.org/10.3390/cells10112896.

[12] Sireci G, Espinosa E, Sano CD, Dieli F, Fournié J-J, Salerno A. Differential activation of human γ δ cells by nonpeptide phosphoantigens. Eur J Immunol 2001;31:1628–35. https://doi.org/10.1002/1521-4141(200105)31:5<1628::AID-IMMU1628>3.0.CO;2-T.

[13] Parrales A, Thoenen E, Iwakuma T. The interplay between mutant P53 and the mevalonate pathway. Cell Death Differ 2018;25:460–70. https://doi.org/10.1038/s41418-017-0026-y.

[14] Sorrentino G, Ruggeri N, Specchia V, Cordenonsi M, Mano M, Dupont S, Manfrin A, Ingallina E, Sommaggio R, Piazza S, et al. Metabolic control of YAP and TAZ by the mevalonate pathway. Nat Cell Biol 2014;16:357–66. https://doi.org/10.1038/ncb2936.

[15] Mu X, Xiang Z, Xu Y, He J, Lu J, Chen Y, Wang X, Tu CR, Zhang Y, Zhang W, et al. Glucose metabolism controls human Γδ T-cell-mediated tumor immunosurveillance in diabetes. Cell Mol Immunol 2022;19:944–56. https://doi.org/10.1038/s41423-022-00894-x.

[16] Lou W, Gong C, Ye Z, Hu Y, Zhu M, Fang Z, Xu H. Lipid metabolic features of T cells in the tumor microenvironment. Lipids Health Dis 2022;21:94. https://doi.org/10.1186/s12944-022-01705-y.

[17] Scotet E, Martinez LO, Grant E, Barbaras R, Jenö P, Guiraud M, Monsarrat B, Saulquin X, Maillet S, Estève J-P, et al. Tumor recognition following Vγ9Vδ2 T cell receptor interactions with a surface F1-ATPase-related structure and apolipoprotein A-I. Immunity 2005;22:71–80. https://doi.org/10.1016/j.immuni.2004.11.012.

[18] Wang X, Lin X, Zheng Z, Lu B, Wang J, Tan AH-M, Zhao M, Loh JT, Ng SW, Chen Q, et al. Host-derived lipids orchestrate pulmonary Γδ T cell response to provide early protection against influenza virus infection. Nat Commun 1914;2021:12. https://doi.org/10.1038/s41467-021-22242-9.

[19] Zhai L, Bell A, Ladomersky E, Lauing KL, Bollu L, Sosman JA, Zhang B, Wu JD, Miller SD, Meeks JJ, et al. Immunosuppressive IDO in cancer: mechanisms of action, animal models, and targeting strategies. Front Immunol 2020;11:1185. https://doi.org/10.3389/fimmu.2020.01185.

[20] Jonescheit H, Oberg H-H, Gonnermann D, Hermes M, Sulaj V, Peters C, Kabelitz D, Wesch D. Influence of indoleamine-2,3-dioxygenase and its metabolite kynurenine on Γδ T cell cytotoxicity against ductal pancreatic adenocarcinoma cells. Cells 2020;9:1140. https://doi.org/10.3390/cells9051140.

[21] Gonnermann D, Oberg H-H, Lettau M, Peipp M, Bauerschlag D, Sebens S, Kabelitz D, Wesch D. Galectin-3 released by pancreatic ductal adenocarcinoma suppresses Γδ T cell proliferation but not their cytotoxicity. Front Immunol 2020;11:1328. https://doi.org/10.3389/fimmu.2020.01328.

[22] Rodrigues NV, Correia DV, Mensurado S, Nóbrega-Pereira S, deBarros A, Kyle-Cezar F, Tutt A, Hayday AC, Norell H, Silva-Santos B, et al. Low-density lipoprotein uptake inhibits the activation and antitumor functions of human Vγ9Vδ2 T cells. Cancer Immunol Res 2018;6:448–57. https://doi.org/10.1158/2326-6066.CIR-17-0327.

[23] Zhuang Y, Liu K, He Q, Gu X, Jiang C, Wu J. Hypoxia signaling in cancer: implications for therapeutic interventions. MedComm 2023;4, e203. https://doi.org/10.1002/mco2.203.

[24] Liao C, Liu X, Zhang C, Zhang Q. Tumor hypoxia: from basic knowledge to therapeutic implications. Semin Cancer Biol 2023;88:172–86. https://doi.org/10.1016/j.semcancer.2022.12.011.

[25] Parodi M, Raggi F, Cangelosi D, Manzini C, Balsamo M, Blengio F, Eva A, Varesio L, Pietra G, Moretta L, Mingari MC, Vitale M, Bosco MC. Hypoxia modifies the transcriptome of human NK cells, modulates their immunoregulatory profile, and influences NK cell subset migration. Front Immunol 2018;9:2358. https://doi.org/10.3389/fimmu.2018.02358.

[26] Boulefour W, Rowinski E, Louati S, Sotton S, Wozny A-S, Moreno-Acosta P, Mery B, Rodriguez-Lafrasse C, Magne N. A review of the role of hypoxia in radioresistance in cancer therapy. Med Sci Monit Int Med J Exp Clin Res 2021;27, e934116-1-e934116-7. https://doi.org/10.12659/MSM.934116.

[27] Jiao Y, Geng R, Zhong Z, Ni S, Liu W, He Z, Gan S, Huang Q, Liu J, Bai J. A hypoxia molecular signature-based prognostic model for endometrial cancer patients. Int J Mol Sci 2023;24:1675. https://doi.org/10.3390/ijms24021675.

[28] Sureshbabu SK, Chaukar D, Chiplunkar SV. Hypoxia regulates the differentiation and anti-tumor effector functions of ΓδT cells in oral cancer. Clin Exp Immunol 2020;201:40–57. https://doi.org/10.1111/cei.13436.

[29] Park JH, Kim H-J, Kim CW, Kim HC, Jung Y, Lee H-S, Lee Y, Ju YS, Oh JE, Park S-H, et al. Tumor hypoxia represses Γδ T cell-mediated antitumor immunity against brain tumors. Nat Immunol 2021;22:336–46. https://doi.org/10.1038/s41590-020-00860-7.

[30] Siegers GM, Dutta I, Lai R, Postovit L-M. Functional plasticity of gamma delta T cells and breast tumor targets in hypoxia. Front Immunol 2018;9.

[31] Sano Y, Kuwabara N, Nakagawa S, Toda Y, Hosogi S, Sato S, Ashihara E. Hypoxia-adapted multiple myeloma stem cells resist Γδ-T-cell-mediated killing by modulating the mevalonate pathway. Anticancer Res 2023;43:547–55. https://doi.org/10.21873/anticanres.16191.

[32] Chen GG, Woo PYM, Ng SCP, Wong GKC, Chan DTM, van Hasselt CA, Tong MCF, Poon WS. Impact of metformin on immunological markers: implication in its anti-tumor mechanism. Pharmacol Ther 2020;213, 107585. https://doi.org/10.1016/j.pharmthera.2020.107585.

[33] Li L, Cao B, Liang X, Lu S, Luo H, Wang Z, Wang S, Jiang J, Lang J, Zhu G. Microenvironmental oxygen pressure orchestrates an anti- and pro-tumoral Γδ T cell equilibrium via tumor-derived exosomes. Oncogene 2019;38:2830–43. https://doi.org/10.1038/s41388-018-0627-z.

[34] Li Y, Wang Y, Shi F, Zhang X, Zhang Y, Bi K, Chen X, Li L, Diao H. Phospholipid metabolites of the gut microbiota promote hypoxia-induced intestinal injury via CD1d-dependent Γδ T cells. Gut Microbes 2022;14:2096994. https://doi.org/10.1080/19490976.2022.2096994.

[35] Castella B, Foglietta M, Sciancalepore P, Rigoni M, Coscia M, Griggio V, Vitale C, Ferracini R, Saraci E, Omedé P, et al. Anergic bone marrow Vγ9Vδ2 T cells as early and long-lasting markers of PD-1-targetable microenvironment-induced immune suppression in human myeloma. Onco Targets Ther 2015;4, e1047580. https://doi.org/10.1080/2162402X.2015.1047580.
[36] Castella B, Foglietta M, Riganti C, Massaia M. Vγ9Vδ2 T cells in the bone marrow of myeloma patients: a paradigm of microenvironment-induced immune suppression. Front Immunol 2018;9:1492. https://doi.org/10.3389/fimmu.2018.01492.
[37] Cazzetta V, Bruni E, Terzoli S, Carenza C, Franzese S, Piazza R, Marzano P, Donadon M, Torzilli G, Cimino M, et al. NKG2A expression identifies a subset of human Vδ2 T cells exerting the highest antitumor effector functions. Cell Rep 2021;37, 109871. https://doi.org/10.1016/j.celrep.2021.109871.
[38] De Vries NL, Van De Haar J, Veninga V, Chalabi M, Ijsselsteijn ME, Van Der Ploeg M, Van Den Bulk J, Ruano D, Van Den Berg JG, Haanen JB, et al. Γδ T cells are effectors of immunotherapy in cancers with HLA class I defects. Nature 2023;613:743–50. https://doi.org/10.1038/s41586-022-05593-1.
[39] Wu Y, Kyle-Cezar F, Woolf RT, Naceur-Lombardelli C, Owen J, Biswas D, Lorenc A, Vantourout P, Gazinska P, Grigoriadis A, et al. An innate-like Vδ1 + Γδ T cell compartment in the human breast is associated with remission in triple-negative breast cancer. Sci Transl Med 2019;11, eaax9364. https://doi.org/10.1126/scitranslmed.aax9364.
[40] Foord E, Arruda LCM, Gaballa A, Klynning C, Uhlin M. Characterization of ascites- and tumor-infiltrating Γδ T cells reveals distinct repertoires and a beneficial role in ovarian cancer. Sci Transl Med 2021;13, eabb0192. https://doi.org/10.1126/scitranslmed.abb0192.
[41] Rancan C, Arias-Badia M, Dogra P, Chen B, Aran D, Yang H, Luong D, Ilano A, Li J, Chang H, et al. Exhausted intratumoral $V\delta2^-$ Γδ T cells in human kidney cancer retain effector function. Nat Immunol 2023;24:612–24. https://doi.org/10.1038/s41590-023-01448-7.
[42] Brauneck F, Weimer P, Schulze Zur Wiesch J, Weisel K, Leypoldt L, Vohwinkel G, Fritzsche B, Bokemeyer C, Wellbrock J, Fiedler W. Bone marrow-resident Vδ1 T cells co-express TIGIT with PD-1, TIM-3 or CD39 in AML and myeloma. Front Med 2021;8, 763773. https://doi.org/10.3389/fmed.2021.763773.
[43] Zheng J, Qiu D, Jiang X, Zhao Y, Zhao H, Wu X, Chen J, Lai J, Zhang W, Li X, et al. Increased PD-1+Foxp3+ Γδ T cells associate with poor overall survival for patients with acute myeloid leukemia. Front Oncol 2022;12:1007565. https://doi.org/10.3389/fonc.2022.1007565.
[44] Jin Z, Lan T, Zhao Y, Du J, Chen J, Lai J, Xu L, Chen S, Zhong X, Wu X, et al. Higher $TIGIT^+$ $CD226^-$ Γδ T cells in patients with acute myeloid leukemia. Immunol Investig 2022;51:40–50. https://doi.org/10.1080/08820139.2020.1806868.
[45] Guo Q, Zhao P, Zhang Z, Zhang J, Zhang Z, Hua Y, Han B, Li N, Zhao X, Hou L. TIM-3 blockade combined with bispecific antibody MT110 enhances the anti-tumor effect of Γδ T cells. Cancer Immunol Immunother 2020;69:2571–87. https://doi.org/10.1007/s00262-020-02638-0.
[46] Wu K, Feng J, Xiu Y, Li Z, Lin Z, Zhao H, Zeng H, Xia W, Yu L, Xu B. Vδ2 T cell subsets, defined by PD-1 and TIM-3 expression, present varied cytokine responses in acute myeloid leukemia patients. Int Immunopharmacol 2020;80, 106122. https://doi.org/10.1016/j.intimp.2019.106122.
[47] Li X, Lu H, Gu Y, Zhang X, Zhang G, Shi T, Chen W. Tim-3 suppresses the killing effect of Vγ9Vδ2 T cells on colon cancer cells by reducing perforin and granzyme B expression. Exp Cell Res 2020;386, 111719. https://doi.org/10.1016/j.yexcr.2019.111719.
[48] Gertner-Dardenne J, Fauriat C, Orlanducci F, Thibult M-L, Pastor S, Fitzgibbon J, Bouabdallah R, Xerri L, Olive D. The co-receptor BTLA negatively regulates human Vγ9Vδ2 T-cell proliferation: a potential way of immune escape for lymphoma cells. Blood 2013;122:922–31. https://doi.org/10.1182/blood-2012-11-464685.
[49] Hu G, Wu P, Cheng P, Zhang Z, Wang Z, Yu X, Shao X, Wu D, Ye J, Zhang T, et al. Tumor-infiltrating $CD39^+$ γδTregs are novel immunosuppressive T cells in human colorectal cancer. Onco Targets Ther 2017;6, e1277305. https://doi.org/10.1080/2162402X.2016.1277305.
[50] Hoeres T, Holzmann E, Smetak M, Birkmann J, Wilhelm M. PD-1 signaling modulates interferon-γ production by gamma delta (Γδ) T-cells in response to leukemia. Onco Targets Ther 2019;8:1550618. https://doi.org/10.1080/2162402X.2018.1550618.
[51] Girard P, Charles J, Cluzel C, Degeorges E, Manches O, Plumas J, De Fraipont F, Leccia M-T, Mouret S, Chaperot L, et al. The features of circulating and tumor-infiltrating Γδ T cells in melanoma patients display critical perturbations with prognostic impact on clinical outcome. Onco Targets Ther 2019;8:1601483. https://doi.org/10.1080/2162402X.2019.1601483.

[52] Wu K, Zhao H, Xiu Y, Li Z, Zhao J, Xie S, Zeng H, Zhang H, Yu L, Xu B. IL-21-mediated expansion of Vγ9Vδ2 T cells is limited by the Tim-3 pathway. Int Immunopharmacol 2019;69:136–42. https://doi.org/10.1016/j.intimp.2019.01.027.
[53] Cerapio J-P, Perrier M, Balança C-C, Gravelle P, Pont F, Devaud C, Franchini D-M, Féliu V, Tosolini M, Valle C, et al. Phased differentiation of Γδ T and T CD8 tumor-infiltrating lymphocytes revealed by single-cell transcriptomics of human cancers. Onco Targets Ther 2021;10:1939518. https://doi.org/10.1080/2162402X.2021.1939518.
[54] Tang L, Wu J, Li C-G, Jiang H-W, Xu M, Du M, Yin Z, Mei H, Hu Y. Characterization of immune dysfunction and identification of prognostic immune-related risk factors in acute myeloid leukemia. Clin Cancer Res 2020;26:1763–72. https://doi.org/10.1158/1078-0432.CCR-19-3003.
[55] Fattori S, Gorvel L, Granjeaud S, Rochigneux P, Rouvière M-S, Ben Amara A, Boucherit N, Paul M, Dauplat MM, Thomassin-Piana J, et al. Quantification of immune variables from liquid biopsy in breast cancer patients links $V\delta2^+$ ΓδT cell alterations with lymph node invasion. Cancer 2021;13:441. https://doi.org/10.3390/cancers13030441.
[56] Reis BS, Darcy PW, Khan IZ, Moon CS, Kornberg AE, Schneider VS, Alvarez Y, Eleso O, Zhu C, Schernthanner M, et al. TCR-Vγδ usage distinguishes protumor from antitumor intestinal Γδ T cell subsets. Science 2022;377:276–84. https://doi.org/10.1126/science.abj8695.
[57] Cazzetta V, Depierreux D, Colucci F, Mikulak J, Mavilio D. NKG2A immune checkpoint in Vδ2 T cells: emerging application in cancer immunotherapy. Cancer 2023;15:1264. https://doi.org/10.3390/cancers15041264.
[58] Zakeri N, Hall A, Swadling L, Pallett LJ, Schmidt NM, Diniz MO, Kucykowicz S, Amin OE, Gander A, Pinzani M, et al. Characterisation and induction of tissue-resident gamma delta T-cells to target hepatocellular carcinoma. Nat Commun 2022;13:1372. https://doi.org/10.1038/s41467-022-29012-1.
[59] Clarke J, Panwar B, Madrigal A, Singh D, Gujar R, Wood O, Chee SJ, Eschweiler S, King EV, Awad AS, et al. Single-cell transcriptomic analysis of tissue-resident memory T cells in human lung cancer. J Exp Med 2019;216:2128–49. https://doi.org/10.1084/jem.20190249.
[60] Rossi C, Gravelle P, Decaup E, Bordenave J, Poupot M, Tosolini M, Franchini D-M, Laurent C, Morin R, Lagarde J-M, et al. Boosting Γδ T cell-mediated antibody-dependent cellular cytotoxicity by PD-1 blockade in follicular lymphoma. Onco Targets Ther 2019;8:1554175. https://doi.org/10.1080/2162402X.2018.1554175.
[61] Xiong D, Wang Y, You M. A gene expression signature of TREM2hi macrophages and Γδ T cells predicts immunotherapy response. Nat Commun 2020;11:5084. https://doi.org/10.1038/s41467-020-18546-x.
[62] Yang R, Shen S, Gong C, Wang X, Luo F, Luo F, Lei Y, Wang Z, Xu S, Ni Q, et al. Bispecific antibody PD-L1 x CD3 boosts the anti-tumor potency of the expanded Vγ2Vδ2 T cells. Front Immunol 2021;12, 654080. https://doi.org/10.3389/fimmu.2021.654080.
[63] Tomogane M, Sano Y, Shimizu D, Shimizu T, Miyashita M, Toda Y, Hosogi S, Tanaka Y, Kimura S, Ashihara E. Human Vγ9Vδ2 T cells exert anti-tumor activity independently of PD-L1 expression in tumor cells. Biochem Biophys Res Commun 2021;573:132–9. https://doi.org/10.1016/j.bbrc.2021.08.005.
[64] Hwang HJ, Lee JJ, Kang SH, Suh JK, Choi ES, Jang S, Hwang S, Koh K, Im HJ, Kim N. The BTLA and PD-1 signaling pathways independently regulate the proliferation and cytotoxicity of human peripheral blood Γδ T cells. Immun Inflamm Dis 2021;9:274–87. https://doi.org/10.1002/iid3.390.
[65] Cui J, Li L, Wang C, Jin H, Yao C, Wang Y, Li D, Tian H, Niu C, Wang G, et al. Combined cellular immunotherapy and chemotherapy improves clinical outcome in patients with gastric carcinoma. Cytotherapy 2015;17:979–88. https://doi.org/10.1016/j.jcyt.2015.03.605.
[66] Pan Y, Chiu Y-H, Chiu S-C, Cho D-Y, Lee L-M, Wen Y-C, Whang-Peng J, Hsiao C-H, Shih P-H. Gamma/delta T-cells enhance carboplatin-induced cytotoxicity towards advanced bladder cancer cells. Anticancer Res 2020;40:5221–7. https://doi.org/10.21873/anticanres.14525.
[67] Zheng Y, Wang P, Fu Y, Chen Y, Ding Z-Y. Zoledronic acid enhances the efficacy of immunotherapy in non-small cell lung cancer. Int Immunopharmacol 2022;110, 109030. https://doi.org/10.1016/j.intimp.2022.109030.
[68] Imai Y, Ayithan N, Wu X, Yuan Y, Wang L, Hwang ST. Cutting edge: PD-1 regulates imiquimod-induced psoriasiform dermatitis through inhibition of IL-17A expression by innate Γδ-low T cells. J Immunol 2015;195:421–5. https://doi.org/10.4049/jimmunol.1500448.
[69] Li N, Xu W, Yuan Y, Ayithan N, Imai Y, Wu X, Miller H, Olson M, Feng Y, Huang YH, et al. Immune-checkpoint protein VISTA critically regulates the IL-23/IL-17 inflammatory axis. Sci Rep 2017;7:1485. https://doi.org/10.1038/s41598-017-01411-1.

[70] Galon J, Costes A, Sanchez-Cabo F, Kirilovsky A, Mlecnik B, Lagorce-Pagès C, Tosolini M, Camus M, Berger A, Wind P, et al. Type, density, and location of immune cells within human colorectal tumors predict clinical outcome. Science 2006;313:1960–4. https://doi.org/10.1126/science.1129139.
[71] Todaro M, Orlando V, Cicero G, Caccamo N, Meraviglia S, Stassi G, Dieli F. Chemotherapy sensitizes colon cancer initiating cells to Vγ9Vδ2 T cell-mediated cytotoxicity. PLoS One 2013;8, e65145. https://doi.org/10.1371/journal.pone.0065145.
[72] Mattarollo SR, Kenna T, Nieda M, Nicol AJ. Chemotherapy and zoledronate sensitize solid tumour cells to Vγ9Vδ2 T cell cytotoxicity. Cancer Immunol Immunother 2007;56:1285–97. https://doi.org/10.1007/s00262-007-0279-2.

CHAPTER

4

Appraising γδ T cell exhaustion and differentiation in the context of synthetic engineering for cancer immunotherapy

John Anderson

UCL Great Ormond Street Institute of Child Health, University College London, London, United Kingdom

Abstract

Successes with cellular immunotherapy using chimeric antigen receptor (CAR) αβ T cells have delivered a paradigm shift in leukemia and multiple myeloma treatment. The failure to date of iterative adaptations of this approach to provide similar benefits in relapsed and refractory solid tumor indications has prompted a search for alternatives to αβ T cell-derived adaptive immunotherapy products. Exorbitant costs and complex logistics have further motivated the examination of allogeneic options to deliver off-the-shelf, as opposed to autologous, products. γδ T cells offer one such alternative chassis, with known tissue-homing properties, favorable prognostic value upon solid tumor infiltration and a lack of alloreactivity. Nonetheless, a great deal about γδ T cell biology remains unknown. Even basic parameters of immunotherapeutically relevant γδ T cell physiology is poorly understood, including their ability to persist, their sensitivity to activation-induced exhaustion and their unique pathways of functional differentiation. A firm understanding of these phenomena has been and continues to be a cornerstone of intelligent αβ T cell adoptive immunotherapy design. The paucity of similar understanding in the context of γδ T cells represents a substantial bottleneck for γδ T cell immunotherapy clinical translation. The herein review summarizes the state-of-the-art knowledge about αβ T cell exhaustion and differentiation and discusses what is—and what isn't—known about γδ T cells in the same context. It then examines the known γδ T cell biology about synthetic engineering and therapeutic combinations aimed at enhancing antitumor efficacy, summarizing the cutting-edge considerations for gene-modified γδ T cell immunotherapy design.

γδ T Cell Cancer Immunotherapy
https://doi.org/10.1016/B978-0-443-21766-1.00009-6

Abbreviations

αβ T	alpha beta T cells
γδ T	gamma delta T cells
Ag	antigen
BCMA	B-cell maturation antigen
BTLA	B and T cell attenuator
BTN3A1 and BTN2A1	butyrophilins 3A1 and 2A1
CARs	chimeric antigen receptors
CCR	chimeric co-stimulatory receptor
CDE	complementarity determining region
CRISPR	clustered regularly interspaced short palindromic repeats
CTLA4	cytotoxic T-lymphocyte associated protein 4
DGK	diacylglycerol kinase
EPCR	endothelial cell protein C receptor
FcγR	Fc gamma receptors
FDA	US Food and Drug Administration
GvHD	graft versus host disease
HIV	human immunodeficiency virus
HVEM	herpes virus entry mediator
IPP	isoprenyl pyrophosphate
KIR	killer-immunoglobulin-like receptors
LAG3	lymphocyte activation gene 3
MHC	major histocompatibility complex
NFAT	nuclear factor of activated T cells
NK cells	natural killer cells
pAg	phosphoantigen
pAPC	professional antigen-presenting cells
PD1	programmed cell death protein 1
SARS-COV2	severe acute respiratory syndrome coronavirus 2
ScFv	single chain variable fragment
TAC	T cell antigen coupler
TALEN	transcription activator-like effector nucleases
TB	tuberculosis
TCR	T cell receptor
TIGIT	T-cell immunoglobulin and immunoreceptor tyrosine-based inhibitory motif domain
TIL	tumor infiltrating lymphocytes
TIM3	T cell immunoglobulin and mucin-domain containing-3
TOX	thymocyte selection-associated HMG box

Conflict of interest

MB and JA hold patents that pertain to the development of T cell immunotherapy. JA holds founder stock in Autolus Ltd.

Introduction

The field of cancer immunotherapy in the first two decades of the 21st century has witnessed stunning exponential expansion. Remarkable too is the evolutionary biology of the gamma delta subset of T lymphocytes (γδ T), a population that still remains relatively overlooked regarding their potential therapeutic capabilities. γδ T cells are present in all jawed vertebrates, attesting to their importance in immunity throughout 500 million years of evolution [1]. Their evolutionary resilience implies a vital role in host defense that complements the adaptive immunity conferred by alpha beta T cells (αβ T). Like in αβ T cells the γδ T cell receptor interacts with CD3 to form a functional complex and its diversity is generated by the rearrangement of variable (V), joining (J), and diversity (D) segments during T cell ontogeny. The integration of the human T cell receptor delta chain genes within the same genetic locus as the alpha gene segments is of significance; not only in terms of the co-evolution of the subsets but also in terms of early T cell fate decisions. T cell ontogeny is driven by the sequential rearrangement of delta, gamma, beta, and alpha gene segments. The stochastic nature of gene segment rearrangements serves dual purposes: to provide huge diversification within the TCR repertoires and to determine whether a nascent T cell progresses down the alpha-beta or γδ T cell route. Delta segments rearrange first within the alpha/delta locus, and alpha segment genes only rearrange if delta rearrangements fail to produce a stable in-frame protein. Hence the quip that αβ T cells can be considered as failed γδ T cells. This coordinated regulation of the respective TCR gene segments to determine subsequent cell identity is a telling example of their complementary and interdependent roles that have maintained the basic γδ T identical over 500 million years of evolution [2,3]. But what exactly are these complementary roles and how can the understanding of them contribute to further refinement of cancer immunotherapy? That is the subject of the herein review.

The power of T cell adoptive transfer to eradicate cancers

As a primer to consider the different potential tools and solutions for exploiting the cellular chassis of γδ T cells in cancer immunotherapy, it is instructive to review progress made in the adoptive T cell field exploiting the inherent anticancer properties of conventional αβ T cells. Before the advent of cellular engineering, the capability of T cells to contribute powerfully to cancer control had been established through two distinct disciplines: allogeneic haemopoietic stem cell transplantation for hematological cancers, and adoptive transfer of expanded tumor-infiltrating lymphocytes, most markedly in patients with melanoma [4–6].

The concept of the graft versus leukemia effect as an important component of the therapeutic benefits of allogeneic stem cell transplantation in high-risk leukemia patients emerged in the late 1980s through several observations from clinical trials that leukemia relapse is inversely correlated with graft versus host disease (GvHD) and is greater in transplants with decreased risk of GVHD, for example, identical twin graft and T cell-depleted grafts, as elegantly summarized by Sosman and Sondel in a 1991 review [7]. That concept has not been challenged; yet for many years, it was broadly thought that T cell-mediated immunotherapy represented a specific sensitivity of leukemia, linked to general alloreactivity of adoptively transferred donor-derived T cells. The concept that autologous T cells are capable of recognition of nonself on cancer cells was somewhat dismissed: the main cited evidence in defense of this immuno-surveillance skepticism was lack of increased spontaneous tumor incidence in the context of immunodeficiency; both human (e.g., solid tumor transplants and inherited severe immunodeficiency) and mice (genetic knock-outs) [6].

Although some of the earliest proof of concept studies demonstrating the capacity of adaptive immunity to reject cancer was derived from dendritic cell vaccine studies in melanoma and renal cell cancer [8,9], the most dramatic demonstration of the capacity of autologous tumor-reactive adaptive immunity to shrink established solid cancers came from the field of tumor infiltrating lymphocytes in melanoma. Pioneering work, most notably from the National Cancer Institute generated protocols for large-scale ex vivo expansion of tumor infiltrating T cells derived from melanoma biopsy samples followed by reinfusion into patients after lymphodepleting chemotherapy with or without total body irradiation [10,11]. Although limited by the labor-intensive nature of bespoke cell product manufacture and the applicability of the approach being restricted to "immune-hot" cancers, such as melanoma, these pioneering studies demonstrated the key proof of concept for the potential of adoptively transferred tumor-reactive T cells to effect long term disease remission and paved the way for the development of engineered T cells with increase the scope of immunotherapies toward colder cancer types.

Engineered αβ T cells in adoptive transfer: State of the art

Engineering with synthetic receptors to redirect T cell specificity toward antigens on the cancer cell surface has been built on two basic concepts; engineering of αβ T cells with a new T cell receptor, and engineering with chimeric antigen receptors (CARs), which comprise specificity for a cell surface target conferred by an antigen-binding ectodomain, combined with a T cell's signaling endodomains, within a single membrane-straddling molecule [12–15]. A comprehensive comparison of TCR versus CAR-engineered autologous αβ T cell adoptive transfer is beyond the scope of this review. However, some key principles have emerged from the clinical experience to date, which are highly relevant for the consideration of engineering approaches for γδ T cell therapeutics (Fig. 1). The following are four key lessons learned and associated pointers to the γδ T cell field:

Lesson 1: CAR engineering has delivered a broader toolbox of cancer antigens to target than the TCR field.

αβ T cell TCRs target peptide antigens presented on cancer cells as complexes with MHC class I or II molecules and are therefore restricted by the MHC type of the cancer itself. Furthermore, the success of genetic modification with new TCRs is dependent on two main factors: (1) the target peptide neoantigen or tumor-associated antigen must be very cancer-specific, and (2) the target cancer must not prove itself capable of avoiding TCR-based recognition by downregulation of antigen expression, antigen processing and/or MHC expression pathways. Virtually all cell surface and cancer-specific molecules can in theory be targeted by a monoclonal antibody of appropriate binding characteristics after engineering into a single-chain CAR-T format. Hence, the number of cancer specificities that have been evaluated and have demonstrated proof of concept in a CAR-T platform surpasses the range of available reagents that have thus far been evaluated using TCR engineering.

Lesson 2: CAR-T approaches have shown greater clinical successes in liquid malignancies than in solid cancers.

The clinical successes of CAR-T products targeting antigens expressed on cancer cells in B cell malignancies have resulted now in six FDA approvals: four for products targeting CD19 and two targeting BCMA. In contrast, there have been no approvals for solid

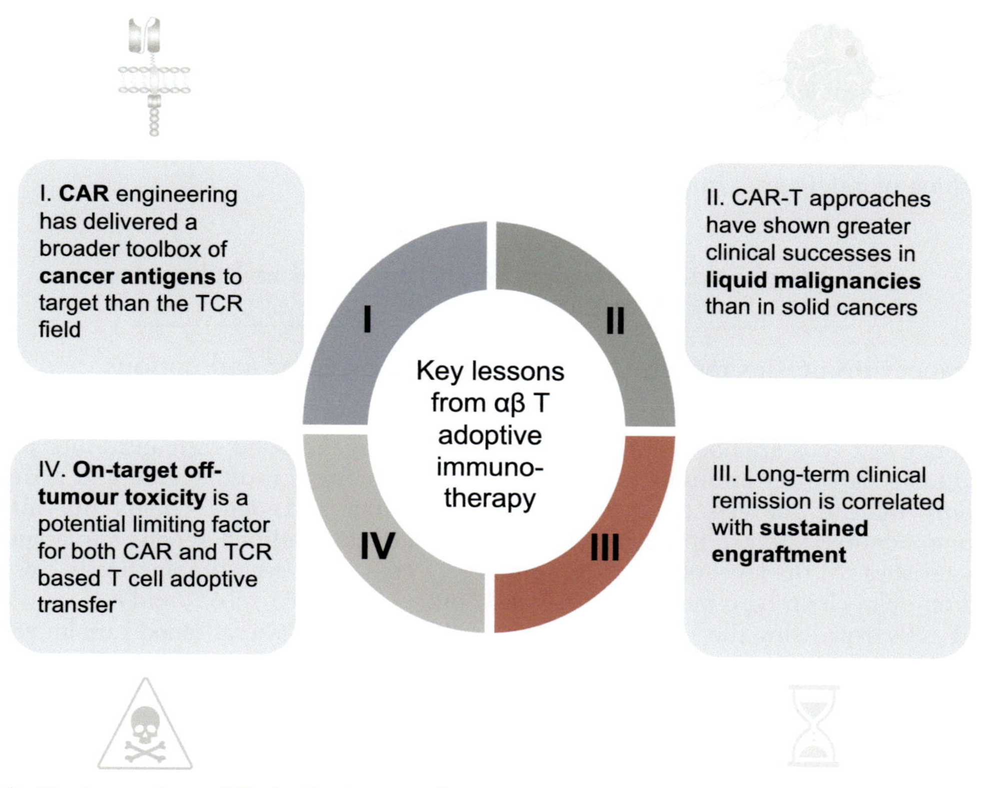

FIG. 1 Key lessons from αβ T adoptive immuno-therapy.

nonhaemopoietic cancers, and early phase academic trials, while demonstrating important proof of concept of clinical response and patient benefit, for example in neuroblastoma [16–18] have overall demonstrated significantly lower best response rates and shorter duration of response than for hematological cancers.

Lesson 3: Long-term clinical remission is correlated with sustained engraftment, which is affected by both T cell biology as impacted by engineering, and cancer-specific factors.

In the haemato-oncology field, the duration and depth of CAR-T cell engraftment are correlated with the duration of clinical response, and a similar pattern is observed in solid cancers both in clinical trials and preclinical models. Many mechanisms of engraftment failure have been demonstrated experimentally. These can be summarized as T cell exhaustion due to excessive antigenic stimulation, inadequate development of memory response, and failure of a proliferative burst of in vivo expansion due to suppressive solid tumor microenvironment [12,13].

Lesson 4: On-target off-tumor toxicity is a potential limiting factor for both CAR and TCR based T cell adoptive transfer.

The targeting of CD19 by CAR-T cells induces a profound degree of toxicity through the killing of CD19-expressing healthy B cells, a toxicity that is unavoidable and is broadly considered manageable: Regular infusions of immunoglobulin to treat B cell aplasia is considered an acceptable price for sustained cancer remission. Identification of new target antigens that are entirely cancer-specific but expressed at high enough levels to be targeted remains something of a holy grail in the discovery science field.

γδ T lymphocytes show significant functional and phenotypic differences from αβ T cells

Anticancer properties mediated by the γδ T cell receptor and options for exploitation in synthetic biology

Human γδ T cells are not a single distinctive cell lineage imbued with antitumor properties suitable for cancer immunotherapy. Their heterogeneity is multilayered and is defined not only by the subtype of T cell receptors gamma and delta chains but also by immunophenotype, transcriptional programming and yet undefined genetic and epigenetic characteristics. At the level of the TCR, human γδ T cells have been pragmatically classified based on delta chain type into Vδ-1, Vδ-2 and "others," i.e., Vδ3,4,5,6,7, and 8.

γδ T cells expressing the Vδ2 TCR are the predominant peripheral blood human γδT cell population, most typically pairing with a Vγ9-chain. Vγ9Vδ2 cells are effector cells primed for innate immune responses capable of robust expansion following acute infection or challenge with cancer targets.

Vγ9Vδ2 cells share a common antigenic target that engages with the TCR to induce signaling and activation, the molecular characterization of which has been elucidated with greater clarity in recent years. Specifically, nonpeptidic pyrophosphates, or phosphoantigens (pAgs), that are intermediates in microbial metabolic pathways and in the mammalian mevalonate pathway [19,20] lead to the formation of an immune synapse with the target cell through a mechanism that involves Vγ9Vδ2 TCR interactions with butyrophilin molecules BTN3A1 and BTN2A1 [21–23]. Studies have identified a key interaction to be between a germline (non-CDR) epitope on the Vγ9 chain with BTN2A1. At least one additional ligand interaction with a separate TCR domain is required, and BTN3A1 interaction with Vδ2 CDR3 motifs is a lead candidate [21,22].

Recent studies indicate that BTN3A1 and TCR both interact with BTN2A1 in a mutually exclusive manner but that BTN3A1/BTN2A1 dimerization is not necessary, implicating a model in which germline-encoded weak TCR/BTN2A1 interaction is converted to a strong interaction through CDR3 encoded interactions with BTN3A1, which is enhanced by phospophantigen [24]. The pAgs that drive this molecular interaction are upregulated during either cellular transformation or microbial infection, and this sensitivity provides a useful clinical tool for in vivo or in vitro expansion specific for the Vγ9Vδ2 subtype. Hence, there is an opportunity for clinical translation through the use of the clinically approved amino bisphosphonate drugs (e.g., zoledronate, pamidronate) that induce accumulation of isoprenyl pyrophosphate (IPP) phosphoantigen through inhibition of farnesyl pyrophosphate synthase, immediately downstream from IPP in the cholesterol biosynthesis pathway [25].

Moreover, the emerging understanding of the key molecular interactions between the Vγ9Vδ2 TCR with its complex cognate ligand/s has equipped researchers with synthetic biology tools for refined chimeric receptors combining antigen recognition with optimal stimulatory and co-stimulatory signals.

Vδ1 cells, in contrast, are more limited in number in human peripheral blood and reside principally in adult peripheral tissues such as the gut [26] and skin [27]. While the more common Vδ2 subtype is rarely paired with any chain apart from Vγ9, Vδ1 cells are more promiscuous for γ-chain pairing, which is reflected in the greater range of cognate ligands of the Vδ1 TCR that include stress-induced self-antigens and CD1c-presented antigens [28,29]. Moreover, in contrast with the blood-borne Vγ9Vδ2 population, Vδ1 cells are more prevalent in peripheral tissues and epithelial surfaces while being relatively sparse in blood.

An exception is in the context of infection where certain pathogens including CMV and malaria are associated with very high circulating levels of blood Vδ1 cells. It is an intriguing hypothesis that the intrinsic tissue-residency properties of Vδ1 cells will be mirrored in superior tropism toward and survival within solid tumors in the context of cancer immunotherapy when compared with the more naturally blood-resident Vγ9Vδ2 T cells. Vδ1 cells in adult tissues and following acute infection have a more focused TCR repertoire than the diverse repertoire in cord blood, consistent with an adaptive-like response to infection [30–32]. Whether this TCR diversity and heterogeneity will result in a more heterogeneous anticancer response of Vδ1 cell products dependent on TCR specificity is currently unknown.

In contrast with Vγ9Vδ2 and Vδ1 T cells, the other human γδT cell subsets are rarer in blood and tissues and are relatively poorly characterized, the molecular ligands for their TCRs are largely unknown, and their anticancer properties are more enigmatic. One clue arises from γδ T TCR chain usage in tumor infiltrating lymphocytes (TILs). Few studies have included single-cell TCR sequencing analysis of the γδ TCR genes cells even though they are a relatively enriched population in TILs in several cancers [33], and a notable study of pan-cancer analysis has identified the presence of γδ T TILs to be the most infiltrating immune cell most significantly associated with favorable prognosis [34].

In the limited number of TIL studies reporting on γδ T TCR repertoire, there is a strong indication for expression of the full range of TCR clonotypes indicating a heterogeneous tumor response [33,35,36]. Whether specific TCR chain types and clonotypes are specifically associated with tumor reactivity or regulatory roles remains unanswered but is an important translational question. Specifically, the identification of TCRs similar to Vg9Vδ2, which appear to have a pan-cancer reactivity, could lead to increased repertoire of off-the-shelf cancer-specific biotherapeutics.

There are, however, several characterized non-Vδ1, non-Vγ9Vδ2 TCRs for which cognate antigens have been discovered, notable examples being Vγ4Vδ5 interaction with EPCR [37], Vγ8Vδ3 interaction with Annexin-A2 [38], poorly-understood Vγ9Vδ2 interaction with tRNA synthetase [39] and Vγ9Vδ1 interaction with Ephrin-A2 [40]. A common feature of these TCRs is direct and MHC-independent antigen binding, and antigens which are upregulated on the cell surface in response to stress conditions of infection or cancer. Hence, for EPCR and EphA2, the TCR was initially derived from analysis of response to cytomegalovirus but subsequently identified as upregulated on tumor cells in stress conditions. In contrast, other non-Vγ9Vδ2 TCRs bind alternate butyrophilin family members [41] or MHC-like molecules CD1c [42] or CD1d, the latter complexed to glycolipid sulfatide [43].

Together these studies of human γδ T TCR specificities complemented by research efforts to identify specificities in mouse and other jawed vertebrate species have shown that γδ T cells are a heterogeneous population of lymphocytes, the phenotype and function of which is only in part dictated by the gamma and delta chain specificities of the TCR. Although there is circumstantial evidence for adaptive behavior as exemplified by focusing of TCR specificities during expansion in response to stress or infection, γδ T can justifiably also be referred to as innate by virtue of (1) overlapping target antigenic targets for TCR that overlaps with other innate lymphocytes such as NK cells, and (2) the expression of an array of NK receptors and Toll-like receptors [44,45].

Natural killer type receptors and innate anticancer reactivity

In addition to the γδ T TCR, which is assumed to play a critical role in the direct recognition of cancer cells to induce anticancer reactivity, the plethora of innate recognition receptors on γδ T cells are reminiscent of the receptor repertoire of NK and NK-T cells [2,45]. Within this family of co-receptors are NK-type receptors such as NKG2D, Fc-gamma receptors (FcγR), ligands for death receptors such as TRAIL ligand and FAS-L, and receptors for cytokine signaling. γδ T also express inhibitory receptors such as NKG2A and the KIR receptors. In equipping γ T with engineered receptors for enhanced function attention must be paid to the existing expression of multiple innate effector receptors and the importance of synthetic receptors being complementary to existing signaling pathways.

Co-stimulation of the γδ T cell

The presence and active function of NK receptors in γδ T cells is well documented but less well established is the dependency of γδ T for co-stimulatory signaling to provide classical co-stimulation to complement TCR engagement. Whereas in classical naïve αβ T cells, co-stimulation (signal 2) through, e.g., CD28, CD27, CD137 and others, is essential to prevent activation-induced cell death on the engagement of TCR, this requirement is not proven in the context of TCR stimulation in naïve γδ T. It is known that γδ T of both Vδ1 and Vδ2 subtypes express both CD27 and CD28, the expression of which changes during activation and expansion [30,46–51] but their functional role is less clear. This issue is of fundamental importance in the design of engineered receptors of γδ T cells since fine-tuning of signal strength for maximal effector function and minimizing exhaustion is a major theme in the adoptive αβ T cell transfer field and may also be highly relevant to γδ T health during expansion. Moreover, the relationship between the signal pathways induced by co-stimulation receptors, NK receptors and Fcγ receptors may prove to be a major emerging theme in fine-tuning of engineered γδ T therapeutics.

Professional antigen presentation by γδ T cells

γδ T cells have the capability for differentiation toward professional-like antigen-presenting cells (pAPCs) following antigen stimulation [52–54]. Although the role of γδT pAPC function to induce robust T cell responses has been proven in vitro, the dependency on this γδT pAPC property for successful in vivo responses to infection or cancer is not established. Mice lacking dendritic cells have a more profound phenotype of immune deficiency than reported in γδT loss-of-function mutants, implying redundancy of the γδT pAPC function [55,56]. Nevertheless, this additional property is often overlooked and might significantly strengthen the depth and duration of responses within the context of adoptive γδT cell transfer studies [53].

Choice of differentiation state for γδ T cell adoptive transfer

Current knowledge on optimal T memory populations for adoptive transfer in the αβ T cells field

Conventional αβ T cells in humans have a range of differentiation fates, which are dependent on the strength of TCR stimulation/co-stimulation, and the cytokine milieu [57]. The use of optimal conditions to enrich less differentiated T cells in the final CAR-T cell product has been shown to improve their antitumor efficacy in vivo [58].

Moreover, distinct differentiation pathways exist for the two main subtypes of αβ T cells; CD4 and CD8. Differentiation of CD4 αβ T cells driven by the specific inflammatory nature of the local environment results in cells with specialist functions; for example Th1 cells that promote proinflammatory immune responses (driven by Interferon-gamma and IL-12), Th2 that promote humoral and helminth immunity and allergy (driven by IL-4), Th9 cells [59] with a role in helminth immunity and allergy (driven by IL-4 and TGF-β), Th17 that plays roles at mucosal barriers and can be either regulatory or pro-inflammatory [60] (driven by IL-1, IL-6, IL-23, TGF-β), Th22 with tissue-specific anti-infective protective roles (driven by IL-6 and TNF-α), and Treg that inhibit Th1 responses and promote cancer growth (driven by high concentration IL-2 and TGF-β). These differentiated populations driven by specific cytokine conditions have distinct transcriptional signatures and lineage-defining transcription factor drivers; for example, Tbet in Th1 cells, FOXP3 in Treg and RORγT in Th17 cells, although there is overlap between transcription factors in some subsets such as Th22.

The similar specialized function has been less fully characterized in CD8 αβ T cells. Independent of specialized functional differentiation, human effector cells of CD4 helper and CD8 cytotoxic lineages are broadly recognized as differentiating from naïve ($CD45RA^+$ $CCR7^+$ $CD27^+$) to central memory ($CD45RA^-$ $CCR7^+$ $CD27^+$) to effector memory ($CD45RA^-$ $CCR7^-$ $CD27^-$) to terminal effector ($CD45RA^+$ $CCR7^-$ $CD27^-$) phenotypes. In commonly used classifications, CD28 or CD62L are frequently substituted for CCR7 or CD27. Naïve and central memory cells expressing the lymph node homing receptors CCR7 and CD62L are enriched in self-renewing capacity while the loss of these markers as cells differentiate is associated with enhanced effector function but diminished replicative capacity. A T memory stem cell population (T_{SCM}: CD45RA CD27 CD28 CCR7 CD62L CD95 CD122 CXCR3 phenotype in humans) with greater stem cell-like properties than conventional central memory cells has been described in both mice and humans and shows particular capacity for long-term engraftment [61–63].

Following on from these flow cytometry marker-based descriptive phenotypes, there have been many studies using unbiased approaches to describe the stem population capable of supporting long-term engraftment and effector populations in CAR-T products [64]. Indeed, a fundamental limitation in the adoptive transfer field is that common transduction protocols require T cell activation, which by its nature promotes differentiation yet is a prerequisite for gamma-retroviral/lentiviral transduction.

Preclinical studies have broadly shown that a CAR-T cell product that is enriched for less differentiated states, i.e., naïve, T_{SCM}, and central memory, will have improved tumor control in long-term in vivo and in vitro assays [65]. Recent studies have demonstrated that high CAR-Treg numbers postinfusion is associated with poor response [66]. Less differentiated T cells have increased oxidative metabolism and mitochondrial biogenesis, and both of these are features of CAR with 41BB endodomains and are associated with enhanced sustained engraftment in preclinical models [67], which is linked with improved outcomes in clinical studies [68].

Recent advances have shown that transduction of nonactivated T cells (at relatively low efficiency) followed by infusion of low CAR-T cell numbers of naïve cells and allowing expansion to occur in vivo on encounter with tumor antigen shows promise for enhanced efficacy compared with conventional protocols, at least within the context of CD19 targeted CARs where encounter with antigen to drive proliferation occurs immediately following infusion [69].

Evidence of optimal γδ T memory/stem populations for adoptive transfer

The body of literature describing putative differentiation subsets of γδ T cells in mice and humans ascribing functional properties to distinct immunophenotypes is significantly smaller than that of the αβ T cell field. Similarities between γδ T marker expression with canonical αβ T memory differentiation have been demonstrated [50,70,71]. However, there are also some emerging differences, and there may be differences between mice and humans as well as between Vδ2 and Vδ1 subsets.

Similar to αβ T cells, for example, human Vδ2 cells—if primed under Th1 conditions—are induced to generate IFN-γ and TNF-α, to produce IL-4 under control of the GATA-3 transcription factor in IL-4 conditions, or to be polarized toward IL-17 production [70]. Indeed, in mice, there is evidence of distinct metabolic programs between Interferon gamma-producing (glycolysis) and IL-17 producing (oxidative metabolism) lineages which are fixed in early γδ T cell ontogeny [72]. In Vδ2 cells conventional αβ T markers CD45RA and CD27 have been used to define naïve, central memory and effector memory and terminal effector cells with similar patterns of differentiation described for αβ T cells. Using this conventional phenotyping, Vδ2 central memory γδ T cells were described to express chemokine receptor markers CXCR3 and CCR5 and have high proliferative capacity but limited effector function while more differentiated subsets have enhanced effector function but reduced proliferative capacity [50,73].

In contrast, other studies in Vδ2 cells have shown significant heterogeneity between individuals paired with retention of phenotype over time. These studies have also noted that CD45RA does not demarcate distinct populations, unlike CD27, CD28, and CD16. It is suggested that CD27 and CD28 broadly identify cells with replicative capacity, while $CD27^-$ cells are quiescent but with enhanced cytotoxic function [74]. Starting with the assumption that cord blood γδ T cells represent truly naïve cells, studies in both Vδ1 cells [30,31] and atypical Vδ2 cells ($Vg9^-V\delta2^+$) [31] indicate that, as a marker, CD45RA is heterogenous and nondistinct, while these naïve cells are uniformly consistently CD27 positive with unfocussed TCR repertoire. In cord blood, naive Vδ1 cells that are $CD27^+$ further co-express CD28, CD62L and CCR7 [30,32]. Following the expansion of cord blood Vδ1 cells the TCR repertoire becomes more focussed and associated with loss of CD27 which is also a feature of the adult $CD27^-$ Vδ1 repertoires [30].

Similarly, highly proliferative Vδ1 cells emerging following haemopoietic stem cell transplantation is noted to have an oligoclonal TCR repertoire in response to viral infection, which is consistent with a more adaptive-type response in the Vδ1 repertoire [75]. Moreover, unsupervised single cell analysis of peripheral blood Vδ1 cells identifies two main populations, a predominantly $CD27^{hi}$ population containing stem/naïve markers and with unfocussed TCR repertoire, and a predominantly CD27low population enriched in genes of effector function [76]. Taken together, the published literature suggests that canonical naïve and central memory marker phenotypes based on CD45RA and CD27 cannot be assumed to have the same stem properties and favorable characteristics for adoptive transfer when applied to γδ T. Moreover, a T_{SCM} phenotype has not been described in γδ T. Nonetheless, the co-expression of CD27, CD28, CD62L and CCR7 in cord blood γδ T and the identification of effector cells expressing these markers in mature peripheral blood γδ T with effector function [32,46,51,76,77] hints that these populations are candidate cells with combined tissue homing and stem-like properties.

Another challenge and area with potential for translation lies in understanding whether differentiation phenotypes are fixed since this has implications for the selection of optimal memory populations for sustained engraftment as has been proposed in the αβ T CAR field [58]. Of note, longitudinal studies in human Vγ9Vδ2 cells have demonstrated a relatively fixed memory state albeit with significant variability between individuals [74]. Insights from mouse γδ T biology suggest that studies of plasticity might be fruitful since the study of mouse thymic and peripheral γδ T indicates that $CD27^+$ cells are fixed for interferon-gamma production while peripheral $CD27^-$ cells can be induced to induce either IFN-γ or IL-17 through epigenetic reprogramming depending on environmental context [78,79].

In conclusion, the current understanding of stem-like phenotypes and behaviors in γδ T cell subsets is incomplete. Since the identification of cell populations with the capacity for long-term engraftment and maintenance of cancer-reactive populations can be assumed to be a prerequisite for successful adoptive transfer, it is clear that further research in this area is the priority.

γδ T cell exhaustion

In the alpha beta CAR-T field, significant effort has been devoted to understanding and addressing the biology of T cell exhaustion. In chronic viral infection, repeated and ongoing antigenic stimulation through microbial persistence has the potential to lead to immune pathology as a result of unrestrained T cell activation. As a homeostatic mechanism for avoidance of immune pathology, T cells enter into a pathway of exhaustion which may be defined as functional hypo-responsiveness in response to ongoing antigenic signaling [80]. What chronic viral infection and growing cancer have in common is a persisting antigenic load, and it is therefore unsurprising that homeostatic pathways to prevent T cell overactivity should come into play in the context of tumor-reactive T cells. The emergence of checkpoint inhibitor therapeutics has followed the discovery of many of the key cell surface molecules called checkpoint receptors, upregulated in response to ongoing T cell activation. Clinical responses using antibodies blocking CTLA4 and PD1 pathways paved the way for a growing field investigating other surface markers of T cell activation/exhaustion including TIM3, LAG3, TIGIT etc. Checkpoint molecule upregulation is one biochemical consequence of initiation of the exhaustion pathway. However, the capacity of checkpoint inhibitor-blocking antibodies that break these inhibitory signals demonstrates that at least some exhausted cells are capable of re-invigoration, i.e. that exhaustion can be reversed.

More recently, the T cell exhaustion field has distinguished exhaustion precursor cells that are hypofunctional but capable of reverting to effector cells following blockade of the exhaustion checkpoints, from terminally exhausted cells considered to be insensitive to checkpoint inhibition [80,81]. Gene-modified T cells are likewise known to be prone to exhaustion as a result of repeated antigenic stimulation and signaling [82]. A significant research effort has been directed toward engineering strategies to prevent or reverse exhaustion. Examples are in three main areas (Fig. 2).

Firstly, targeting the exhaustion checkpoints. For example, CAR-T cells can be engineered to produce checkpoint-blocking secreted molecules, or the checkpoint receptors such as PD1 can be deleted by, e.g., CRISPR knock-out, or the cells can be induced to co-express chimeric switch-receptors (or "decoy"-receptors) whereby the ectodomain of a checkpoint receptor such as PD1 can be fused with a positive signaling endodomain such as CD28 [83,84].

A second general approach is to target the transcriptional pathway driving the exhaustion phenotype. Several transcription factor exhaustion driver genes have been identified in recent years with particular insightful observations focussed on understanding transcriptional complexes around effector and exhaustion genes downstream from TCR and co-stimulation signaling [85,86]. Key proteins are TOX [87–89] (driving the establishment of epigenetic imprinting of exhaustion states), NR4A [90], and NFAT/AP1 transcription factor occupation of promoters of effector function genes. Indeed, both TOX and NR4A are targets of calcium influx/calcineurin pathway-induced NFAT1 transcription factor, in the presence or absence of AP1-mediated co-stimulatory signals [91].

An important hypothesis to test is that knock-out of exhaustion pathway genes in engineered CAR-T cells will lead to augmentation of long-term effector function through mitigation of exhaustion [92]. Several studies have demonstrated the feasibility of CRISPR/Cas-mediated knockdown of genes that are markers of exhaustion such as CTLA4, PD1, TIM3, LAG3 and TIGIT, in CAR-T [93,94]. Other genes implicated in the exhaustion pathway such as DGK have been similarly inhibited by CRISPR/CAS to enhance function in CAR-T cells [95] but to date, no studies have demonstrated that targeting transcriptional drivers of exhaustion such as TOX or NR4A is effective.

Several approaches to augment CAR-T cells with engineering strategies that either block or compete with these pro-exhaustion pathways have shown promise [82,90,91,96]. Of particular note, enhanced CAR-T function associated with reduced exhaustion signature was observed following overexpression in CAR-T cells of c-jun, which drives canonical AP1 transcription complexes at NFAT promoter [12,97]. Co-expressed c-jun and CAR have shown sufficient preclinical promise to lead to clinical translation.

The concept of "transient rest" to reverse exhaustion has been developed in cancer models through the use of drug treatments or molecular switches to prevent CAR signaling during

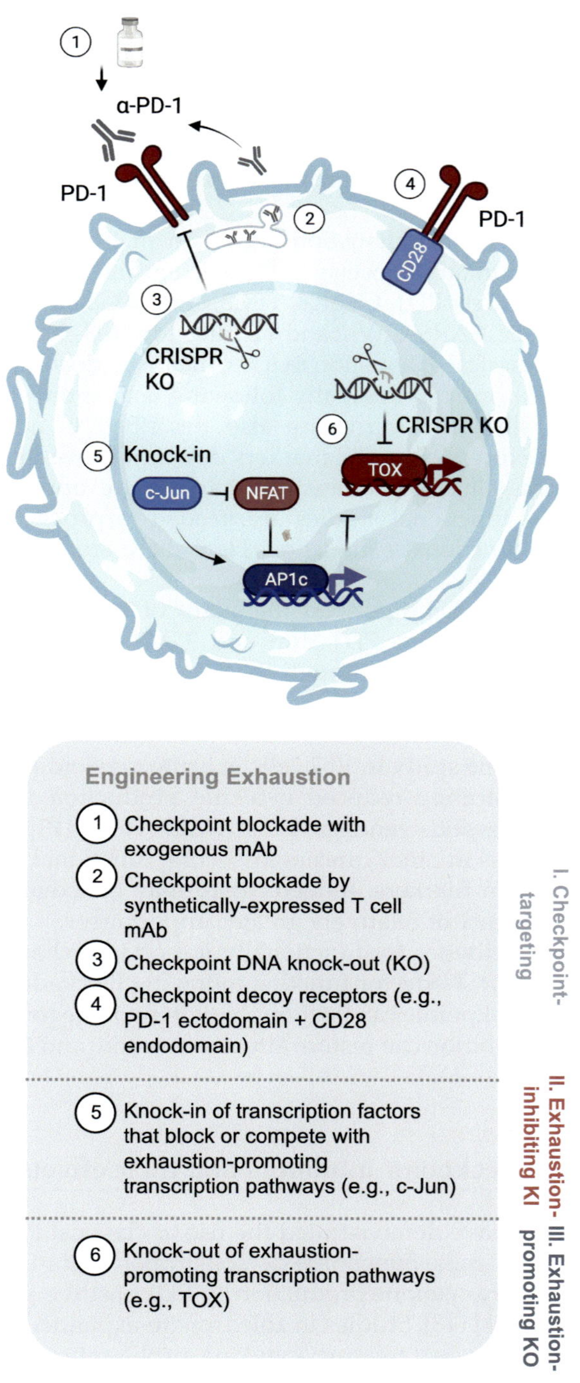

FIG. 2 Strategies to prevent or reverse γδ T cell exhaustion.

manufacture or following adoptive transfer through several synthetic biology approaches [98,99]. One attractive approach makes use of drugs to switch CAR-T cells between on and off states through the use of synth drugs or chemical tools to effect reversible degradation or activation; known as off-switch and on-switch respectively [100–102]. Determination of engineering approaches to promote prolonged in vivo activity of adoptively transferred γδ T will first require a deepened understanding of exhaustion pathways in γδ T and their differences from αβ T cells.

Data demonstrating functional exhaustion in human γδ T cells has been documented in experimental models and human subjects in both infection and cancer settings. Upregulation of canonical exhaustion markers PD1, CTLA4, TIM3, and LAG3 have all been documented in studies of plasmodium, HIV, SARS-CoV-2 and TB infection [103–106]. In human malignancy, solid cancer elevation of classical exhaustion markers has been documented in tumor infiltrating lymphocytes [107,108] or experimentally following activation—by tumor cells in vitro [109,110]. Interestingly, similar patterns are also described in acute myeloid leukemia [111–113]. However, the expression of the markers does not prove the phenomenon of functional hyporesponsiveness since all exhaustion markers are upregulated naturally in response to activation, and functional studies are needed to prove that marker expression correlates with a state of functional exhaustion. Here there are fewer published functional confirmatory studies.

The αβ T exhaustion field has described the states of function exhaustion in terms of exhaustion precursor cells that are PD1 dim and TIM3 negative and are amenable to reinvigoration of full effector function through checkpoint blockade, while $PD1^{hi}$ $TIM3^{+}$ cells are terminally exhausted and insensitive to checkpoint inhibitors [80,114]. Moreover, the exhaustion precursor cells express markers associated with stem or memory states such as TCf7 and BLIPM1 [86,115–117]. One study in Vδ2 cells in acute myeloid leukemia is confirmatory for this paradigm, demonstrating reduced cytokine production in $PD1^{+}TIM3^{+}$ cells but relatively normal cytokine responsiveness in $PD1^{+}$ $TIM3^{-}$ cells [113]. However, there are relatively few published studies in other cancers dissecting functional status about the expression of canonical exhaustion markers. Indeed, there may be exhaustion markers that are unique for γδ T cells or at least of relatively greater importance.

Probably the strongest evidence for functionally important exhaustion pathways in γδ T cells comes from studies of γδ T effector function following blockade of those putative pathways. The γδ T putative checkpoints can be thought of in two categories; those with evidence from αβ T cells that their inhibition can restore effector function, and those that appear unique to γδ T cells.

Classical αβ T checkpoint inhibitors and their effects on γδ T cells

Various studies in vitro have demonstrated the use of classical PD1-blocking reagents to enhance the function of γδ T cells. Functional demonstration that PDL1 expression in tumor cells can inhibit inflammatory cytokine production by PD1 positive γδ T cells was established in a Daudi lymphoma model [118]. Studies in zoledronate-expanded/activated Vγ9Vδ2 cells show that following PD1 pathway targeting, cytokine, proliferation or cytotoxicity responses to TCR/IL-2 stimulation can be restored [119–122].

Much interest has been directed to the blockade of TIM3, given its association with terminally-exhausted T cells. In γδ T cells, it has been observed that malaria exposure increases TIM3, the expression of which is associated with loss of inflammatory cytokine production, conforming to the exhaustion paradigm of functional hyporesponsiveness to chronic infection [106,123]. Similar to studies in αβ T cells, the addition of LAG3 blockade to PD1 can be synergistic or additive for blockade of pro-inflammatory cytokine response by γδ T cells [120,121,124].

Putative innate-specific checkpoint pathways and their inhibition

B and T cell attenuator (BTLA) is an inhibitory molecule of interest in γδ T by its coexpression with the Vg9Vδ2 TCR [125]. Interestingly, the coinhibitory ligand molecule Herpes virus entry mediator (HVEM) is known to interact with BTLA, and this pathway has been identified as functionally important in γδ T cells for blocking their proliferation response. Several experimental approaches to block the HVEM-BTLA axis have successfully increased Vg9Vδ2 responsiveness, specifically regarding proliferation in several cancer models [119,126–128].

TIGIT checkpoint inhibition is an emerging field with a particular focus on its potential role in the control of adoptively transferred or endogenous natural killer cells in the context of cancer [129,130]. TIGIT has been described in γδ T cells and interestingly is noted to have higher expression in Vδ1 than Vδ2 cells [74,131].

Taken together, these data on responses to existing checkpoint inhibitor clinical agents provide insight into exhaustion as a real phenomenon in γδ T cells and hence encourage further mechanistic research into γδ T-specific exhaustion pathways as well as the further exploration of existing checkpoint inhibitors in combination, and γδ T cellular engineering to prolong in vivo function following adoptive transfer. However, it is not possible to assume that strategies showing success in αβ T will be the most favorable approaches in γδ T. For example, it is yet to be established if γδ T become epigenetically hardwired into exhaustion, whether TIM3-positive $PD1^{hi}$ cells represent terminally and irreversibly exhausted γδ T as has been described in αβT, and whether the checkpoint inhibitors that have shown the greatest clinical successes in αβ T (i.e., CTLA4 and PD1) are the logical first choice pathways for engineering or combination strategies within γδ T.

It is not yet known if the genes most linked to epigenetic hardwiring of exhaustion (TOX, NR4A, ID2, etc.) have similar pivotal function roles in exhausted γδ T cells and therefore represent lead targets for engineering approaches. Encouragingly, NR4A and ID2 have been reported to be co-expressed with classical inhibitor receptors including TIGIT in γδ T [132]. An unbiased screening approach to uncover the key exhaustion mitigation pathways through functional screens is surely justified to inform the field regarding the most fruitful avenues of investigation.

Current progress in γδ T cell engineering

There are a limited number of published preclinical studies on engineering γδ T with synthetic receptors as recently reviewed [133]. Interestingly, the first described γδ CAR-T product was inspired by the observation that putative large-scale CAR-T cell products generated by transduction of induced pluripotential stem cells by chimeric antigen receptors clustered transcriptionally with γδ T cells [134]. Subsequently, a range of methods has been used to generate large-scale CAR-expressing γδ T cell products by combining expansion methods from blood with gamma-retroviral and lentiviral transduction. Several different stimulation and expansion protocols have been described including artificial antigen-presenting cells combined with transposons [135,136], Zoledronic acid and IL-2 (for Vg9Vδ2 specific product) with gamma retrovirus [109,137] lentivirus [138] or RNA electroporation [139,140], Concanavalin-A with IL-7 [110], and anti CD3 stimulation with lentiviral transduction for Vδ1 specific product [141].

The first reported clinical translation of CAR expressing γδ T involves a CD20-directed second-generation chimeric antigen receptor expressed in a Vδ1 cell product and infused into patients with CD20-expressing lymphomas. Of note this first trial is evaluating an allogeneic CAR-T cell product and the published phase I safety data demonstrated it to be well tolerated with no significant graft versus host disease [142].

Considerations of optimal CAR structure for evaluation in γδ T lymphocytes

Broadly, successful expansion of adoptively transferred engineered T cells requires signaling from immune receptors on encountering target cells that is qualitatively and quantitatively balanced to permit immediate cytotoxicity, proliferation and the generation of a memory population with capacity for re-expansion on re-encounter of tumor antigen. However, these positive attributes and consequences of signaling must be balanced against the imperative of not inducing limiting T cell exhaustion through excessive T cell signaling. Here the field for γδ T cell CARs might be even more challenging than in αβ-CAR T cells, since unknown and unpredictable exhaustion-inducing signaling may be derived from the γδ T cell receptor, Fc-gamma receptors and NK receptors in a potentially uncontrollable and unpredictable manner. Moreover, the heterogeneity of TCR ligands for a Vδ1-CAR cell product may result in some populations being overactivated on encountering cognate TCR ligand expressed by the cancer target, while other subclones might be insufficiently activated.

Two innovative approaches to mitigate this potential problem are the use of chimeric co-stimulatory receptor (CCR), and nonsignaling CAR (∅-CAR) (Fig. 3). In the former approach, the chimeric receptor provides only co-stimulation on encountering a target antigen through the antibody component of the CCR thereby relying on the natural cancer-sensing properties of the TCR to provide signal-1. In the nonsignaling CAR [138], its purpose is simply that the ScFv component of the receptor enhances specificity and affinity of recognition of cancer targets by enhancing immune synapse formation with cancer targets. A further variant of this approach is the T cell antigen coupler (TAC) by which a membrane-tethered bispecific antibody expressed on the surface of the γδ T cell forms a bridge between TCR and a target antigen [143].

In brief, a signal that does not lead to a high level of immediate effector function risks the CAR-T cell being overwhelmed by tumor growth in the tumor environment and never establishing sufficient expansion for tumor control. Conversely, an explosive initial T cell response to a tumor might generate overwhelming cytokine release syndrome, which has the potential for fatal immune activation [144] or might lead to early T cell exhaustion before complete tumor control can be achieved. In reality, a perfect CAR-T product probably

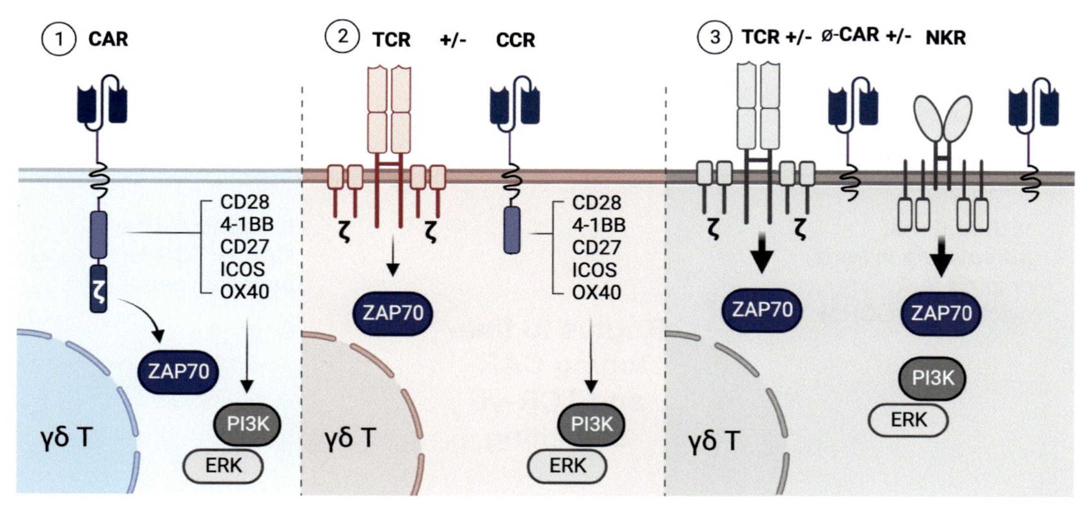

FIG. 3 Strategies to tune γδ T cell activation with chimeric antigen receptors.

incorporates a balance of cells with immediate effect or function and cells also capable of memory and stem cell characteristics. Such balanced signaling for optimal expansion/persistence is determined in turn by each component of primary immune receptors, either CAR or TCR as well as by the range of antigen densities on the target cells.

Variables that can be fine-tuned to achieve optimal signaling are therefore (1) expression levels of the immune receptor, (2) affinity of the CAR or TCR for antigen, (3) the number, nature, and arrangement of stimulatory endodomains, (4) the CAR hinge and transmembrane and their capacity to dimerise or interact with endogenous co-stimulatory molecules, and (5) the addition of further co-stimulatory or cytokine signaling either *in cis* or *in trans* (Fig. 4).

Several paradigms on balanced signaling emerging from the αβ T CAR field are yet to be systematically tested in the γδ T CAR field. For example, whether 41BB co-stimulation incorporation into the CAR promotes increased oxidative metabolism, mitochondrial biogenesis, and longer in vivo survival of effector CAR T cells than second-generation constructs incorporating CD28 co-stimulation.

Fine-tuning for differential antigen sensing

The broad conceptual approaches that have been developed for avoiding toxicity have been (i) fine-tuning of activation thresholds and (ii) the use of dual receptors in Boolean logic-gated configurations (the principle of Boolean logic-gating is illustrated in Fig. 5). The activation threshold concept is best understood in terms of the ability of an engineered T cell to distinguish between a relatively low antigen density of a particular antigen in a healthy tissue versus a higher antigen density in the cancer cells of an individual. In this simple model of a T cell with a single CAR against one antigen, the mechanism of toxicity

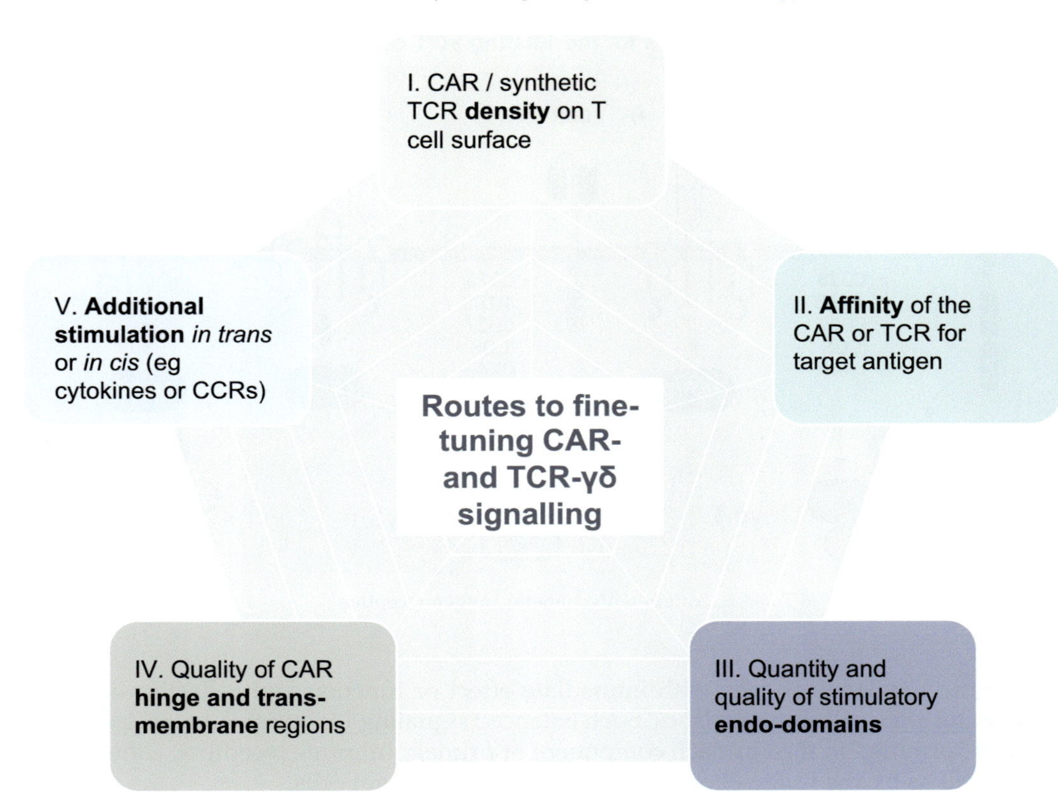

FIG. 4 Factors to consider when designing synthetic γδ T cell signaling circuits.

can be thought of as direct cytolysis of the healthy cell after the formation of an immune synapse and/or inflammatory damage caused by cytokine release. For each of these mechanisms, the likelihood of toxicity is dependent on the capacity of the engineered T cell to form a functional synaptic interaction with the target, which in turn is dependent on the number and affinity of the immune receptor to ligand interactions; i.e., the functional avidity of interaction. Hence by fine-tuning the expression levels and affinity of, for example, the ScFv component of a CAR it is in theory possible to fine-tune responsiveness for a threshold level of antigen density that can distinguish healthy from tumor cells [145].

For some cancer antigens, however, there is no clear cut-off between expression on healthy and transformed cells. In these situations, the concept of logic gates can be exploited in manners that can harness the natural cancer-sensing innate properties of γδ T cells, as described above. This can be done through the use of chimeric co-stimulatory receptors, whereby the engineered γδ T cell elegantly requires only one additional and relatively small synthetic receptor [109,137]. In conventional αβ T CAR-T cells, Boolean AND-gate engineering requires two antigens to be present for full activation and cytotoxicity and, therefore, two synthetic receptors or a more complex bispecific single receptor. By selecting antigens that never or

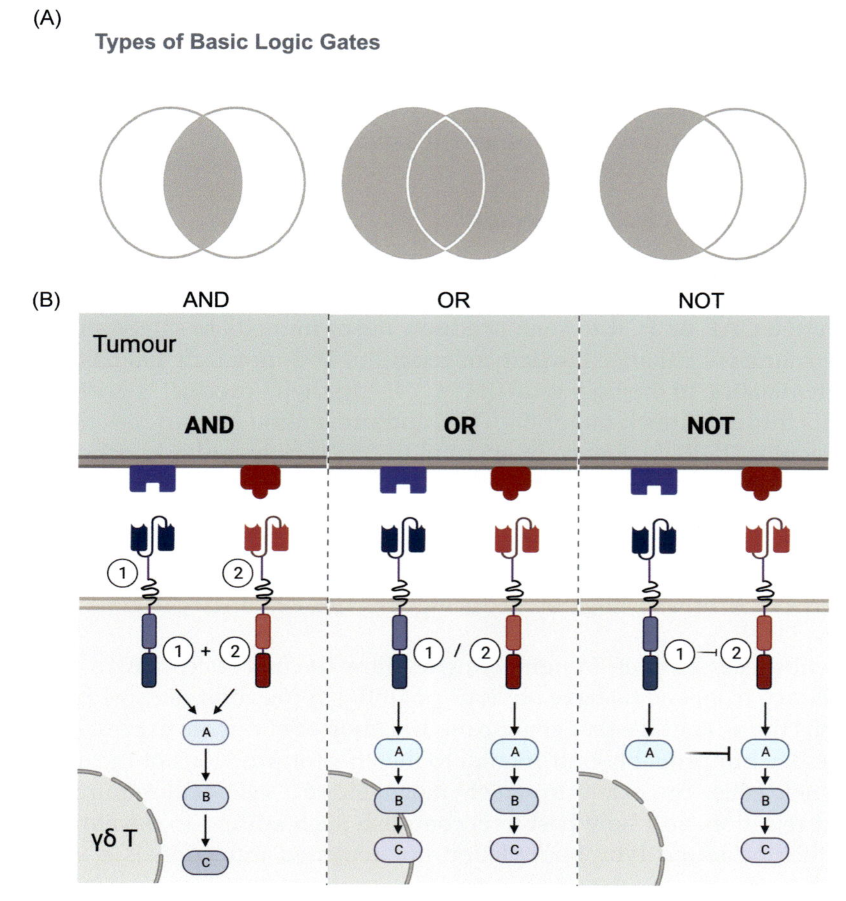

FIG. 5 Illustration of (A) Boolean Logic, and (B) how it can be applied to synthetic signaling circuit design. An "AND" gate will enable signaling only when both receptors are engaged, an "OR" gate will signal when either of the receptors is engaged, while a "NOT" gate will propagate a signal when one—but not the other—receptor is engaged. These designs can be used to achieve various immunotherapeutic goals, including avoidance of on-target off-tumor toxicity (with an "AND" or "NOT" gate) or enhanced CAR signaling in the presence of low or varied target antigen density (with an "OR" gate).

rarely co-express on healthy cells and employing immune receptor affinities that are insufficient for activation against single antigen-positive targets, an additional level of specificity has been introduced. This AND-gate concept have recently been extended to incorporate approaches such as the sequential AND-gate whereby activation of one chimeric receptor leads to the transcription of a second, through the use of activation-dependent promoters or synthetic chemistry such as *syn*-notch [146–148].

A recently described concept involves an overhaul of conventional endodomain arrangements such that both antigens are required to form a single immune receptor by bringing together two essential interacting components of the proximal signaling complex [149]. Since the CD3 component and its interaction with proximal signaling moieties is considered largely identical between αβ T and γδ T, the capability of synthetic proximal signaling machinery to enhance sensitivity to low antigen density may also have applications within the γδ T cell field, although this remains to be investigated experimentally.

A platform for allogeneic cell therapy?

Adoptively transferred unmodified allogeneic conventional T cells can effect graft versus host disease (GVHD) if their immediate rejection is limited by systemic immune-suppression. The addition of a CAR or TCR to such products raises immediate safety and efficacy objections on the grounds of enhanced toxicity or rejection. The approach that has been developed to allow the realization of the holy grail of the "off-the-shelf" product is genetic modification to limit GvHD through knock-out of the TCR, and prevention of graft rejection by knock-out of MHC molecules [150]. Further refinement of the concept includes knock-out of genes that sensitize to immunosuppressive drugs, and overexpression of cell surface proteins such as HLA-G that limit NK-mediated rejection. The first clinical demonstrations of such gene-edited CAR-T cells made use of TALENs technology [151] and this has been superseded by CRISPR base editing using CAS protein variants that can knock out without recourse to double-strand DNA breaks, thereby reducing the risk of off-target gene editing such as translocations [152].

These recent successes in αβ T engineering to allow the first successful infusions of allogeneic products free from GvHD have obvious potential applicability in γδ T cell immunotherapy, which has the advantage that genetic modification to eliminate expression of TCR is not only unnecessary but probably detrimental to the effector functions of the adoptively transferred allogeneic cells. The extent to which allogeneic γδ T cells will require gene editing to prevent graft rejection, and how best to accomplish such editing to develop a product that requires neither profound lymphodepletion nor repeated infusions, is a key objective for the field.

Concluding remarks

The remarkable growth of engineered cell therapy product development in αβ T cell adoptive transfer field especially in CAR-T cells has not yet been translated into large numbers of trial-ready novel concepts in γδ T cells. However, the potential of γδ T for long-term memory responses, additional innate sensing that could mitigate antigen loss variants, and natural capacity to avoid graft versus host disease in the allogeneic setting mark out this novel cellular chassis as a high priority for research and development.

Acknowledgment

I thank Dr. Marta Barisa for her contribution to the generation of figures that accompany this review.

References

[1] Hirano M, et al. Evolutionary implications of a third lymphocyte lineage in lampreys. Nature 2013;501:435–8.
[2] Hayday AC. Gammadelta T cells and the lymphoid stress-surveillance response. Immunity 2009;31:184–96.
[3] Hayday AC, Vantourout P. The innate biologies of adaptive antigen receptors. Annu Rev Immunol 2020;38:487–510.
[4] Hinrichs CS, Rosenberg SA. Exploiting the curative potential of adoptive T-cell therapy for cancer. Immunol Rev 2014;257:56–71.
[5] Stevanović S, et al. Complete regression of metastatic cervical cancer after treatment with human papillomavirus-targeted tumor-infiltrating T cells. J Clin Oncol 2015;33:1543–50.
[6] Kim R, Emi M, Tanabe K. Cancer immunoediting from immune surveillance to immune escape. Immunology 2007;121:1–14.
[7] Sosman JA, Sondel PM. The graft versus leukemia effect: possible mechanisms and clinical significance to the biologic therapy of leukemia. Bone Marrow Transplant 1991;7(Suppl. 1):33–7.
[8] Nestle FO, et al. Vaccination of melanoma patients with peptide- or tumorlysate-pulsed dendritic cells. Nat Med 1998;4:328–32.
[9] Kugler A, et al. Regression of human metastatic renal cell carcinoma after vaccination with tumor cell–dendritic cell hybrids. Nat Med 2000;6:332–6.
[10] Wang R, Rosenberg SA. Human tumor antigens for cancer vaccine development. Immunol Rev 1999;170:85–100.
[11] Rosenberg SA, Restifo NP. Adoptive cell transfer as personalized immunotherapy for human cancer. Science 2015;348:62–8.
[12] Labanieh L, Mackall CL. CAR immune cells: design principles, resistance and the next generation. Nature 2023;614:635–48.
[13] Brown CE, Mackall CL. CAR T cell therapy: inroads to response and resistance. Nat Rev Immunol 2019;19:73–4.
[14] Majzner RG, Mackall CL. Clinical lessons learned from the first leg of the CAR T cell journey. Nat Med 2019;25:1341–55.
[15] Bosse KR, Majzner RG, Mackall CL, Maris JM. Immune-based approaches for the treatment of pediatric malignancies. Annu Rev Cancer Biol 2020;4:353–70.
[16] Bufalo FD, et al. GD2-CART01 for relapsed or refractory high-risk neuroblastoma. New Engl J Med 2023;388:1284–95.
[17] Straathof K, et al. Antitumor activity without on-target off-tumor toxicity of GD2–chimeric antigen receptor T cells in patients with neuroblastoma. Sci Transl Med 2020;12:eabd6169.
[18] Pule MA, et al. Virus-specific T cells engineered to coexpress tumor-specific receptors: persistence and antitumor activity in individuals with neuroblastoma. Nat Med 2008;14:1264–70.
[19] Morita CT, Jin C, Sarikonda G, Wang H. Nonpeptide antigens, presentation mechanisms, and immunological memory of human Vgamma2Vdelta2 T cells: discriminating friend from foe through the recognition of prenyl pyrophosphate antigens. Immunol Rev 2007;215:59–76.
[20] Gober H-J, et al. Human T cell receptor gammadelta cells recognize endogenous mevalonate metabolites in tumor cells. J Exp Med 2003;197:163–8.
[21] Karunakaran MM, et al. Butyrophilin-2A1 directly binds germline-encoded regions of the Vγ9Vδ2 TCR and is essential for phosphoantigen sensing. Immunity 2020;52:487–498.e6.
[22] Rigau M, et al. Butyrophilin 2A1 is essential for phosphoantigen reactivity by γδ T cells. Science 2020;367, eaay5516.
[23] Sandstrom A, et al. The intracellular B30.2 domain of butyrophilin 3A1 binds phosphoantigens to mediate activation of human Vγ9Vδ2 T cells. Immunity 2014;40:490–500.
[24] Willcox CR, et al. Phosphoantigen sensing combines TCR-dependent recognition of the BTN3A IgV domain and germline interaction with BTN2A1. Cell Rep 2023;42, 112321.
[25] Bertaina A, et al. Zoledronic acid boosts γδ T-cell activity in children receiving αβ+ T and CD19+ cell-depleted grafts from an HLA-haplo-identical donor. Onco Targets Ther 2017;6, e1216291.

[26] Halstensen TS, Scott H, Brandtzaeg P. Intraepithelial T cells of the TcRγ/δ+CD8− and Vδ1/Jδ1+ phenotypes are increased in coeliac disease. Scand J Immunol 1989;30:665–72.
[27] Bos JD, et al. T-cell receptor γδ bearing cells in normal human skin. J Invest Dermatol 1990;94:37–42.
[28] Spada FM, et al. Self-recognition of CD1 by gamma/delta T cells: implications for innate immunity. J Exp Med 2000;191:937–48.
[29] Reijneveld JF, et al. Human γδ T cells recognize CD1b by two distinct mechanisms. Proc Natl Acad Sci USA 2020;117:22944–52.
[30] Davey MS, et al. Clonal selection in the human Vδ1 T cell repertoire indicates γδ TCR-dependent adaptive immune surveillance. Nat Commun 2017;8:14760.
[31] Davey MS, et al. The human Vδ2^{+} T-cell compartment comprises distinct innate-like Vγ9^{+} and adaptive Vγ9^{-} subsets. Nat Commun 2018;9:1760.
[32] Davey MS, Willcox CR, Baker AT, Hunter S, Willcox BE. Recasting human Vδ1 lymphocytes in an adaptive role. Trends Immunol 2018;39:446–59.
[33] Hurtado MO, et al. Tumor infiltrating lymphocytes expanded from pediatric neuroblastoma display heterogeneity of phenotype and function. PLoS One 2019;14, e0216373.
[34] Gentles AJ, et al. The prognostic landscape of genes and infiltrating immune cells across human cancers. Nat Med 2015;21:938–45.
[35] Raverdeau M, Cunningham SP, Harmon C, Lynch L. γδ T cells in cancer: a small population of lymphocytes with big implications. Clin Transl Immunol 2019;8, e01080.
[36] Thorsson V, et al. The immune landscape of cancer. Immunity 2019;51:411–2.
[37] Willcox CR, et al. Cytomegalovirus and tumor stress surveillance by binding of a human γδ T cell antigen receptor to endothelial protein C receptor. Nat Immunol 2012;13:872–9.
[38] Marlin R, et al. Sensing of cell stress by human γδ TCR-dependent recognition of annexin A2. Proc Natl Acad Sci 2017;114:3163–8.
[39] Bruder J, et al. Target specificity of an autoreactive pathogenic human γδ-T cell receptor in myositis. J Biol Chem 2012;287:20986–95.
[40] Harly C, et al. Human γδ T cell sensing of AMPK-dependent metabolic tumor reprogramming through TCR recognition of EphA2. Sci Immunol 2021;6(61):eaba9010. https://doi.org/10.1126/sciimmunol.aba9010.
[41] Melandri D, et al. The γδTCR combines innate immunity with adaptive immunity by utilizing spatially distinct regions for agonist selection and antigen responsiveness. Nat Immunol 2018;19:1352–65.
[42] Roy S, et al. Molecular analysis of lipid-reactive Vδ1 γδ T cells identified by CD1c tetramers. J Immunol 2016;196:1933–42.
[43] Uldrich AP, et al. CD1d-lipid antigen recognition by the γδ TCR. Nat Immunol 2013;14:1137–45.
[44] Hayday A, Vantourout P. A long-playing CD about the γδ TCR repertoire. Immunity 2013;39:994–6.
[45] Vantourout P, Hayday A. Six-of-the-best: unique contributions of γδ T cells to immunology. Nat Rev Immunol 2013;13:88–100.
[46] Guerra-Maupome M, Palmer MV, Waters WR, McGill JL. Characterization of γδ T cell effector/memory subsets based on CD27 and CD45R expression in response to Mycobacterium bovis infection. Immunohorizons 2019;3:208–18.
[47] Wu Y, et al. A local human Vδ1 T cell population is associated with survival in nonsmall-cell lung cancer. Nat Cancer 2022;3:696–709. https://doi.org/10.1038/s43018-022-00376-z.
[48] Ribot JC, et al. CD27 is a thymic determinant of the balance between interferon-gamma- and interleukin 17-producing gammadelta T cell subsets. Nat Immunol 2009;10:427–36.
[49] Odaira K, et al. CD27-CD45+ γδ T cells can be divided into two populations, CD27-CD45int and CD27-CD45hi with little proliferation potential. Biochem Biophys Res Commun 2016;478:1298–303.
[50] Dieli F, et al. Differentiation of effector/memory Vdelta2 T cells and migratory routes in lymph nodes or inflammatory sites. J Exp Med 2003;198:391–7.
[51] Berglund S, Gaballa A, Sawaisorn P, Sundberg B, Uhlin M. Expansion of gammadelta T cells from cord blood: a therapeutical possibility. Stem Cells Int 2018;2018:1–15.
[52] Brandes M, et al. Cross-presenting human gammadelta T cells induce robust CD8+ alphabeta T cell responses. Proc Natl Acad Sci USA 2009;106:2307–12.
[53] Brandes M, Willimann K, Moser B. Professional antigen-presentation function by human gammadelta T cells. Science 2005;309:264–8.

[54] Himoudi N, et al. Human γδ T lymphocytes are licensed for professional antigen presentation by interaction with opsonized target cells. J Immunol 2012;188:1708–16.
[55] Wang L, Kang F-B, Shan B-E. B7-H3-mediated tumor immunology: friend or foe? Int J Cancer 2014;134:2764–71.
[56] Cabeza-Cabrerizo M, Cardoso A, Minutti CM, da Costa MP, Sousa CR, e. Dendritic cells revisited. Annu Rev Immunol 2021;39:131–66.
[57] Dutton RW, Bradley LM, Swain SL. T cell memory. Annu Rev Immunol 1998;16:201–23.
[58] Riddell SR, et al. Adoptive therapy with chimeric antigen receptor-modified T cells of defined subset composition. Cancer J 2014;20:141–4.
[59] Veldhoen M, et al. Transforming growth factor-beta "reprograms" the differentiation of T helper 2 cells and promotes an interleukin 9-producing subset. Nat Immunol 2008;9:1341–6.
[60] Ivanov II, et al. The orphan nuclear receptor RORγt directs the differentiation program of proinflammatory IL-17+ T helper cells. Cell 2006;126:1121–33.
[61] Gattinoni L, et al. A human memory T cell subset with stem cell-like properties. Nat Med 2011;17:1290–7.
[62] Gattinoni L, et al. Acquisition of full effector function in vitro paradoxically impairs the in vivo antitumor efficacy of adoptively transferred CD8+ T cells. J Clin Invest 2005;115:1616–26.
[63] Gattinoni L, Klebanoff CA, Restifo NP. Paths to stemness: building the ultimate antitumour T cell. Nat Rev Cancer 2012;12:671–84.
[64] Deng Q, et al. Characteristics of anti-CD19 CAR T cell infusion products associated with efficacy and toxicity in patients with large B cell lymphomas. Nat Med 2020;26:1878–87.
[65] Hinrichs CS, et al. Human effector CD8+ T cells derived from naive rather than memory subsets possess superior traits for adoptive immunotherapy. Blood 2011;117:808–14.
[66] Good Z, et al. Post-infusion CAR TReg cells identify patients resistant to CD19-CAR therapy. Nat Med 2022;28:1860–71.
[67] Kawalekar OU, et al. Distinct signalling of coreceptors regulates specific metabolism pathways and impacts memory development in CAR T cells. Immunity 2016;44:380–90.
[68] Wittibschlager V, et al. CAR T-cell persistence correlates with improved outcome in patients with B-cell lymphoma. Int J Mol Sci 2023;24:5688.
[69] Ghassemi S, et al. Rapid manufacturing of non-activated potent CAR T cells. Nat Biomed Eng 2022;6:118–28.
[70] Caccamo N, et al. Differentiation, phenotype, and function of interleukin-17-producing human Vγ9Vδ2 T cells. Blood 2011;118:129–38.
[71] Fiala GJ, Gomes AQ, Silva-Santos B. From thymus to periphery: molecular basis of effector γδ-T cell differentiation. Immunol Rev 2020;298:47–60.
[72] Lopes N, et al. Distinct metabolic programs established in the thymus control effector functions of γδ T cell subsets in tumor microenvironments. Nat Immunol 2021;22:179–92.
[73] Gioia C, et al. Lack of $CD27^{-}CD45RA^{-}V\gamma9V\delta2^{+}$ T cell effectors in immunocompromised hosts and during active pulmonary tuberculosis. J Immunol 2002;168:1484–9.
[74] Ryan PL, et al. Heterogeneous yet stable Vδ2(+) T-cell profiles define distinct cytotoxic effector potentials in healthy human individuals. Proc Natl Acad Sci USA 2016;113:14378–83.
[75] Ravens S, et al. Human γδ T cells are quickly reconstituted after stem-cell transplantation and show adaptive clonal expansion in response to viral infection. Nat Immunol 2017;18:393–401.
[76] McMurray JL, et al. Transcriptional profiling of human Vδ1 T cells reveals a pathogen-driven adaptive differentiation program. Cell Rep 2022;39, 110858.
[77] Wang Q, Sun Q, Chen Q, Li H, Liu D. Expression of CD27 and CD28 on γδ T cells from the peripheral blood of patients with allergic rhinitis. Exp Ther Med 2020;20:224.
[78] Schmolka N, et al. Epigenetic and transcriptional signatures of stable versus plastic differentiation of proinflammatory γδ T cell subsets. Nat Immunol 2013;14:1093–100.
[79] Schmolka N, et al. MicroRNA-146a controls functional plasticity in γδ T cells by targeting NOD1. Sci Immunol 2018;3, eaao1392.
[80] Blank CU, et al. Defining 'T cell exhaustion'. Nat Rev Immunol 2019;19:665–74.
[81] Mann TH, Kaech SM. Tick-TOX, it's time for T cell exhaustion. Nat Immunol 2019;20:1092–4.
[82] Gumber D, Wang LD. Improving CAR-T immunotherapy: overcoming the challenges of T cell exhaustion. EBioMedicine 2022;77, 103941.

[83] Liu X, et al. A chimeric switch-receptor targeting PD1 augments the efficacy of second-generation CAR T cells in advanced solid tumors. Cancer Res 2016;76:1578–90.

[84] Hoogi S, et al. A TIGIT-based chimeric co-stimulatory switch receptor improves T-cell anti-tumor function. J Immunother Cancer 2019;7:243.

[85] McLane LM, Abdel-Hakeem MS, Wherry EJ. CD8 T cell exhaustion during chronic viral infection and cancer. Annu Rev Immunol 2019;37:457–95.

[86] Beltra J-C, et al. Developmental relationships of four exhausted CD8+ T cell subsets reveals underlying transcriptional and epigenetic landscape control mechanisms. Immunity 2020. https://doi.org/10.1016/j.immuni.2020.04.014.

[87] Khan O, et al. TOX transcriptionally and epigenetically programs CD8+ T cell exhaustion. Nature 2019;571:211–8.

[88] Alfei F, et al. TOX reinforces the phenotype and longevity of exhausted T cells in chronic viral infection. Nature 2019;571:265–9.

[89] Scott AC, et al. TOX is a critical regulator of tumour-specific T cell differentiation. Nature 2019;571:1–5.

[90] Chen J, et al. NR4A transcription factors limit CAR T cell function in solid tumours. Nature 2019;6, 224ra25.

[91] Seo H, et al. TOX and TOX2 transcription factors cooperate with NR4A transcription factors to impose CD8+ T cell exhaustion. Proc Natl Acad Sci USA 2019;116:12410–5.

[92] Wellhausen N, Agarwal S, Rommel PC, Gill SI, June CH. Better living through chemistry: CRISPR/Cas engineered T cells for cancer immunotherapy. Curr Opin Immunol 2022;74:76–84.

[93] Ren J, et al. Multiplex genome editing to generate universal CAR T cells resistant to PD1 inhibition. Clin Cancer Res 2016;23:2255–66.

[94] Rupp LJ, et al. CRISPR/Cas9-mediated PD-1 disruption enhances anti-tumor efficacy of human chimeric antigen receptor T cells. Sci Rep 2017;7:737.

[95] Jung I-Y, et al. CRISPR/Cas9-mediated knockout of DGK improves antitumor activities of human T cells. Cancer Res 2018;78:4692–703.

[96] Li F, Zhang Y. Targeting NR4As, a new strategy to fine-tune CAR-T cells against solid tumors. Signal Transduct Target Ther 2019;4:7.

[97] Lynn RC, et al. C-Jun overexpression in CAR T cells induces exhaustion resistance. Nature 2019;576:293–300.

[98] Weber EW, et al. Transient rest restores functionality in exhausted CAR-T cells through epigenetic remodeling. Science 2021;372, eaba1786.

[99] Mestermann K, et al. The tyrosine kinase inhibitor dasatinib acts as a pharmacologic on/off switch for CAR T cells. Sci Transl Med 2019;11, eaau5907.

[100] Zajc CU, et al. A conformation-specific ON-switch for controlling CAR T cells with an orally available drug. Proc Natl Acad Sci USA 2020;117:14926–35.

[101] Jan M, et al. Reversible ON- and OFF-switch chimeric antigen receptors controlled by lenalidomide. Sci Transl Med 2021;13, eabb6295.

[102] Labanieh L, et al. Enhanced safety and efficacy of protease-regulated CAR-T cell receptors. Cell 2022;185:1745–1763.e22.

[103] Cerapio JP, et al. Single-cell RNAseq profiling of human γδ T lymphocytes in virus-related cancers and COVID-19 disease. Viruses 2021;13:2212.

[104] Shen H, et al. Selective destruction of interleukin 23-induced expansion of a major antigen-specific γδ T-cell subset in patients with tuberculosis. J Infect Dis 2016;215:420–30.

[105] Dunne PJ, et al. CD3ε expression defines functionally distinct subsets of Vδ1 T cells in patients with human immunodeficiency virus infection. Front Immunol 2018;9:940.

[106] Gogoi D, Biswas D, Borkakoty B, Mahanta J. Exposure to plasmodium vivax is associated with the increased expression of exhaustion markers on γδ T lymphocytes. Parasite Immunol 2018;40, e12594.

[107] Cerapio J-P, et al. Phased differentiation of γδ T and T CD8 tumor-infiltrating lymphocytes revealed by single-cell transcriptomics of human cancers. Onco Targets Ther 2021;10, 1939518.

[108] Weimer P, et al. Tissue-specific expression of TIGIT, PD-1, TIM-3, and CD39 by γδ T cells in ovarian cancer. Cells 2022;11:964.

[109] Fisher J, et al. Engineering γδT cells limits tonic signalling associated with chimeric antigen receptors. Sci Signal 2019;12, eaax1872.

[110] Capsomidis A, et al. Chimeric antigen receptor-engineered human γδ T cells: enhanced cytotoxicity with retention of cross presentation. Mol Ther 2018;26:354–65.
[111] Tang L, et al. Characterization of immune dysfunction and identification of prognostic immune-related risk factors in acute myeloid leukemia. Clin Cancer Res 2020;26:1763–72.
[112] Jin Z, et al. Higher $TIGIT^{+}CD226^{-}$ γδ T cells in patients with acute myeloid leukemia. Immunol Investig 2022;51:40–50.
[113] Wu K, et al. Vδ2 T cell subsets, defined by PD-1 and TIM-3 expression, present varied cytokine responses in acute myeloid leukemia patients. Int Immunopharmacol 2020;80, 106122.
[114] Im SJ, et al. Defining CD8+ T cells that provide the proliferative burst after PD-1 therapy. Nature 2016;537:417–21.
[115] Zehn D, Thimme R, Lugli E, de Almeida GP, Oxenius A. 'Stem-like' precursors are the fount to sustain persistent CD8+ T cell responses. Nat Immunol 2022;23:836–47.
[116] Chen Z, et al. TCF-1-centered transcriptional network drives an effector versus exhausted CD8 T cell-fate decision. Immunity 2019;51:840–855.e5.
[117] Shin H, et al. A role for the transcriptional repressor Blimp-1 in CD8+ T cell exhaustion during chronic viral infection. Immunity 2009;31:309–20.
[118] Iwasaki M, et al. Expression and function of PD-1 in human γδ T cells that recognize phosphoantigens. Eur J Immunol 2011;41:345–55.
[119] Castella B, et al. Anergic bone marrow Vγ9Vδ2 T cells as early and long-lasting markers of PD-1-targetable microenvironment-induced immune suppression in human myeloma. Onco Targets Ther 2015;4, e1047580.
[120] Yang R, et al. Bispecific antibody PD-L1 x CD3 boosts the anti-tumor potency of the expanded Vγ2Vδ2 T cells. Front Immunol 2021;12, 654080.
[121] Hoeres T, Smetak M, Pretscher D, Wilhelm M. Improving the efficiency of Vγ9Vδ2 T-cell immunotherapy in cancer. Front Immunol 2018;9:800.
[122] Tomogane M, et al. Human Vγ9Vδ2 T cells exert anti-tumor activity independently of PD-L1 expression in tumor cells. Biochem Biophys Res Commun 2021;573:132–9.
[123] Pamplona A, Silva-Santos B. γδ T cells in malaria: a double-edged sword. FEBS J 2021;288:1118–29.
[124] Yang Z-Z, et al. Expression of LAG-3 defines exhaustion of intratumoral PD-1+ T cells and correlates with poor outcome in follicular lymphoma. Oncotarget 2017;8:61425–39.
[125] Tani-ichi S, et al. Innate-like CD27+CD45RBhigh γδ T cells require TCR signalling for homeostasis in peripheral lymphoid organs. J Immunol 2020;204:2671–84.
[126] Gertner-Dardenne J, et al. The co-receptor BTLA negatively regulates human Vγ9Vδ2 T-cell proliferation: a potential way of immune escape for lymphoma cells. Blood 2013;122:922–31.
[127] Boice M, et al. Loss of the HVEM tumor suppressor in lymphoma and restoration by modified CAR-T cells. Cell 2016;167:405–418.e13.
[128] Catafal-Tardos E, Baglioni MV, Bekiaris V. Inhibiting the unconventionals: importance of immune checkpoint receptors in γδ T, MAIT, and NKT cells. Cancer 2021;13:4647.
[129] Yu X, et al. The surface protein TIGIT suppresses T cell activation by promoting the generation of mature immunoregulatory dendritic cells. Nat Immunol 2009;10:48–57.
[130] Wienke J, et al. Integrative analysis of neuroblastoma by single-cell RNA sequencing identifies the NECTIN2-TIGIT axis as a target for immunotherapy. Biorxiv 2022. https://doi.org/10.1101/2022.07.15.499859. 2022.07.15.499859.
[131] Paquin-Proulx D, et al. Inversion of the Vδ1 to Vδ2 γδ T cell ratio in CVID is not restored by IVIg and is associated with immune activation and exhaustion. Medicine 2016;95, e4304.
[132] Tirier SM, et al. Subclone-specific microenvironmental impact and drug response in refractory multiple myeloma revealed by single-cell transcriptomics. Nat Commun 2021;12:6960.
[133] Ferry GM, Anderson J. Augmenting human γδ lymphocytes for cancer therapy with chimeric antigen receptors. Explor Immunol 2022;2:168–79.
[134] Themeli M, et al. Generation of tumor-targeted human T lymphocytes from induced pluripotent stem cells for cancer therapy. Nat Biotechnol 2013;31:928–33.
[135] Deniger DC, et al. Activating and propagating polyclonal γδ T cells with broad specificity for malignancies. Clin Cancer Res 2014;20:5708–19.

[136] Deniger DC, et al. Bispecific T-cells expressing polyclonal repertoire of endogenous γδ T-cell receptors and introduced CD19-specific chimeric antigen receptor. Mol Ther 2013;21:638–47.
[137] Fisher J, et al. Avoidance of on-target off-tumor activation using a co-stimulation-only chimeric antigen receptor. Mol Ther 2017;25:1234–47.
[138] Fleischer LC, et al. Non-signalling chimeric antigen receptors enhance antigen-directed killing by γδ T cells in contrast to αβ T cells. Mol Ther Oncolytics 2020;18:149–60.
[139] Harrer DC, et al. RNA-transfection of γ/δ T cells with a chimeric antigen receptor or an α/β T-cell receptor: a safer alternative to genetically engineered α/β T cells for the immunotherapy of melanoma. BMC Cancer 2017;17:551.
[140] Ang WX, et al. Electroporation of NKG2D RNA CAR improves Vγ9Vδ2 T cell responses against human solid tumor xenografts. Mol Ther Oncolytics 2020;17:421–30.
[141] Ferry GM, et al. A simple and robust single-step method for CAR-Vδ1 γδT cell expansion and transduction for cancer immunotherapy. Front Immunol 2022;13, 863155.
[142] Neelapu SS, et al. A phase 1 safety and efficacy study of ADI-001 anti-CD20 CAR-engineered allogeneic γδ (γδ) T cells in adults with B cell malignancies, in monotherapy and combination with IL-2. Blood 2021;138:2834.
[143] Helsen CW, et al. The chimeric TAC receptor co-opts the T cell receptor yielding robust anti-tumor activity without toxicity. Nat Commun 2018;9:3049.
[144] Morgan RA, et al. Case report of a serious adverse event following the administration of T cells transduced with a chimeric antigen receptor recognizing ERBB2. Mol Ther 2010;18:843–51.
[145] Hotblack A, Straathof K. Fine-tuning CARs for best performance. Cancer Cell 2022;40:11–3.
[146] Zhu I, et al. Modular design of synthetic receptors for programmed gene regulation in cell therapies. Cell 2022;185:1431–1443.e16.
[147] Choe JH, et al. SynNotch-CAR T cells overcome challenges of specificity, heterogeneity, and persistence in treating glioblastoma. Sci Transl Med 2021;13:eabe7378.
[148] Roybal KT, et al. Precision tumor recognition by T cells with combinatorial antigen-sensing circuits. Cell 2016;164:770–9.
[149] Tousley AM, et al. Co-opting signalling molecules enables logic-gated control of CAR T cells. Nature 2023;615 (7952):507–16. https://doi.org/10.1038/s41586-023-05778-2.
[150] Depil S, Duchateau P, Grupp SA, Mufti G, Poirot L. 'Off-the-shelf' allogeneic CAR T cells: development and challenges. Nat Rev Drug Discov 2020;19:185–99.
[151] Qasim W, et al. Molecular remission of infant B-ALL after infusion of universal TALEN gene-edited CAR T cells. Sci Transl Med 2017;9, eaaj2013.
[152] Dimitri A, Herbst F, Fraietta JA. Engineering the next-generation of CAR T-cells with CRISPR-Cas9 gene editing. Mol Cancer 2022;21:78.

CHAPTER

5

γδ T cells for cancer immunotherapy: A 2024 comprehensive systematic review of clinical trials

Marta Barisa[a], *Callum Nattress*[b], *Daniel Fowler*[a], *John Anderson*[a], *and Jonathan Fisher*[a]

[a]UCL Great Ormond Street Institute of Child Health, University College London, London, United Kingdom [b]UCL Cancer Institute, University College London, London, United Kingdom

Abstract

Despite a substantial increase in funding, activity, and interest, the overall picture regarding outcomes of γδ T cell immunotherapeutic trials for cancer remains ambiguous. We set out to catalogue and discuss the range of different approaches that have been evaluated in the clinic over two decades, from the first trial in 2003 to the latest trial registered in June 2024. We identified 90 γδ T cell intervention-based oncology clinical studies, with Europe, China, and the United States being the busiest sponsor sites. The majority of clinical trials have evaluated therapeutic efficacy in the solid tumor context, with adoptive cell therapy being the most common type of intervention examined. While the first 10 years of studies were dominated by autologous γδ T cell adoptive transfer, the latter decade has been largely allogeneic. Our particular areas of interest were documented trial outcomes and cell therapeutic persistence in the solid and hematological malignancy setting, the range of preconditioning regimens (or lack, thereof) that have been utilized, and the types of γδ T cell genetic modifications that have been progressed to the clinic.

Abbreviations

7PEBL PersonGen Biotherapeutics anti-CD7 protein expression blocker
aAPC artificial antigen presenting cell
AML acute myeloid leukemia
ASCO American Society for Clinical Oncology
B-ALL B cell acute lymphoblastic leukemia

γδ T Cell Cancer Immunotherapy
https://doi.org/10.1016/B978-0-443-21766-1.00002-3

BCG	*M. bovis* bacillus Calmette-Guérin
BCMA	B cell maturation antigen
BiTE	bi-specific T cell engager
BTN	butyrophilin
CAR	chimeric antigen receptor
CLL	chronic lymphocytic leukemia
CML	chronic myelocytic leukemia
CRS	cytokine release syndrome
CTCAE	Common Terminology Criteria for Adverse Events
CTL	cytotoxic lymphocyte
CTM-N2D	CytoMed NKG2DL-redirected γδ CAR-T
DC	dendritic cell
DC-CIK	dendritic cell-cultured cytokine-induced killer cells
DL1/2/3/4	dose levels 1, 2, 3, and 4
DOT	delta one T cells
DRI	drug-resistant immunotherapy
EBV	Epstein-Barr virus
ECOG	Eastern Cooperative Oncology Group (ECOG) Performance Status
ELN 2017	European LeukemiaNet
FDA	Food and Drug Administration (United States)
GD2	disialoganglioside GD2
GPC3	glypican 3
GvHD	graft versus host disease
HCC	hepatocellular carcinoma
HCT	hematopoietic cell transplantation
HLA	human leukocyte antigen
IFN-γ	interferon-γ
IP	intellectual property
IV	intravenous
KO	knockout
LBCL	large B cell lymphoma
mAb	monoclonal antibody
MDS	myelodysplastic syndrome
MGMT	methylguanine-DNA methyltransferase
MHC	major histocompatibility complex
NHL	non-Hodgin's lymphoma
NK	natural killer (cell)
NSCLC	nonsmall cell lung carcinoma
OS	overall survival
PBMC	peripheral blood mononuclear cells
PSCA	prostate stem cell antigen.
Pt	patient
PTCL	peripheral T cell leukemia
Q1/Q2/Q3/Q4	calendar quarters of the year
QC	quality control
RECIST	response evaluation criteria in solid tumors
RP2D	recommended phase II dose
scFv	single chain variable fragment
SFP	scFv-Fc fusion protein
TCR	T cell receptor
TEG	GaDeta "T cells engineered to express a defined γδ TCR"
Th1	type 1 T helper cell
TIL	tumor infiltrating lymphocyte

T-LL	T cell acute lymphoblastic leukemia
TME	tumor microenvironment
TMZ	temozolomide
UCL	University College London

Conflict of interest

CN declares no conflict of interest. MB, DF, JA, and JF hold patents that pertain to the development T cell immunotherapy. JA further holds founder stock in Autolus Ltd.

Introduction

Investment into and exploration of cancer immunotherapy using noncanonical lymphocytes has grown steadily since the early 2000s [1,2]. Much of this interest derives from perceived limitations of more widely adopted gene-modified αβ T cell therapies [3,4], which have produced paradigm shifts in the treatment of refractory B cell malignancies. These perceived limitations can be grouped into two broad categories: (i) logistics and cost of therapeutic manufacture; and (ii) the biological properties of peripheral αβ T cells compared to innate lymphocytes, such as γδ T cells and natural killer cells (NK cells).

In the allogeneic setting, canonical αβ T cell receptors (TCRs) recognize nonself-peptides loaded onto host major histocompatibility complex (MHC) molecules and can mediate potent responses against healthy recipient tissue (graft-versus-host disease or GvHD). For this reason, adoptively transferred αβ T cell drug development has largely focused on autologous therapeutic application, necessitating personalized manufacture for each patient. Associated with a range of logistical difficulties, including complex reagent equipment and specialist staffing resources, per-patient autologous product manufacture incurs an estimated 10-fold cost increase compared to matched allogeneic cell therapy products [5]. Per-patient manufacture further complicates the setup of stringent quality control (QC) screening, since batch-to-batch variability is directly impacted by the quality of individual donor material available. Autologous manufacture often entails production from the apheresate of heavily pretreated and often lymphopenic patients, whose T cell content—relative to healthy donor material—can be poorer quantitatively and qualitatively. Finally, having to execute a complete manufacturing cycle for each patient can increase the time to treatment following decision to treat, with a potentially negative impact on patient outcomes. This can be particularly pronounced when the patient condition is poor or unstable. It can be further compounded by protracted and technically complex manufacturing protocols, as, for example, can be the case with ultra-personalized manufacturing of therapeutics from patient biopsy-derived infiltrating lymphocytes (TIL) [6–8].

In response to the logistical difficulties associated with autologous product manufacture, a range of alternative approaches are being evaluated clinically and preclinically. These include strategies that aim to reduce the risk of alloreactive αβ-TCRs by TCR knockout (KO) or repertoire-skewing, such as selection of Epstein-Barr virus-derived cytotoxic lymphocytes (EBV-CTLs). Other strategies look to explore naturally nonalloreactive innate lymphocyte subsets, such as γδ T cells and NK cells. All of the above have been explored in the clinic for cancer indications, but none have yet been approved. Of note is that Atara's EBV-CTL therapy against posttransplant lymphoproliferative disease became the first approved allogeneic adoptive T cell therapy in December 2022 [9].

Considerations around the biological properties of an optimal cell type for allogeneic cellular immunotherapy (other than lack of alloreactivity) are the key determinants of their prioritization for noncanonical lymphocyte therapeutic development. It may be the case that αβ T cells can be rendered nonalloreactive through various means, but multiple lines of evidence—beyond the lack of alloreactivity—point to γδ T cells and NK cells as being highly suited for use as anticancer immunotherapies. These include decades of in vitro data (and some animal modeling data), which demonstrate broad and potent antitumor reactivity of unmodified γδ T cells and NK cells [2,10], in contrast to αβ T cells which require significant manipulation to acquire therapeutic antitumor responsiveness. Perhaps the strongest biological rationale for the therapeutic exploration of γδ T and NK cell immunotherapy exploration derives from analysis of large cohort tumor transcriptomic data, which correlates patient outcomes with the presence of particular lymphocyte subsets. Indeed, such studies have consistently found infiltrating proinflammatory innate lymphocytes to correlate with improved patient outcomes over and above αβ T cell subsets (including CD8+ αβ CTLs) [11–13]. Unexpectedly, the superior correlation of innate lymphocyte infiltrate with better outcomes held not only for different solid tumor indications but also for hematological malignancies, against which a range of αβ T cell chimeric antigen receptor (CAR) therapies have now been approved. This positive correlation is particularly strong for γδ T cells, the presence of which has been linked by different groups to the improved outcomes in patients with solid as well as liquid tumors [11,12,14–16]. Additional evidence for the physiological importance of γδ T cells for antitumor responses has emerged from retrospectively analyzed bone marrow transplantation clinical data, which indicates that patient event-free survival is enhanced upon higher graft γδ T cell expansion [17–21].

Given the relative wealth of evidence stretching back decades that innate immune lymphocyte, and especially γδ T cell, presence is associated with better outcomes for cancer patients, it is noteworthy that there are still no approved autologous or allogeneic innate lymphocyte tumor immunotherapies. Meanwhile, as of 2024, six different autologous αβ T cell CAR-T products have been FDA-approved for the treatment of hematological malignancies (Kymriah, Yescarta, Tecartus, and Breyanzi targeting CD19, and Abecma and Carvykti targeting BCMA) [22], a highly promising interim report has been published for autologous GD2-targeting αβ T CAR for the treatment of neuroblastoma [23], and Atara's allogeneic αβ-CTL product has been approved for the treatment of lymphoproliferative disease.

The reasons for this discrepancy are likely multiple. First (and, perhaps, foremost), αβ T cell biology is far better understood than that of γδ T cells and other innate lymphocytes. Even basic questions about Vγ9Vδ2 cell biology—the most common γδ T cell subset in easily accessible peripheral blood—remain largely unanswered. It is unclear whether Vγ9Vδ2 cells form immunological memory, whether they become exhausted in a manner similar to canonical αβ T cells, how their responsiveness is regulated peripherally, or even where their primary physiological function lies. Such questions are less clear still for other subsets, with the possible exception of immunological memory, given that large peripheral and tissue-resident clonal expansions upon exposure to viral and other infectious antigens have been observed with Vδ1, Vδ3, and other γδ T cell subsets [24–26]. Whether these protect against future reinfection, however, remains to be firmly established. Such paucity of knowledge, relative even to NK cells, is predominantly attributable to the lack of easily accessible animal models for the study of human γδ T cells. Human-like γδ T cells appear to have evolved relatively recently and appear only in primates [27]. This makes the study of their biology substantially more difficult, ruling out immune competent murine models of chronic viral infection or, indeed, cancer, that have been seminal for our understanding of acute, chronic, protumorigenic, and antitumorigenic αβ T cell responses. Hence, in vitro assays, immunocompromised xenograft murine models and, more recently, human organoid platforms are the mainstay of γδ T cell immunotherapeutic modeling. Even in these systems it is difficult to benchmark their functionality to matched αβ T cell products, given tumor nonspecific αβ TCR-mediated xeno- and alloreactivity that is naturally absent in γδ T cells.

This paucity of high-throughput, translationally relevant models has delayed the development of even basic aspects of the γδ T cell immunotherapy design. The exact engagement mechanism of even the peripherally enriched Vγ9Vδ2 cell TCR remains a questions of some debate [28], and the biological insights that produced Vγ9Vδ2 cell-specific expansion protocols from peripheral blood mononuclear cells (PBMC) were published as recently as 2012 [29]. Understanding of alternative γδ T cell subsets remains less developed still. This lack of clarity is partly explained by the apparent diversity of γδ T cell types, relative even to the highly polyclonal adaptive αβ T cell pool. Due to enhanced numbers of V and D gene segments, human γδ T cells possess a significantly greater theoretical TCR CDR3 diversity than that of αβ T cells, theoretically encompassing 10^{15} unique clones [27]. Different γδ TCR cell clones have been shown to engage butyrophilins (BTN), classical and nonclassical human leukocyte antigens (HLAs), different types of CD1 molecules, and even to bind some antigens directly and in the absence of presentation [27,30]. Some γδ TCR clones appear to engage antigens using multiple modalities simultaneously. The most studied example of this unusual phenomenon is the oligoclonal Vγ9Vδ2 TCR, which is broadly activated by phosphoantigen-driven butyrophilin changes during states of dysregulated cholesterol synthesis pathway-induced cellular "stress" in a manner that is CDR3-independent, but has also been documented to engage a number of protein antigens directly or via the CDR3 through mechanism that are not well understood [28]. Our understanding of the γδ T cell therapeutic design is compounded further by the observation that some γδ T cells can be activated in the apparent absence of direct TCR stimulation, relying instead, wholly or partially, on cytokine stimulation or innate cytotoxicity receptors, such as NKG2D, DNAM-1, NKp30, NKp44, and others [2,31]. For current insights into the knowns and unknowns of translationally relevant γδ T cell biology, we refer you to detailed recent reviews [2,32].

The heterogeneity of immune receptors and effector functions among γδ T cells creates particular problems for quality assessment of a γδ T therapeutic product. An unmodified, activated peripheral αβ T cell can be predicted with some certainty to be poorly cytotoxically responsive to any given tumor target. An αβ T cell engineered to stably express an optimized and validated second-generation CAR is likely to be strongly responsive to tumor targets, in a manner that is influenced by target positivity for the CAR target antigen. This renders αβ T cells a system that is both predictable and malleable—ideal as the cellular substrate for synthetic engineering. The same is not true for γδ T cells. Unmodified γδ T cells are often already responsive to tumor targets in terms of cytotoxicity and Th1-type cytokine production. This responsiveness is dependent on the array of TCR and/or natural cytotoxicity receptors in a manner that has a high degree of functional redundancy and association with cognate ligand expression on the tumor targets. Hence, there can be high interdonor variability of γδ T cell responses against the same target cells. Conversely, activatory ligand expression on tumor target cells is not sufficient to guarantee γδ T cell cytotoxicity against the target. Instead of a binary, αβ TCR or CAR-like on/off switch, γδ T cell activation is hypothesized to be regulated by a range of stimulatory and inhibitory inputs, which produces a cytotoxic or cytokine output only when a particular (and, as of yet, poorly understood) balance of inputs is achieved. This complicates optimal synthetic activatory signal engineering, such as may be derived from a classic CD28-CD3ζ (28ζ) containing second-generation CAR. The amount of synthetic signal provided may be just enough to tip the balance of signaling into optimal cytotoxic functionality against one tumor target with its particular stimulating ligand repertoire, but not another that comes with a distinct set of stimulating ligands. It is conceivable, for example, that the same amount of synthetic CAR-derived signal upon encountering an antigen-bright tumor is optimally activating against a naturally nonimmunogenic target, but excessive against a target that is naturally strongly immunogenic. Unlike αβ T cells, γδ T cells are poorly predictable and, therefore, basic therapeutic deliverables like the provision of an optimal synthetic signal, donor selection, or choice of target present a considerable challenge.

Several established observations about γδ T cell biology have, nonetheless, enabled immunotherapeutic development to proceed. These are: (i) most blood-derived human γδ T cell subsets are nonalloreactive; (ii) most, given appropriate stimulus, are capable of antitumor cytotoxicity via direct (e.g., TCR or natural cytotoxicity receptors) or indirect (antibody-mediated) means, and (iii) most can be a prolific source of Th1-type proinflammatory cytokines, such as IFN-γ [2]. Recent literature further describes that, as with αβ T cells, some naturally occurring γδ T cell subsets, in particular those from nonhematopoietic tissues, can acquire apparently protumorigenic roles, and act as a source of IL-17 or IL-10 and TGF-β in the tumor microenvironment (TME) [33,34].

We set out to evaluate the different approaches that have sought to exploit γδ T cells as antineoplastic therapeutics in the clinic. We have aspired to gather a broad understanding of the clinical activity that has taken place in the first instance, by recording every registered trial since 2014 of adoptive cell transfer, avoiding geographical bias. Our staged search methodology is illustrated in Fig. 1. For a systematic review of γδ T cell trials prior to 2014, please refer to our 2014 systematic clinical trial review by Fisher et al. [35], and the 2012 review by Fournié and colleagues [36]. Our areas of particular interest for this review were trials targeting solid tumors, allogeneic adoptive transfer trials, and gene-engineered γδ T cell trials.

Results

Broad overview

Prior to delving into granular analysis of different trial specifics, we evaluated the macro trends in the use of γδ T cell interventions for antineoplastic clinical trials, including trends in trial sponsorship location, types of indications targeted, and types of interventions utilized. An in-depth discussion of solid tumor, allogeneic and gene-engineered γδ T cell adoptive transfer trials is presented in subsequent sections.

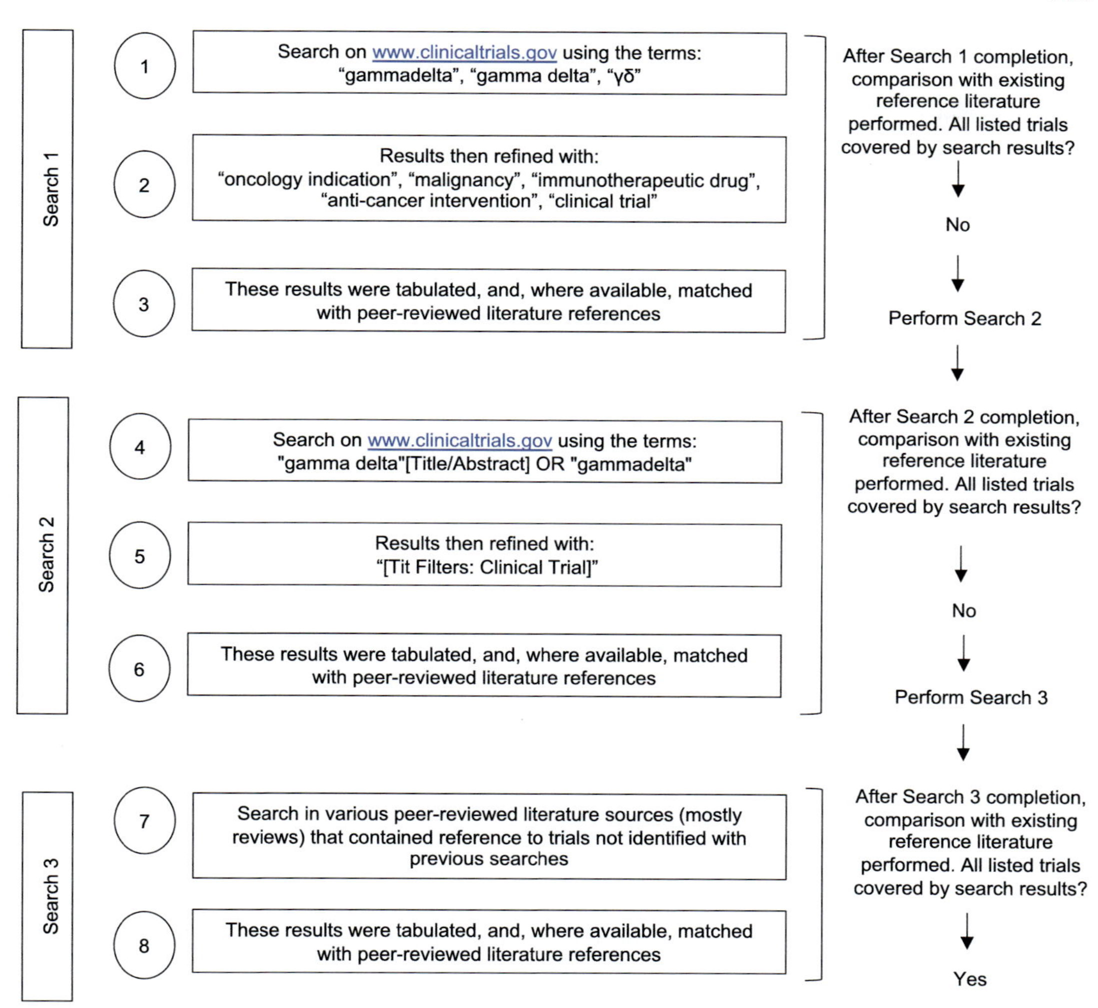

FIG. 1 Search methodology. The search was an iterative process, performed in three stages. Search 1 consisted of a highly refined clinicaltrials.gov search, followed by Search 2, which was a less stringent version of the first search. The final search ventured beyond clinicaltrials.gov, and included screening of literature, primarily peer-reviewed original research and review articles, that referenced γδ T cell clinical trial outcomes. All identified relevant trials were included. The search was carried out during June 2024.

Trial location

The search identified a total of 90 clinical trials, listed in the Appendix. While new γδ T cell trial registration has seen numerous periods of increased activity since beginning in the early 2000s, the rate of new trial registration has been at its highest since 2019, despite a likely SARS-CoV-2-driven slowdown of new trial registration in 2020 (Fig. 2A). Since the earliest γδ T trials, Europe has consistently been the busiest location in terms of new trial sponsors, but this has changes with a rapid growth of the number of γδ T clinical trials sponsored in

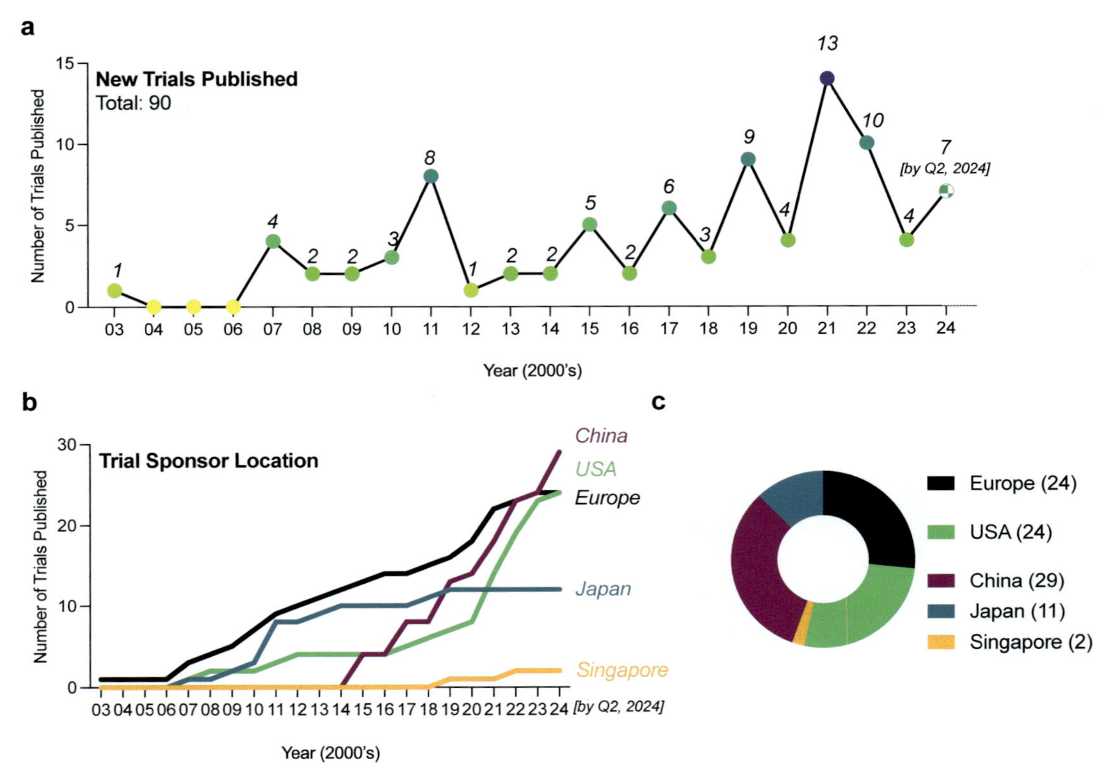

FIG. 2 γδ T cell-based oncology clinical trials over time, sorted by location. (A) We identified a total of 90 γδ T cell oncology clinical trials, and (B) categorized these by location of the trial sponsor and time. Panel (C) indicates the total number of trials that have been sponsored in a given location.

China and the United States (Fig. 2B and C). If current trends persist, the United States and China will become the leading locations for γδ T cell oncoimmunotherapy trials over the next decade, with Europe maintaining a steady but overall slower rate of new trial registration.

In contrast, Japan, an early global leader in γδ T cell oncology trial sponsorship, has substantially reduced new trial activity since the early 2010s. An active newcomer to the γδ T clinical trial space is Singapore, where much of the trial growth is driven by the cell therapy company, CytoMed Therapeutics.

The global trends for busy cellular immunotherapy locales are broadly consistent for CAR-T studies in general, where the United States has now been overtaken by China as the most active study sponsor [37]. It is worth noting, however, that Europe has played a proportionally greater role in γδ T clinical translation compared to CAR-T trials at large.

Trial indications

With respect to targeted disease indication, γδ T cell clinical trials stand in contrast to mainstream CAR-αβ T cells. The majority of CAR-αβ T trials have examined efficacy against hematological cancer indications [38], while almost two-thirds of γδ T oncology trials have

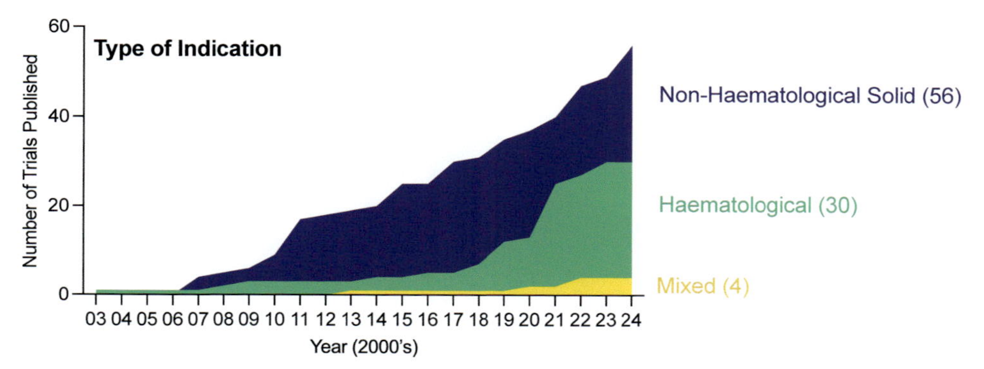

FIG. 3 γδ T cell-based oncology interventional clinical trials over time, sorted by type of tumor targeted. γδ T cell clinical studies were grouped according to what type of malignancy was targeted.

targeted nonhematological solid tumor indications (Fig. 3). Many of the trials have been basket-indication studies in some contrast to CAR-αβ trials, where specific disease indication and strong target antigen expression are often a prerequisite for trial enrolment.

The rarity of γδ T hematological trials was rapidly reversed upon the first approvals of CAR-αβ therapies for acute lymphoblastic leukemia in 2017, and hematological γδ T trials have been on the rise since, to the extent that 2021 saw the registration of 11 hematological indication-targeting γδ T clinical trials and only three trials for nonhematological solid tumors. This trend looks set to continue, as allogeneic alternatives are sought for the clinically proven autologous CD19 and BCMA CAR-αβ T products.

Types of γδ T interventions

The vast majority (~80%) of γδ T cell oncology clinical trials have employed adoptive transfer of ex vivo-activated γδ T cells as the mode of therapeutic intervention. This trend has accelerated since 2017 and is again likely influenced by the concurrent successes of CAR-αβ T adoptive transfer therapies for blood cancers (Fig. 4A). Prior to 2014, the more common trial intervention consisted of the administration of Vγ9Vδ2 subset-activating bisphosphonate drugs [35] rather than adoptive cell therapy, with or without T cell-supporting cytokines, such as IL-2. While an attractive therapeutic approach due to low cost and straightforward scalability, disappointing clinical results [35,39] have meant that new clinical trials based on bisphosphonate drug administration without further combination treatments, such as the co-administration of activated γδ T cells, have been largely halted. A relatively recent and expanding trend for γδ T cell therapeutic interventions utilizes T cell engagers, that aim to bind to tumor antigens with one binder and to the γδ TCR with a linked second binder [40–42]. A recent trial has also sought to examine the use of γδ T cell adoptive transfer in combination with already approved tumor-specific antibodies [43].

The earliest γδ T adoptive trials were all autologous, though allogeneic γδ T trials have been taking place since 2014 (Fig. 4B). While new γδ T trial registration since 2017 is for predominantly allogeneic adoptive transfer products, autologous γδ T products are still being explored. The most striking trend concerns the rapid proliferation of gene-modified γδ T cell trials since 2019. Interestingly, the trend for genetic modification is not spread evenly over

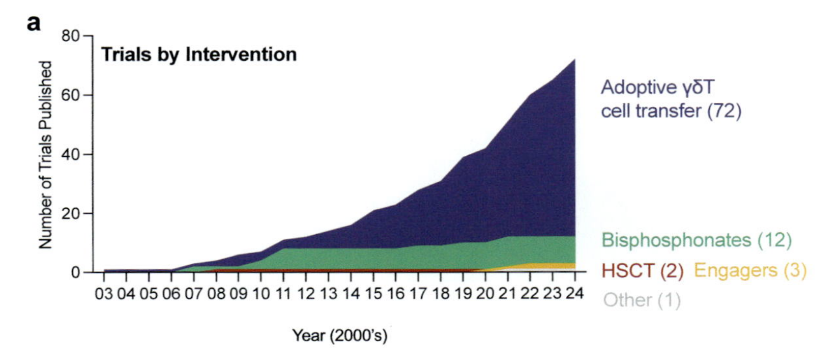

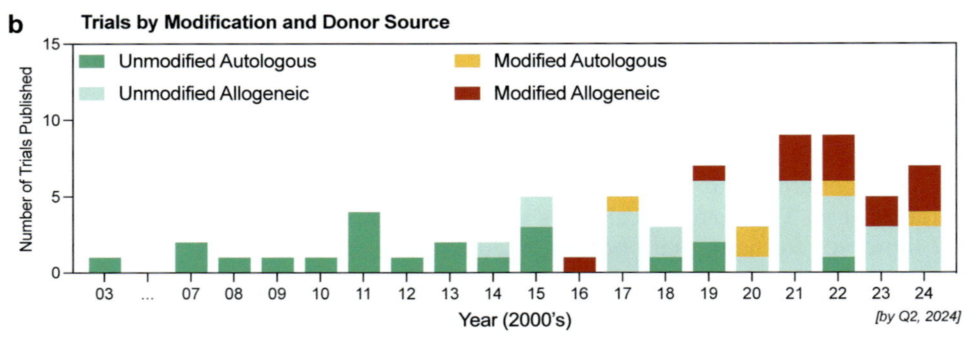

FIG. 4 γδ T cell-based oncology clinical trials over time, sorted by the type of γδ T cell-based intervention. (A) Studies were tracked over time and sorted according to whether adoptive cell transfer, bisphosphonates, hematopoietic stem cell transplantation (HSCT), T cell engagers or other modalities formed the primary basis of the therapeutic intervention. The single "Other" study in this context was trial NCT02753309, described further by Mukherjee et al., which evaluated the use of rapamycin combined with intravesical BCG in patients with nonmuscle invasive bladder cancer, and tracked endogenous γδ T cell responses. (B) The adoptive cell transfer trials were further subcategorized into autologous and allogeneic trials, which were either modified or unmodified.

autologous and allogeneic products. Since 2020, all three new autologous γδ T cell trials were gene-modified, while the majority of allogeneic γδ T trials were still for unmodified γδ T cells over that same time period.

Solid tumor targeting

Nearly two-thirds of γδ T cell-based clinical trials for cancer have sought to achieve therapeutic efficacy specifically in the solid, as opposed to hematological, tumor setting (Fig. 3). The range of solid tumor indications targeted with γδ T cell-based interventions since 2014 is summarized in Fig. 5 and covers 38 total trials registered. Trial indications were distributed relatively evenly across common solid tumor indications, including breast, lung, gastric, and prostate cancers. Of 38 trials, 30 targeted a specific indication, and only 8 of the trials aimed to

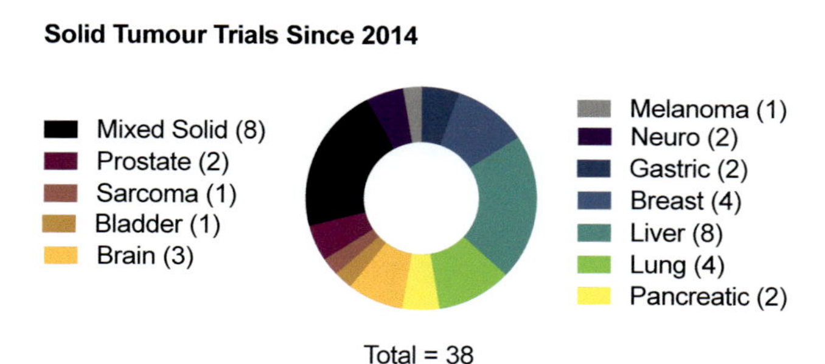

FIG. 5 Solid tumor-targeting γδ T cell clinical studies since 2014, grouped by tumor type.

evaluate γδ T intervention in basket indication clinical trials. Interestingly, the most commonly targeted indication with eight trials was liver cancer, an area that remains of substantial therapeutic interest for γδ T cell immunotherapy development [44]. The next most common single indication was breast cancer with four trials. A vast majority of trials evaluated unmodified autologous or allogeneic γδ T cell adoptive cell transfer, followed by trials employing gene-modified autologous and allogeneic γδ T cell interventions, and several based treatment on cell-free bisphosphonate interventions and T cell engagers.

To evaluate the outcomes of the solid tumor-targeting trials, we tabulated the trials for which the results have been published on PubMed (see Table 1), covering 12/38 trials. None of the trials reported dose-limiting toxicity, presenting a universally favorable safety profile. Objective response rates were collated as reported by the trial and publication authors. Where available, objective response criteria were included. Two of the trials were based on bisphosphonate interventions [54,55], including one that examined the effect of adding zoledronic acid to letrozole for the treatment of breast cancer [54]. Neither of the trials reported objective responses attributable to zoledronic acid. One of the trials sought to evaluate the efficacy of BTN3A-targeting mAb, ICT01, in engaging endogenous Vγ9Vδ2 cells in a basket solid tumor trial, but reported no objective responses [40]. Interestingly, however, the trial did report a decrease in peripheral γδ T cell levels, potentially indicative of enhanced ICT01-mediated recruitment to tumor sites. The remaining nine trials utilized unmodified adoptive Vγ9Vδ2 cell transfer (three autologous [56–58], six allogeneic [55–59]) to treat gastric, liver, pancreatic, nonsmall cell lung cancer, and cholangiocarcinoma. Summarizing data from these trials, there are no reported instances of objective radiological response attributable to γδ T cell infusion or activation, although not all trials defined objective response criteria.

In three of the nine reported adoptive transfer trials, mixed radiological response was reported. While one adoptive transfer trial (UMIN000004130) delivered T cells locoregionally to treat malignant intraperitoneal ascites and is, therefore, noninformative for T cell homing, eight other trials involved intravenous delivery of γδ T cells, of which two trials reported responses in lymph nodes (NCT02425735) and lung lesions (UMIN000006128). This suggests that the differential responses observed may be dependent on transferred Vγ9Vδ2 cell tropism or sensitivity to direct and indirect tumor microenvironmental (TME) suppression that may vary between different tumor sites.

TABLE 1 γδ T cell solid tumor clinical trial publications and outcomes since 2014.

Year (Refs.)	Publication title	Trial ID	Indication	Well-being	Type of γδ T intervention	Adjunct treatment	ORRC	ORR	Findings of note
In vivo activation									
2016 [45]	In vivo expansion and activation of γδ T cells as immunotherapy for refractory neuroblastoma	NCT01404702	Neuroblastoma	≥ 50% Karnofsky	Intravenous zoledronic acid and IL-2	Not specified	Not specified	0/4	Well tolerated; trial terminated due to slow accrual (4 pts)
2018 [46]	Combined effects of neoadjuvant letrozole and zoledronic acid on γδ T cells in postmenopausal women with early-stage breast cancer	UMIN000008701	Breast cancer	ECOG: 0–2	Zoledronic acid, used as a neoadjuvant	Letrozole	RECIST		No notable benefit conferred by adding zoledronic acid to regimen (55 pts)
2021 [47]	Development of ICT01, a first-in-class, anti-BTN3A antibody for activating Vγ9Vδ2 T cell-mediated antitumor immune response	NCT04243499	Basket solid and hematological tumor trial	ECOG: 0–1	Anti-BTN3A agonist antibody	Not specified	Not specified	0/6	Decrease in peripheral Vγ9Vδ2 cells, potentially indicative of tumor recruitment; trial ongoing (6 pts)
Adoptive transfer									
2014 [48]	Intraperitoneal injection of in vitro-expanded Vγ9Vδ2 T cells together with zoledronate for the treatment of malignant ascites due to gastric cancer	UMIN000004130	Gastric cancer	ECOG: 0–2	Autologous Vγ9Vδ2 enriched, zoledronic acid and IL-2-activated PBMCs delivered intraperitoneally	Zoledronic acid (IV and IP)	Not specified	0/7	Reduction of ascites in 2 patients, but progression at distal sites (7 pts)

2017 [49]	Adjuvant combination therapy with gemcitabine and autologous γδ T-cell transfer in patients with curatively resected pancreatic cancer	UMIN000000931	Pancreatic cancer	ECOG: 0–1	Autologous Vγ9Vδ2-enriched, zoledronic acid and IL-2-activated T cells, delivered IV	Gemcitabine	Not specified		No notable benefit conferred by adding γδ T cells to regimen (28 pts)
2019 [50]	Allogenic Vγ9Vδ2 T cell as new potential immunotherapy drug for solid tumor: a case study for cholangiocarcinoma	NCT02425735	Cholangiocarcinoma	>50%: Karnofsky	Allogeneic Vγ9Vδ2 enriched, zoledronic acid and undisclosed cytokine-activated T cells, delivered IV	Not specified	Not specified		Reduction in lymph node size, well tolerated (1 pt)
2020 [51]	Adoptive transfer of zoledronate-expanded autologous Vγ9Vδ2 T-cells in patients with treatment-refractory non-small cell lung cancer: a multicenter, open-label, single-arm, phase 2 study	UMIN000006128	Nonsmall cell lung carcinoma	ECOG: 0–1	Autologous Vγ9Vδ2 enriched, zoledronic acid and IL-2-activated T cells, delivered IV	Not specified	RECIST	0/25	Lung lesion response in 1 patient, but with liver metastasis progression (25 pts. altogether)
2020 [52]	Irreversible electroporation plus allogenic Vγ9Vδ2 T cells enhances antitumor effect for locally advanced pancreatic cancer patients	NCT03180437	Pancreatic cancer	Not specified	Allogeneic Vγ9Vδ2 enriched, zoledronic acid and IL-2-activated T cells, delivered IV	Not specified	Not specified	0/30	Modest survival improvement Vγ9Vδ2 + IRE arm (30 pts)
2021 [53]	Allogeneic Vγ9Vδ2 T-cell immunotherapy exhibits promising clinical safety and prolongs the survival of patients with late-stage lung or liver cancer	NCT03183206 NCT03183219 NCT03183232 NCT03180437	Different for different trials; lung and liver cancer	>50%: Karnofsky	Allogeneic Vγ9Vδ2 enriched, zoledronic acid and IL-2/IL-15/vitamin C-activated T cells, delivered IV	IRE, Iodine-125 and/or cryoablation	RECIST	1/132	One complete response in a patient that was receiving concurrent iodine-125 (132 pts)

Abbreviations: *PBMC*, peripheral blood mononuclear cells; *IV*, intravenous; *IP*, intraperitoneal; *pts*, patients; *ECOG*, Eastern Cooperative Oncology Group; *RECIST*, response evaluation criteria in solid tumors; *BTN3A*, butyrophilins 3A; *ORRC*, objective response rate criteria; *ORR*, overall response rate.

While definitive conclusions from these trials are difficult to draw due to nonuniform objective response criteria, the studies challenge the perspective that unmodified Vγ9Vδ2 cell adoptive cell transfer is sufficient to effect substantial therapeutic efficacy against solid tumors, in either the allogeneic or autologous transfer setting. Gene modification or therapeutic combinations to enhance Vγ9Vδ2 cell persistence, homing and effector function may be a prerequisite for achieving objective responses for Vγ9Vδ2 cell immunotherapy for solid tumors. While disappointing by some measures, this finding is broadly consistent with clinical experience of adoptive transfer immunotherapies using unmodified αβ T cells and NK cells. That is, most unmodified immune cells in the absence of adjuvant treatment or genetic enhancement are a poor match for the rapid proliferation, immune suppression, and evolutionary phenotypic agility of solid tumors.

Hematological cancer targeting: Mixed results

The range of hematological cancer indications targeted with γδ T cell-based interventions since 2014 is summarized in Fig. 6 and constitutes 28 registered studies. Compared to the solid tumor-targeting γδ T trials where a single indication was the most common type of target, hematological malignancy trials were often for mixed indications with or without targeted antigen expression eligibility criteria. Nine of the 24 trials targeted multiple (>3) hematological malignancies. The most common type of single indication trial focused on targeting acute myeloid leukemia (AML), with eight trials. This may be in part explained by the substantial unmet need represented by AML, given that it is often seen as a difficult indication for antigen-targeted, synthetic αβ T cell-based therapies. Notably, all the AML trials used unmodified γδ T cells, aiming to evaluate the innate recognition and cytotoxic capacity of ex vivo activated γδ T cells—likely driven in part by the difficulty with identifying AML-restricted target antigens that are not expressed on the healthy myeloid compartment. Among trials that focused on antigen-specific targeting, the most common interventions were for CD20-expressing B cell malignancies with three trials, followed by one trial each targeting CD19-expressing chronic lymphocytic leukemia (CLL), B-lineage acute lymphoblastic leukemia (B-ALL) and lymphoma. Two trials targeted CD7-expressing malignancies.

A majority of the trials employed unmodified autologous or allogeneic ex vivo-expanded γδ T cell adoptive cell transfer, often in the context of an allogeneic hematopoietic stem cell

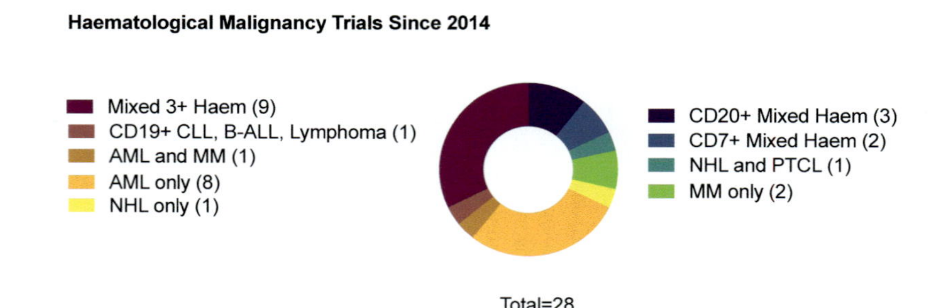

FIG. 6 Hematological malignancy-targeting γδ T cell clinical studies since 2014, grouped by cancer type.

transplantation. A single trial examined the efficacy of TCR-targeting BiTe therapy and eight trials employed gene-modified autologous or allogeneic γδ T cell infusions. Of note—and in some contrast with solid tumor targeting—we could identify no γδ T cell trials for hematological indications since 2014 that sought to utilize cell-free bisphosphonate interventions, which may suggest a belief in the community that such intervention is unlikely to be sufficient to effect meaningful improvements in patient outcomes.

To evaluate the outcomes of the hematological cancer-targeting trials, we tabulated the results of the 4/28 trials for which the results have been published and indexed on PubMed (see Table 2). Of these, one trial evaluated IV infusion of the Vγ9Vδ2 cell activating anti-BTN3A mAb, ICT01 [40], and reported no objective responses. The rest employed cell-based therapies. Of those, γδ T cells were delivered either as an ex vivo bisphosphonate-activated product [63], as a part of an αβ T cell-depleted but otherwise nonactivated peripheral leukapheresate [61], or a bone marrow transplantation combined with zoledronic acid and IL-2 [64]. All three reported a number of complete or partial responses up to 8 months following treatment, although it is uncertain whether responses resulted from γδ T cell adoptive transfer or as a consequence of the preconditioning chemotherapy.

Taken together, the trials have confirmed the essential safety and feasibility of adoptive transfer of autologous and allogeneic γδ T cell products. These studies have generated important early proof-of-concept data that γδ T cell-based adoptive transfer has potential to effect clinical responses in diseases like multiple myeloma and acute myeloid leukemia. Whether these transient responses can be consolidated to produce meaningful and sustained disease remission will be informed by upcoming trials with gene-modified adoptive γδ T cell transfer in combination with treatments beyond IL-2 and zoledronic acid.

Allogeneic γδ T cell adoptive transfer trials

It is likely that the persistence of γδ T cell product in patient circulation will be a crucial determinant of the durability of therapeutic effect. Various approaches are being examined to address this issue, ranging from evaluation of different preconditioning chemotherapy regimens, to opting for autologous products to avoid allo-rejection. Some have sought to address this with repeated infusions of cryopreserved, allogeneic and unmodified γδ T cell products, though it is unclear how effective this approach is in the absence of paired repeated lymphodepleting chemotherapy.

No γδ T cellular immunotherapies that are cytokine armored for therapeutic persistence or cloaked to reduce allo-rejection have yet reached the clinic, though preclinical pipelines remain in development from academic and commercial cell therapy specialists [65,66].

Choice of γδ T cell subset for allogeneic trials

All of the above-discussed and results-reported PubMed-indexed solid and hematological cancer γδ T cell intervention trials have utilized the Vγ9Vδ2 subset. Interestingly, this is not a trend that has held firm as time has gone on for more recent allogeneic γδ T cell adoptive transfer trials, of which we identified a total of 38 (Fig. 7). Vγ9Vδ2 cells are the predominant γδ T cell type in 14/38 allogeneic studies registered since 2014 (Fig. 7 and Table 3—Group A), while the others were based on less explicitly defined TCR identities.

TABLE 2 γδ T cell hematological tumor trial publications and outcomes since 2014.

Year (Refs.)	Publication title	Trial ID	Indication	Well-being	Type of γδ T intervention	Adjunct treatment	ORRC	ORR	Findings of note
In vivo activation									
2021 [47]	Development of ICT01, a first-in-class, anti-BTN3A antibody for activating Vγ9Vδ2 T cell-mediated antitumor immune response	NCT04243499	Basket solid and hem liquid tumor trial	ECOG: 0–1	Anti-BTN3A agonist antibody	Not specified	Not specified	0/6	Decrease in peripheral Vγ9Vδ2 cells, potentially indicative of tumor recruitment (6 pts)
2021 [60]	Phase II trial of maintenance treatment with IL2 and zoledronate in multiple myeloma after bone marrow transplantation: biological and clinical results	2013-001188-22	Multiple myeloma	Not specified	IV zoledronic acid and IL-2 following autologous bone marrow transplantation	Not specified	International uniform response criteria for multiple myeloma	8/44	Treatment resulted in increased peripheral Vγ9Vδ2 numbers that subsequently declined (44 pts)
Adoptive transfer									
2014 [61]	Successful adoptive transfer and in vivo expansion of haplo-identical γδ T cells	Not identified	Basket hematological (lymphoid and myeloid)	Not specified	Allogeneic αβ T cell-depleted nonactivated leukapheresate delivered IV	IV zoledronic acid and IL-2	Not specified	3/4	Transient complete remissions that lasted 2–8 months, which may be attributable in part to Flu/Cy and non-γδ T components; role of γδ T cells unclear (4 pts)
2023 [62]	A phase I trial of allogeneic γδ T lymphocytes from haploidentical donors in patients with refractory or relapsed acute myeloid leukemia	NCT03790072	AML	≥ 50% Karnofsky	Allogeneic, unmodified PBMC and zoledronic acid-activated Vγ9Vδ2 cells, delivered IV	Not specified	ELN 2017	2/3 at 3 at DL 3	One transient complete remission that lasted ~2 months and one MLFS of the same duration

Abbreviations: *PBMC*, peripheral blood mononuclear cells; *IV*, intravenous; *IP*, intraperitoneal; *pts*, patients; *ELN*, European Leukemia Net; *RECIST*, response evaluation criteria in solid tumors; *BTN3A*, butyrophilins 3A; *ORRC*, objective response rate criteria; *ORR*, overall response rate; *AML*, acute myeloid leukemia; *DL*, dose level; *Flu/Cy*, fludarabine/cyclophosphamide; *MLFS*, morphologic leukemia-free state.

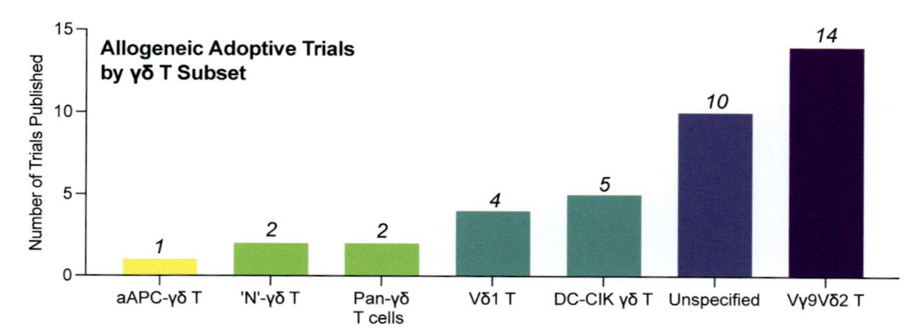

FIG. 7 Adoptive transfer γδ T cell clinical studies since 2014, grouped by γδ T subset used. The six identifiable subtypes were Vγ9Vδ2 cells, dendritic cell-cultured cytokine-induced killer γδ T cells (DC-CIK γδ T), Vδ1 cells, "NK/T-like" γδ T cells ("N"-γδ T), pan-γδ T cells, and artificial antigen presenting cell-expanded γδ T cells (aAPC-γδ T). The remaining trials did not specify the type of γδ T cells used.

TABLE 3 Allogeneic adoptive γδ T cell transfer trials since 2014 by subset.

	Refs.	γδ T intervention	TCR	Manufacture	Trial ID	Sponsor(s)
A	[46,49,51,63]	Vγ9Vδ2 T	Vγ9, Vδ2	Various bisphosphonate stimulation with common-γ chain cytokines (predominantly IL-2)	NCT03183219 NCT05664243 NCT05358808 NCT05400603 NCT05628545 NCT05302037 NCT04107142 NCT03790072 NCT03533816 NCT03183232 NCT03183206 NCT02656147 NCT05686538 NCT06150885	Various
B	[50]	DC-CIK γδ T	Unknown (likely Vγ9, Vδ2)	Unspecified cytokine-induced killer cells co-cultured with dendritic cells (DC-CIK)	NCT02585908 NCT02425748 NCT03180437 NCT02425735 NCT02418481	Fuda Cancer Hospital, Guangzhou (China)
C	[67,68]	Vδ1 T	Vγ-variable, Vδ1	Common-γ chain cytokines and anti-CD3 mAb (OKT-3), or anti-Vδ1 mAb	[a]NCT05001451 [b]NCT04911478 [b]NCT04735471 [c]NCT05377827	[a]GammaDelta Therapeutics (United Kingdom) [b]Adicet Bio (United States) [c]Takeda (United States, Japan)

Continued

TABLE 3 Allogeneic adoptive γδ T cell transfer trials since 2014 by subset—cont'd

	Refs.	γδ T intervention	TCR	Manufacture	Trial ID	Sponsor(s)
D	[69,70]	"N"-γδ T	Vγ-unspecified, Vδ2, NK-T unspecified	Phosphoantigens combined with unspecified "NK-T-stimulating" cytokines	NCT05653271 NCT06415487	Acepodia (United States)
E	[71]	aAPC-γδ T	Unspecified (likely mixed)	K-562 CD3 scFv—CD137L—CD28 scFv—IL15RA quadruplet artificial antigen-presenting cell (aAPC), with zoledronic acid and IL-2	NCT05015426	H. Lee Moffitt Cancer Center and Research Institute (United States)
F	[72,73]	"Pan"—T cells including γδ T	Unspecified (likely mixed)	[d]CRISPR/Cas9 CD7 and TRAC-K/O expanded T cells [e]Pan-γδ mAb-expanded T cells	[d]NCT05377827 [e]NCT06404281	[d]Washington University School of Medicine [e]Changzhou No.2 People's Hospital
G	Unspecified				[1]NCT04765462 [1]NCT04764513 [2]NCT04696705 [3]NCT04518774 [4]NCT04008381 [5]NCT06018363 [6]NCT06069570 [3]NCT06364800 [3]NCT06364787 [7]NCT06212388	[1]Chinese PLA General Hospital [2]Institute of Hematology & Blood Diseases Hospital [3]Beijing 302 Hospital [4]Wuhan Union Hospital [5]Dushu Lake Hospital Affiliated to Soochow University (all China) [6]Beverly Hills Cancer Center (United States) [7]Xijing Hospital (China)

Abbreviations: *mAb*, monoclonal antibody; *TRAC*, T cell receptor α constant locus; *NK-T*, natural killer T cells; *aAPC*, artificial antigen presenting cells; *scFv*, single chain variable fragment; *IL15R*, IL15 receptor; *CRISPR*, Clustered Regularly Interspaced Short Palindromic Repeats.

Of these, the next most common trial type (Group B) described "(DC)-CIK/γδ T cells" or dendritic cell cytokine-induced killer cells combined with ex vivo-expanded γδ T cells (of apparently Vγ9Vδ2 cell phenotype), which have been evaluated at the Fuda Cancer Hospital in Guangzhou, China. While a stringent definition for this cell type is lacking, CIK cells are broadly defined as a "[…] heterogeneous population of effector $CD3^{+}CD56^{+}$ natural killer T cells" [74], traditionally manufactured by PBMC stimulation with an anti-CD3 mAb, IL-2 and IFN-γ, with or without additional stimulation in the presence of DCs [74]. CIK cells have been evaluated in a range of adoptive cell therapy trials for cancer in various formats: on their own, in combination with chemotherapy or surgery, or in combination with other cell therapies. γδ T cells have been one of the most commonly studied CIK therapy cellular combinations [75]. Trial NCT02585908 evaluated CIK+γδ T immunotherapy for gastric cancer, NCT02425735 examined DC-CIK+γδ T for liver cancer, NCT02425748 assessed DC-CIK+γδ T for lung cancer, and NCT02418481—DC-CIK+γδ T for breast cancer. Other than trial NCT02425735, which has been ongoing since 2015 with current status "unknown," all the other DC-CIK+γδ T trials have been completed, and enrolled a total of 120 patients.

Unfortunately, we were unable to identify any PubMed-indexed references to the CIK/γδ T trial outcomes, other than a single-patient case report from trial NCT02425735. This patient was diagnosed with cholangiocarcinoma (stage IV), had received liver transplantation and was subsequently diagnosed with recurrent mediastinal lymph node metastasis [50]. The patient was then enrolled onto trial, received a total of 400×10^{6} ex vivo-expanded allogeneic Vγ9Vδ2 cells across eight infusions, experienced an improvement in quality of life and saw shrinkage of lymph node metastases. This case study is briefly discussed in Table 1. We are hopeful that the results from the remaining 119 patient cohort may yet be published. Operating on the public data currently available, it is impossible to evaluate the safety and efficacy of (DC-)CIK+γδ T cellular immunotherapy.

There are four registered trials (group C) that describe the T cell product as "Vδ1" cells—a population that is defined by V delta chain TCR expression, in which V gamma chain expression is not specified. In interesting contrast to a landscape dominated by Vγ9Vδ2 alone or (DC-)CIK-γδ T academic trials, all three registered Vδ1 trials were commercially sponsored. Trial NCT05001451 was sponsored by Gamma Delta Therapeutics Limited, and evaluated TCR polyclonal Vδ1 "DOT cell" monotherapy for patients with minimal residual disease positive acute myeloid leukemia. The "DOT" cells were batch-manufactured from allogeneic PBMC using anti-CD3 mAb (clone OKT-3) and a cocktail of cytokines, as described in preclinical manuscripts [67,76]. The trial was first posted in 2021, enrolled three patients and was terminated in 2022. The formal reason for trial termination is listed as a "business decision" by the trial sponsor [77]. We were unable to identify further published safety and efficacy data for the trial. Takeda will evaluate a similar approach for relapsed or refractory acute myeloid leukemia in a larger multicenter study, registered in 2023 (NCT05377827).

Trials NCT04735471 and NCT04911478, sponsored by Adicet Bio, both pertain to one and the same immunotherapeutic product and disease indication: allogeneic gene-modified CD20-CAR Vδ1 cells, ADI-001, for the treatment of patients with $CD20^{+}$ B cell malignancies. "ADI-001" cells were batch-manufactured from allogeneic PBMC using anti-Vδ1 TCR mAb stimulation and a cocktail of cytokines, and then transduced with a γ-retroviral vector to express the CD20 CAR, comprising a 3H7 scFv, CD8α hinge and transmembrane domains, and 4-1BBζ signaling domains, as described in a preclinical manuscript [68]. Further discussion of the synthetic construct is contained in a subsequent review section, "Modified γδ T cell adoptive transfer trials." Interventional Phase I trial NCT04735471 evaluates the safety of ADI-001 treatment and aims to enroll a total of 78 patients, while trial NCT04911478 is the associated long-term follow-up observational study. The interventional trial has been ongoing since 2021 but has led to an interim ASCO abstract [78] and press release [79]. The press release reports a 71% ORR and 63% complete response rate across all dose levels in patients with median of four prior lines of therapy. Fifty percent of the patients enrolled had previously progressed on anti-CD19 CAR-T therapy. At 6 months following treatment, the CR rate was 25%, and in patients that had progressed following autologous CD19 CAR-T therapy, the 6-month CR rate was 33%. Adicet Bio reported that their allogeneic CD20 CAR-T expansion and persistence at the recommended Phase II dose level exceeds values reported for "approved autologous CD19 CAR T cell therapy" (presumably referring to tisagenlecleucel (Kymriah) and axicabtagene ciloleucel (Yescarta)); DL4 demonstrated a mean C-max of 483 cells/μL with a mean time-to-peak at approximately day 9 following infusion and persistence through day 28, with mean blood levels of 21 cells/μL. This, however, must be interpreted in the context of ADI-001 DL4 recommended Phase II dose level being 1×10^9 CAR-positive cells—a dose 2–5 times higher than that recommended for Kymriah and Yescarta therapy [80,81]. An important aspect of the reported ADI-001 data is the relative lack of treatment-related toxicities, compared to αβ T cell anti-CD19 CAR-T therapies, including high grade cytokine release syndrome (CRS) and immune effector cell-associated neurotoxicity syndrome (ICANS). Adicet Bio plan to progress their CD20 CAR-Vδ1 product into a pivotal Phase II study in post-CAR-T relapse large B cell lymphoma (LBCL) during 2024 [79].

"N"-γδ T cells (Group D), the development of which has been sponsored by the biotech Acepodia, derive their name from a Vγ9Vδ2-oriented γδ T cell expansion process that includes further undisclosed "NK-T stimulating cytokines." The resulting allogeneic products, named ACE1831 and ACE2016, are not genetically modified, but covalently conjugated to an anti-CD20 and EGFR binders, respectively, via DNA aptamers (described in abstracts [69,70]). ACE1831 and ACE2016 product engineering is illustrated in greater detail in the "Modified γδ T cell adoptive transfer trials" section of the review and Fig. 9. ACE products are reported to express high levels of CD56, CD16, NKG2D, and CD69, and low levels of PD-1 and KIRs [69], though the data have not yet been made public. Ongoing trial NCT05653271 is evaluating ACE1831 in adult subjects with relapsed/refractory CD20-expressing B-cell malignancies, while trial NCT06415487 examines ACE2016 safety and efficacy for locally advanced or metastatic solid tumors expressing EGFR. The CD20-targeted trial estimates a total enrolment of 42 patients and a 2027 completion date. Acepodia announced the dosing of their first patient (with relapsed, refractory non-Hodgkin's lymphoma) in May 2023 [82]. We anticipate an interim trial report in 2024. The EGFR-targeted trial estimates it will enroll 30 patients, and has begun in 2024.

Artificial antigen presenting cell-expanded γδ T cells (aAPC-γδ T cells; Group E), are currently under clinical evaluation at H. Lee Moffitt Cancer Center and Research Institute in the United States. Trial NCT05015426 is evaluating unmodified, allogeneic aAPC-γδ T cells for the prevention of AML relapse in high-risk patients following allogeneic hematopoietic stem cell transplantation [83]. A published preclinical manuscript [71] described aAPC-γδ T cell product manufacture consisting of PBMC stimulation with irradiated K562-based aAPCs that have been engineered to express an anti-CD3 scFv, CD137L, and CD28 scFv co-stimulatory molecules, and IL15Rα, in the presence of zoledronic acid and IL-2. While the protocol describes substantial expansion of γδ T cells overall, we were unable to identify information pertaining to specific γδ T cell subset composition of the resulting product. The addition of zoledronic acid to the protocol implies a substantial (though likely not uniquely so) content of Vγ9Vδ2 cells.

Two 2023 trials (Group F) were registered as utilizing pan-γδ T cell products. One US trial from Washington University School of Medicine in collaboration with Wugen is evaluating CRISPR/Cas9 CD7 and TRAC-K/O αCD7 CAR-T cells for $CD7^+$ hematological malignancies [72]. A Chinese trial by Changzhou No.2 People's Hospital, meanwhile, is evaluating pan-γδ mAb-expanded αPD-1-secreting γδ T cell efficacy for the treatment of advanced solid tumors [73].

Ten further trials (Group G) were run with unspecified and apparently unmodified γδ T cell products; all were sponsored by entities in China. We were unable to identify further information on the manufacturing process for γδ T products for these studies.

Allogeneic trial dosing and preconditioning

A likely significant determinant of adoptively transferred allogeneic (and autologous) cell therapeutic efficacy is the content, dose, and scheduling of the preconditioning or lymphodepletion regimen. In order to describe the relationship of these parameters to trial outcomes, we collated all the allogeneic trials since 2014 for which we could identify information on the preconditioning regimen (Table 4). Trials that specified not having a preconditioning regimen were included. We have shown the depth of information that was available for each trial, which was variable between different trials. Trial efficacy outcomes were included where available; none of the trials reported T cell treatment-related adverse events, so this information is not tabulated. All the adoptive transfer trials delivered T cell product intravenously, except for brain tumor trials NCT05664243 with DeltEx (Vγ9Vδ2) cells for glioblastoma and trial NCT06018363 with QH104 αB7H3 CAR-γδ T cells for malignant brain glioma, both of which deliver T cells locoregionally into the central nervous system.

Four of 26 trials excluded preconditioning prior to T cell infusion. Several did not specify whether preconditioning was carried out. Interestingly, the trials that specify no chemotherapy are either locoregional delivery trials (NCT06018363), or combination trials that also combine γδ T cell immunotherapy with radiotherapy or chemotherapy of one kind or another. NCT04765462 is currently recruiting for an unmodified γδ T cell solid tumor basket study, and does not include preconditioning, but may include "combinations with chemotherapy, targeted therapy, radiotherapy, immune checkpoint inhibitors and other therapies, which are allowed depending on the disease status of the enrolled patients" [88]; further information is not provided. Trial NCT05400603 (Aflac-NBL-2002) [43] is currently recruiting relapsed or refractory neuroblastoma patients, to identify a maximum tolerated dose of unmodified γδ T cell combination with anti-GD2 chemoimmunotherapy (dinutuximab ($17.5\,mg/m^2$)), temozolomide ($100\,mg/m^2$), irinotecan ($50\,mg/m^2$), and zoledronic acid (0.0125 mg/kg/dose). Trial NCT06069570 for NSCLC will combine T cells with repeated doses of low-dose radiotherapy (1.0 Gy/fraction), while trial NCT06018363 for glioma will deliver T cells locoregionally into the central nervous system (CNS) and may presume a lack of product allo-rejection at this site.

TABLE 4 Allogeneic, adoptive transfer γδ T trial preconditioning regimens and dosing.

Year (Refs.)	Indication	Trial ID	γδ T product	Preconditioning	Adjunct treatment	Dosing	Outcomes
2016	CD19+ CLL, ALL, B-cell lymphoma	NCT02656147	CAR-positive; otherwise unspecified	Induction chemotherapy		Dose escalation	Not available
2017 [52]	Pancreatic cancer	NCT03180437	Allogeneic PBMC-derived γδ T cells	Neoadjuvant chemotherapy, incl gemcitabine ± nab-paclitaxel, FOLFI-RINOX or others	Prior pancreatic tumor cryosurgery or irreversible electroporation procedure prior to γδ T cell infusion	One to three courses with 28 day interval; 2 infusions per cycle	γδ T cell-treated group had a 3.5 month median OS benefit ($P = .01$)
2018 [62]	Relapsed or refractory AML	NCT03790072	Haploidentical, allogeneic PBMC-derived γδ T cells	Fludarabine 25 mg/m^2 from day 6 until day 2 (inclusive) and cyclophosphamide 500 mg/m^2 on days 6 and 5	Unspecified	3+3 dose escalation DL1: 1×10^6/kg DL2: 1×10^7/kg DL3: 1×10^8/kg	2 died prior to D14 post treatment (1 from infection unlikely due to product and one from disease progression); 4 had 28-day bone marrow evaluation: 1/4 CR; 1/4 morphologic leukemia-free state; 1/4 SD; 1/4 no evidence of response
2018	AML, CML, ALL, MDS	NCT03533816	Allogeneic PBMC-derived γδ T cells	Hematopoietic stem cell transplantation and posttransplant cyclophosphamide		3+3 dose escalation DL1: 1×10^6/kg DL2: 3×10^6/kg DL3: 1×10^7/kg	Not available

2021	Malignant solid tumors	NCT04765462	Allogeneic, unspecified PBMC-derived γδ T cells	None	Chemotherapy, targeted and radiotherapy, immune checkpoint inhibitors and other therapies "were allowed in this study, depending on the disease status of the enrolled patients"	Dose escalation DL1: 2×10^6/kg DL2: 1×10^7/kg DL3: 5×10^7/kg Phase 2 study at recommended dose for efficacy confirmation	Not available
2021	AML, ALL, MDS, Lymphoma	NCT04764513	Allogeneic PBMC-derived γδ T cells	Hematopoietic stem cell transplantation and posttransplant chemotherapy		3+3 dose escalation DL1: 2×10^6/kg DL2: 1×10^7/kg DL3: 5×10^7/kg Phase 2 study at recommended dose for efficacy confirmation	Not available
2021	Relapsing/remitting NHL, PTCL	NCT04696705	Allogeneic PBMC-derived γδ T cells from blood-related donor	Pretreatment for lymphodepleting chemotherapy	Unspecified	Dose escalation 2 cycles with 14 day intervals. 2 infusions per cycle DL1: 1×10^7/kg DL2: 3×10^7/kg DL3: 9×10^7/kg	Not available
2021	AML	NCT05001451	Allogeneic PBMC-derived Vδ1 γδ T cells	Fludarabine and cyclophosphamide	Unspecified	Unspecified	Not available
2021	Lymphoma (various)	NCT04735471	Allogeneic PBMC-derived Vδ1 γδ T cells	Fludarabine and cyclophosphamide	Unspecified	3+3 dose escalation; dosing unspecified	Not available

Continued

TABLE 4 Allogeneic, adoptive transfer γδ T trial preconditioning regimens and dosing—cont'd

Year (Refs.)	Indication	Trial ID	γδ T product	Preconditioning	Adjunct treatment	Dosing	Outcomes
2021	AML	NCT05015426	Allogeneic, unspecified γδ T cells	Infusion following allogeneic HCT	Unspecified	3+3 dose escalation DL1: 1×10^6/kg DL2: 5×10^6/kg DL3: 2.5×10^7/kg DL4: 1×10^8/kg	Not available
2022	Newly diagnosed or recurrent glioblastoma multiforme	NCT05664243	PBMC-derived Vγ9Vδ2 T cells Phase I: autologous Phase II: allogeneic	Surgical resection, standard of care induction TMZ/radiation therapy followed by 6 cycles of maintenance TMZ, then daily oral TMZ 150 mg/m2 follows for days 2–5		DeltEx cells will be administered on Day 1 of each of 6, 28-day cycles in combination with TMZ maintenance	Not yet recruiting
2022	AML	NCT05358808	PBMC-derived Vγ9Vδ2 T cells	Fludarabine [30 mg/m^2] Day -6 to Day -3 [total 120 mg/m^2] and cyclophosphamide 0.5 g/m^2 [total 1.5 g/m^2) Day -5 to Day -3. This will be followed by a rest day (Day -2).	None	TCB008 treatment (consisting initially of one dose of 7×10^7 or 7×10^8 cells) is administered on Day 0	Not available
2022	Neuroblastoma	NCT05400603	Allogeneic, unspecified γδ T cells. Frozen then thawed for infusion	None	Dinutuximab (17.5 mg/m^2), temozolomide (100 mg/m^2), irinotecan (50 mg/m^2) and zoledronic acid (0.0125 mg/kg/dose) consistent across all dose levels	1 Course = two infusions. Escalate after each course. 1st infusion D6; 2nd D13 3+3 dose escalation DL1: 3×10^6/kg DL2: 1×10^7/kg DL3: 3×10^7/kg	Not available

2022	Solid and hematological tumors	NCT05302037	Allogeneic, unspecified PBMC-derived	Unspecified lymphodepleting chemotherapy	Unspecified	Four infusions of CTM-N2D at escalating doses: 1×10^7, 1×10^8, 3×10^8 or 1×10^9 per infusion at an interval of one infusion every 7 days	Not yet recruiting
2022	Lymphoma (various)	NCT05653271	Allogeneic PBMC-derived γδ T cells	Fludarabine and cyclophosphamide	Obinutuzumab	Unspecified	Not available
2023	AML, myelodysplastic syndrome	NCT05686538	Haplo-matched allogeneic innate donor lymphocyte infusion (iDLI)	Myeloablative chemotherapy	HSCT	N/A	Not yet available
2023	Stage 4 Metastatic NSCLC	NCT06069570	Allogeneic unmodified γδ T cells, Deltacel (unspecified TCR)	None	Low-dose radiotherapy (LDRT; 1.0 Gy/fraction)	0.4×10^9, 0.8×10^9 or 1.6×10^9 cells, pre- and pro-ceded by LDRT	Full report not yet available; informal communication from CEO about first patient response with 13% reduction in tumor size at 6 months [84]
2023	B7H3⁺ malignant brain glioma	NCT06018363	Allogeneic αB7H3 CAR-γδ T cells (unspecified TCR)	None	None	3+3 Dose escalation, but doses unspecified	Not yet available
2023 [85]	AML	NCT05377827	Allogeneic, unmodified Vδ1 cells	Unspecified "chemotherapy"		3 Dose levels, but doses unspecified	Not yet available

Continued

TABLE 4 Allogeneic, adoptive transfer γδ T trial preconditioning regimens and dosing—cont'd

Year (Refs.)	Indication	Trial ID	γδ T product	Preconditioning	Adjunct treatment	Dosing	Outcomes
2023	$CD7^+$ hematological malignancies	NCT05377827	Allogeneic TRAC & CD7-K/O αCD7 CAR-T	"Preparative lymphodepleting chemotherapy"		Accelerated titration 3+3 design, but doses unspecified	Full report not yet available; press release: at median follow-up of 107 days ORR was 57% (4/7), including 2 CR, 1 morphological leukemia-free state, 1 PR [86]
2024 [70]	$EGFR^+$ advanced or metastatic solid tumors	NCT06415487	Allogeneic Vγ-unspecified, Vδ2, NK-T unspecified	Fludarabine and cyclophosphamide lymphodepletion	Pembrolizumab (αPD-1 mAb)	Dose escalation with 1 then 3 doses of cell therapy, then combination with pembrolizumab	Not yet recruiting
2024 [73]	Advanced solid tumors	NCT06404281	Allogeneic αPD-1-secreting pan-γδ T cells	Not specified		Atypical 3+3 dose-escalation (DLs: 3×10^7/kg, 1×10^8/kg, 3×10^8/kg); 6× two-weekly infusions of T cells	Not yet recruiting
2024	αPD-1 therapy-resistant HCC	NCT06364800	Allogeneic unmodified γδ T cells (unspecified TCR)	Unspecified	αPD-1 mAb; multitarget kinase inhibitors that can act on VEGFR-1, VEGFR-2, VEGFR-3, FGFR1, PDGFR, cKit, Ret and other targets	3+3 Dose-escalation with 4× cycles of three-weekly infusions: 1×10^8/kg, 2×10^8/kg, 4×10^8/Kg	Not yet recruiting

2024	First-line treatment of HCC	NCT06364787	Allogeneic unmodified γδ T cells (unspecified TCR)	Unspecified	αPD-1 mAb; multitarget kinase inhibitors that can act on VEGFR-1, VEGFR-2, VEGFR-3, FGFR1, PDGFR, cKit, Ret and other targets	3+3 Dose-escalation with 4× cycles of three-weekly infusions: 1×10^8/kg, 2×10^8/kg, 4×10^8/kg	Not yet recruiting
2024	Melanoma amenable to surgical resection	NCT06212388	Allogeneic unmodified γδ T cells (unspecified TCR)	Unspecified	Surgery, pembrolizumab (αPD-1 mAb); recombinant IFN-α1b	Dose levels unspecified; γδ T cells infused every 3 weeks	Not yet recruiting
2024 [87]	Relapsed/refractory solid tumors	NCT06150885	Allogeneic α HLA-G-CAR BiTE-secreting Vγ9Vδ2 T cells	Unspecified	None	3+3 Design with five cohorts: low dose for single administration, low dose for twice administrations for 2 weeks; low, middle and high dose for 4× repeated administrations for 4 weeks	Not yet recruiting

Abbreviations: *PBMC*, peripheral blood mononuclear cells; *TCR*, T cell receptor; *DL*, dose level; *CAR*, chimeric antigen receptor; *IFN*, interferon; *LDRT*, low-dose radiotherapy; *mAb*, monoclonal antibody; *TRAC*, T cell receptor α constant locus; *NK-T*, natural killer T cells; *CRISPR*, Clustered Regularly Interspaced Short Palindromic Repeats; *K/O*, knockout; *BiTe*, bi-specific T cell engager; *HLA*, human leukocyte antigen; *VEGFR*, vascular endothelial growth factor receptor; *FGFR*, fibroblast growth factor receptor; *PDGFR*, platelet-derived growth factor receptor; *ORR*, overall response rate; *CR*, complete response; *PR*, partial response; *SD*, stable disease; *CEO*, chief executive officer; *TMZ*, temozolomide.

Six [63,77,89–92] of 26 trials have specified that the preconditioning regimen will consist of fludarabine and cyclophosphamide, five [93–97] have listed unspecified preconditioning chemotherapy, and one [98] lists induction chemotherapy for B cell malignancy treatment (both as a preconditioning regimen and adjunct therapy). Four trials [83,95,99,100] list hematopoietic stem cell transplantation as an adjunct/pretreatment, two of which specify additional posttransplant chemotherapy including cyclophosphamide. Completed trial NCT03180437 for locally advanced pancreatic cancer combined T cell infusion with preinfusion irreversible electroporation (IRE) surgery and "neoadjuvant chemotherapy, including gemcitabine with or without Nab-paclitaxel, FOLFIRINOX or others." Locoregional brain tumor trial NCT05664243 combined T cell adoptive transfer with preinfusion surgical resection, standard of care induction temozolomide and radiation therapy, followed by six cycles of maintenance temozolomide, then daily oral temozolomide at 150 mg/m^2 for days two to five. The other locoregional brain tumor trial (NCT06018363) did not specify adjunct treatments.

Of the allogeneic studies listed in Table 4, only NCT03180437 for pancreatic and NCT03790072 for AML have thus far formally reported efficacy and T cell persistence. Evidence for immune modulation of the patient T cell compartment, potentially mediated by the Vγ9Vδ2 product, could be detected at 90 days following treatment in trial NCT03180437 [52]. Patients in the treatment arm who received T cells as part of their therapy had a 3.5 month median overall survival (OS) benefit ($P = .01$), and patients who received multiple Vγ9Vδ2 cell doses had a higher OS than patients who received a single dose ($P < .05$) [52]. Trial NCT03790072 reported T cell persistence following three re-infusions of product, performing lymphodepletion only prior to the first infusion [62]. Evidence for T cell product persistence was detectable for more than a month following each infusion in two examined patients, though the multiple infusions make it impossible to assess the long-term capacity of product engraftment in this context. One of the patients achieved a transient complete remission (CR), while the other achieved stable disease (SD). Both patients relapsed to active disease ~100 days after starting treatment even though, notably, donor Vγ9Vδ2 cells were still detectable in the peripheral blood of at least one of the patients. While generating the proof of concept that unmodified allogeneic Vγ9Vδ2 cells in combination with lymphodepleting preconditioning chemotherapy are capable of mediating a meaningful therapeutic effect against advanced AML, the specific contribution of the Vγ9Vδ2 cells (versus fludarabine and cyclophosphamide) remains elusive. It is noteworthy that trial NCT03790072 reports the most substantial blast reductions following specifically lymphodepleting chemotherapy, in a manner that was not recapitulated following repeat administrations of IV Vγ9Vδ2 cells.

Modified γδ T cell adoptive transfer trials

Overview

The most striking recent trend in γδ T cell oncoimmunotherapy trials has been the growing number of modified (usually, gene-modified) adoptive transfer therapies. We identified a total of 18 modified γδ T products that have been progressed to clinic (Fig. 8). Five of the 18 have been with autologous, and 13 with allogeneic γδ T cells. While intuitively one might assume from a safety and efficacy perspective that the field might evaluate allogeneic approaches after initial trials with autologous products, this has not been the case. The earliest

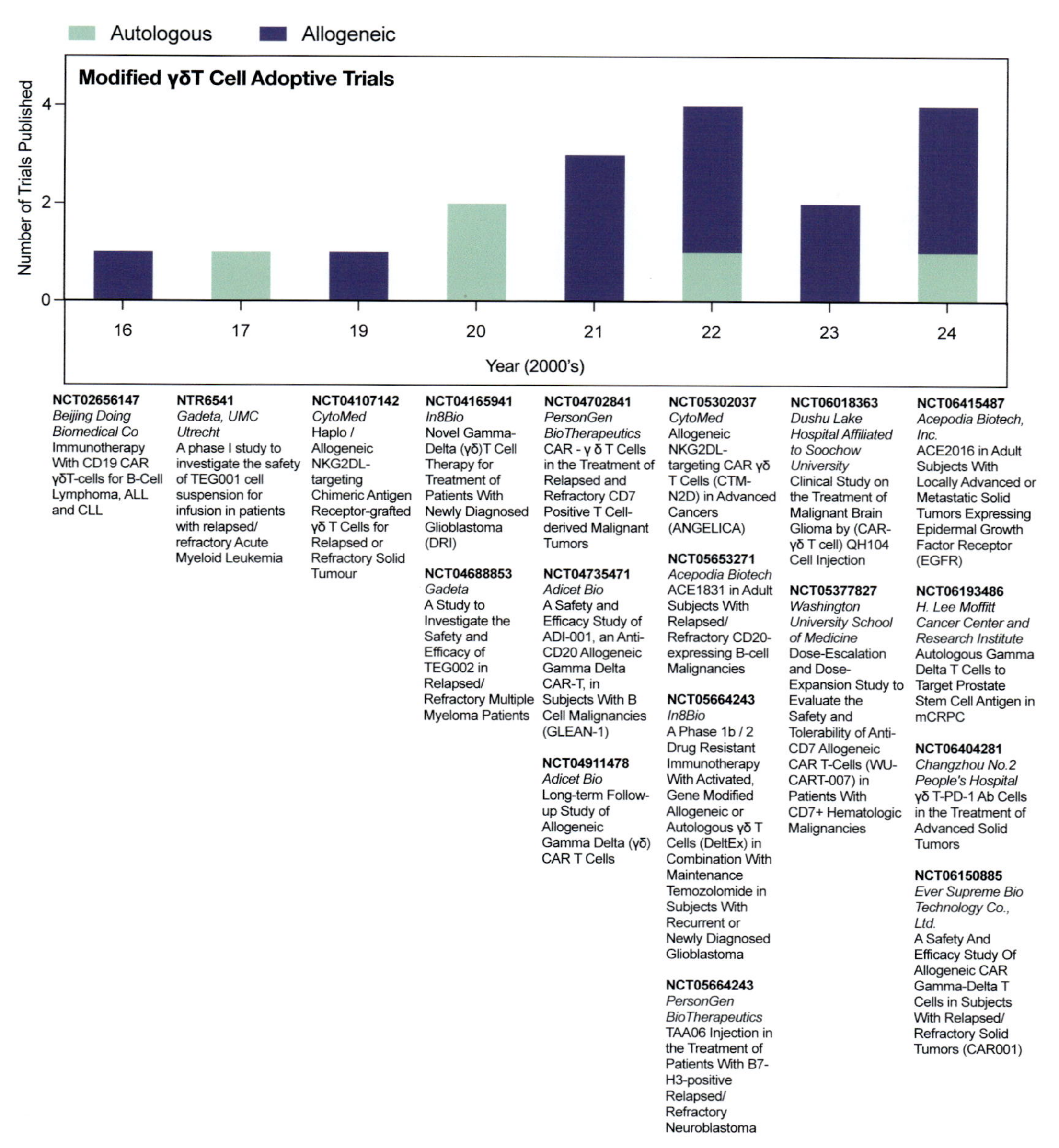

FIG. 8 All modified adoptive transfer γδ T cell clinical studies, grouped by year and autologous versus allogeneic intervention. Trial ID, sponsor and title are listed.

gene-modified γδ T cell trial we could identify (CD19 CAR-γδ by sponsor, Beijing Doing Biomedical Co) was with an allogeneic product, while the latest trial registered in 2024 (αPSCA CAR-γδ, H. Lee Moffitt Cancer Center and Research Institute [101]) was an autologous CAR-T product. An overview of the modified γδ T trials we could identify is summarized in Table 5. The different approaches to modified product design are illustrated in Fig. 9.

TABLE 5 Modified γδ T cell transfer trials.

Year (Refs.)	γδ T subset	Modification method	Modification	Trial ID	Sponsor(s)
2016 (None)	Allogeneic, unspecified PBMC-derived γδ T cells	Unknown	CD19-redirected γδ CAR-T (construct unknown)	NCT02656147	Beijing Doing Biomedical Co
2017 [102,103]	Autologous PBMC-derived αβ T cells	γ-Retrovirus	TEG001, γδ TCR-engineered TCR-T	NTR6541	UMC Utrecht, Gadeta
2019 (None)	Allogeneic, unspecified γδ T cells	mRNA electroporation	NKG2DL-redirected γδ CAR-T (CTM-N2D)	NCT04107142	CytoMed Therapeutics
2020 (None)	Autologous PBMC-derived αβ T cells	γ-Retrovirus	TEG002, γδ TCR-engineered TCR-T	NCT04688853	Gadeta
2020 [104,105]	Autologous, PBMC-derived Vγ9Vδ2 T cells	Unspecified lentivector	Drug (temazolamide, TMZ)-resistance via AGT and MGMT enzyme expression to yield "DeltEx" cells	NCT04165941	University of Alabama at Birmingham, In8Bio
2021 (None)	Autologous γδ T cells, unspecified source	Unspecified	CD7-redirected γδ CAR-T (PA3-17; construct unknown)	NCT04702841	PersonGen BioTherapeutics
2021 [78]	Allogeneic, PBMC-derived Vδ1 γδ T cells	γ-Retrovirus	CD20-redirected γδ 4-1BBζ CAR-T (ADI-001)	NCT04735471	Adicet Bio
2021 (None)	ADI-001 observational trial			NCT04911478	Adicet Bio
2022 (None)	Allogeneic, unspecified PBMC-derived γδ T cells	mRNA electroporation	NKG2DL-redirected γδ CAR-T (CTM-N2D)	NCT05302037	CytoMed Therapeutics
2022 [69]	Allogeneic PBMC Vδ2 "N"-γδ T cells	Covalent DNA aptamers	Gene-unmodified γδ T cells, covalently linked to anti-CD20 mAb via DNA aptamers	NCT05653271	Acepodia Biotech
2022 (None)	Autologous γδ T cells, unspecified source	Not specified	B7H3-redirected γδ CAR-T (TAA06; construct unknown)	NCT05562024	PersonGen BioTherapeutics

TABLE 5 Modified γδ T cell transfer trials—cont'd

Year (Refs.)	γδ T subset	Modification method	Modification	Trial ID	Sponsor(s)
2022 [104,105]	PBMC-derived Vγ9Vδ2 T cells Phase I: autologous Phase II: allogeneic	Unspecified lentivector	Drug (temazolamide, TMZ)-resistance via AGT and MGMT enzyme expression to yield "DeltEx" cells	NCT05664243	In8Bio
2023	Allogeneic unspecified γδ T cells	Unspecified	Unspecified design B7H3-redirected γδ CAR-T	NCT06018363	Dushu Lake Hospital Affiliated to Soochow University
2023 [72,86]	Allogeneic TRAC-K/O mixed T cells	Unspecified CRISPR/Cas9 gene-editing	CD7-redirected TRAC and CD7 CRISPR K/O CAR-T cells (CAR-T design unspecified)	NCT05377827	Washington University School of Medicine in collaboration with Wugen
2024 [70]	Allogeneic PBMC Vδ2 "N"-γδ T cells	Covalent DNA aptamers	Gene-unmodified γδ T cells, covalently linked to anti-EGFR mAb via DNA aptamers	NCT06415487	Acepodia Biotech
2024 [101]	Autologous, PBMC-derived Vγ9Vδ2 T cells	RD114-pseudotyped γ-retrovirus	Anti-PSCA CD8 linker 28ζ CAR-T cells	NCT06193486	H. Lee Moffitt Cancer Center and Research Institute
2024 [73]	Allogeneic pan-γδ T cells	Unspecified lentivirus	Constitutive PGK-promoter αPD-1 mAb-secretion cassette	NCT06404281	Changzhou No.2 People's Hospital
2024 [87]	Allogeneic, PBMC-derived Vγ9Vδ2 T cells	Unspecified lentivirus	EF1α-promoter-driven αHLA-G CAR with a 41BB co-stimulatory domain that includes an inserted Tyk2-binding motif from the interferon-α/β receptor 1 (IFNAR1), and a DAP12-derived CD3ζ chain; this CAR is co-expressed via P2A cleavage sequence with a PD-L1/CD3ε BiTE	NCT06150885	Ever Supreme Bio Technology Co

Abbreviations: *PBMC*, peripheral blood mononuclear cells; *TCR*, T cell receptor; *CRISPR*, Clustered Regularly Interspaced Short Palindromic Repeats; *BiTe*, bi-specific T cell engager; *CAR*, chimeric antigen receptor; *IFN*, interferon; *mAb*, monoclonal antibody; *TRAC*, T cell receptor α constant locus; *K/O*, knockout; *HLA*, human leukocyte antigen; *TMZ*, temozolomide.

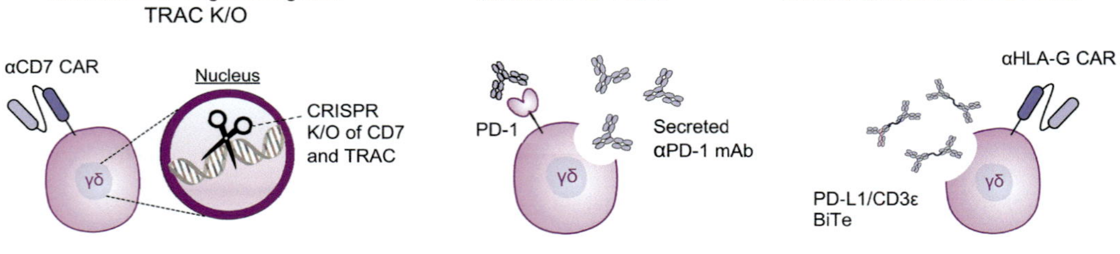

FIG. 9 Illustration of the different types of γδ T cell engineering that has been progressed to clinical trial. (A) TEG cells for hematological indications, and (B) CTM-N2D cells for mixed solid and hematological cancers have been evaluated in two trials each. (C) Classical, ostensibly 2nd generation, CAR-T γδ T cells have been and are being tested in six studies. (D) One trial has examined (anti-CD7) CAR-T γδ T cells, with an additional CD7 Golgi retention module, for T and NK cell malignancies. (E) Two trials are examining genetically unmodified γδ T cells that are covalently linked to anti-CD20 or EGFR mAbs via DNA aptamers for B cell lymphoma or advanced and metastatic solid tumors, while (F) two trials continue to evaluate DeltEx drug-resistant γδ T cells for glioblastoma. The following T cell modification strategies are each being examined in a single trial: (G) a mixed CRISPR/Cas9 gene-edited T cell product that expresses an anti-CD7 CAR, with an endogenous T cell receptor α constant locus (TRAC) and CD7 knockout (K/O); (H) γδ T cells engineered to secrete anti-PD-1 mAb; (I) γδ T cells that are engineered to express not only a CAR that targets the HLA-G antigen, but also bispecific T cell engagers (BiTEs) that engage the PD-L1 checkpoint and CD3.

Modification strategies

Only half (9/18) of modified γδ T therapies were genetically engineered to express a CAR, in some contrast to the αβ T cell immunotherapy space, which is dominated by CAR engineering [86,91,98,106–111], and one of those was an observational trial to track the performance of ADI-001 αCD20 CAR-T long-term [108]. Three of the CAR studies have been registered by Chinese biotech PersonGen Biotherapeutics, of which the 2016 allogeneic trial NCT02656147 which examined an anti-CD19 CAR in patients with B cell lymphoma, ALL and CLL was the earliest gene-modified γδ T cell trial we could identify. The other two PersonGen Biotherapeutics studies were the 2021 autologous trial NCT04702841 evaluating anti-CD7 CAR-T against relapsed, refractory T-ALL and NK/T-ALL and 2022-registered autologous trial (NCT05562024) with an anti-B7H3 CAR-T product, TAA006, for the treatment of relapsed or refractory neuroblastoma. We note the approach that PersonGen have taken to avoid fratricide within their anti-CD7 CAR-T product. CD7-targeting nanobody (CD7Nb) was recombinantly fused with an endoplasmic reticulum/Golgi-retention motif peptide to generate an anti-CD7 protein expression blocker (7PEBL) that mediated intracellular retention of CD7 [112]. Information on the synthetic construct, methodology or trial outcomes for the PersonGen trials have not have not been published in peer-reviewed journals, though PersonGen's website states that their products rely on "γδ T cell receptors to recognise tumour phosphoantigens" [113], implying that the Vγ9Vδ2 subset is the basis of the therapeutic approach. It is further stated that lentiviral approaches and "typical" CAR construct design are employed. Peer-reviewed publication of the outcomes of the above three trials would provide important insights for the field at large.

Another registered interventional CAR-γδ trial was the already-mentioned Adicet Bio trial NCT04735471 to evaluate ADI-001, an anti-CD20 allogeneic CAR-T in subjects with B cell malignancies (GLEAN-1). The ADI-001 CAR comprises a classical second-generation 4-1BBζ design, with a 3H7 anti-CD20 scFv, a CD8α hinge, and transmembrane domains, driven by a synthetic EF1α promoter within a γ-retroviral vector that was used to transduce activated Vδ1 cells. Adicet report that their γδ CAR-T product displays a predominantly $CD27^{+}CD45RA^{+}$ phenotype that is further positive for CD62L, CD95, and CD45RO, but not CCR7 [68]. ADI-001 is an almost entirely Vδ1 product, containing traces of donor NK cells, but essentially no other γδ T subsets. While there are no peer-reviewed publications on the outcomes of the ongoing trial NCT04735471, Adicet Bio have released regular trial updates in the form of press releases. The latest update we identified was posted on June 26, 2023 [79] and reports ADI-001 performance in patients with relapsed or refractory B cell NHL.

According to this nonreviewed trial data, of the 24 efficacy-evaluable patients, three received ADI-001 at DL1 (30×10^6 CAR^+ cells), three received DL2 (100×10^6 CAR^+ cells), six received DL3 (300×10^6 CAR^+ cells), four received two infusions at DL3 (one on the first day of treatment and the second dose 7 days later, following a single lymphodepletion), and eight received DL4 (1×10^9 CAR^+ cells). Patients were heavily pretreated with a median of four prior lines of therapy, had a relatively high tumor burden, and a poor prognostic outlook based on their median International Prognostic Index (IPI) score. Of the patients enrolled in the study, 50% had progressed on prior (presumably, anti-CD19) CAR-T therapy. ADI-001 treatment yielded a 71% ORR across all dose levels. Adicet Bio further specify that ADI-001 displayed an ORR that is consistent with autologous αβ CAR-T, if factoring in the number of prior lines of therapy and percentage of patients enrolled in the study who had previously progressed on (speculatively, anti-CD19) CAR-T. Adicet selected the recommended Phase II dose (RP2D) as 1×10^9 CAR^+ cells (DL4), at which the radiological disease status at 6 months following the first T cell infusion was 25% CR. In patients who had previously progressed on autologous CAR-T therapy, the 6-month CR was 33%. ADI-001 was generally well-tolerated, and there were no occurrences of dose-limiting toxicities or GvHD. Of the 24 patients evaluable for safety, there was one report of Grade 3 or higher CRS and one report of Grade 3 or higher ICANS. ADI-001 is being progressed for a pivotal Phase II study in post-CAR-T LBCL in the first half of 2024.

The press release clinical data from ADI-001 trial NCT04735471 suggest that classical second-generation CAR-γδ T cells can effect meaningful, durable, and safe clinical responses in patients with hematological malignancies. Two areas of particular interest will be (i) the persistence of allogeneic Vδ1 cells, as driven by both allo-rejection and exhaustion of the product, and (ii) the effect of the anti-CD20 CAR on tumor antigen expression. This will shed light on whether the relapses seen in the 6 months following treatment were driven by collapse of product anti-CD20 functionality or by antigen-negative cancer relapse.

Study NCT06018363 by Dushu Lake Hospital Affiliated to Soochow University examines a locoregionally delivered anti-B7H3 CAR-T product for $B7H3^+$ malignant glioma. Similar approaches are being evaluated preclinically and clinically at a range of institutions [114–118], though, unfortunately, we were unable to identify further information on the specific product or engineering for this trial.

Another brain tumor-targeting trial, NCT06150885, examines the efficacy in the wider context of a basket trial for a range of solid tumors, with intravenous product delivery. From Ever Supreme Bio Technology Co in Taiwan, the trial is evaluating the efficacy of a novel approach of bi-specific T cell engager (BiTE)-secreting CAR-Vγ9Vδ2 cells. Specifically, Vγ9Vδ2 cells were mRNA-modified to express a nanobody-based αHLA-G CAR and a secretable nanobody-based αPD-L1-targeting αCD3ε BiTE [87]. The approach, therefore, capitalizes not only on targeting multiple tumor-associated antigens and a checkpoint receptor, but also has the potential to utilize the efficacy of a range of T cell subsets. The CAR structure consists of a 41BB co-stimulatory domain that includes an inserted Tyk2-binding motif from the interferon-α/β receptor 1 (IFNAR1), and a DAP12-derived CD3ζ chain. No results from the trial have yet been made public.

Trial NCT06193486 at H. Lee Moffitt Cancer Center and Research Institute is evaluating a novel and potent approach for targeting bone-metastatic castration-resistant prostate cancer, in more classical CAR format. The CD8 linker and 28ζ endodomain CAR-T targets prostate stem cell antigen (PSCA, binder clone Ha1–4.117 [101,119]). The trial capitalizes on promising preclinical data by targeting patients who are already on zoledronate for bone metastases, whereby the bone-targeting efficacy of Vγ9Vδ2 CAR-T was potentiated using zoledronic acid in preclinical animal models [101]. Unusually for recent and modified γδ T cell immunotherapeutic products, the above trial utilizes autologous material. No data from the trial have yet been published.

WU-CART-007 αCD7 CAR-T cells for CD7+ hematological malignancies have been progressed to trial by the Washington University School of Medicine in collaboration with Wugen. To our knowledge, the product is the first instance of CRISPR-engineered γδ T cells to reach the clinic. The exact details of the product have not been published in-depth, but the approach has been described as consisting of T cell receptor α constant (TRAC) and CD7 loci CRISPR-knockout T cells that hail from a mix of αβ and γδ T cells [87]. While a peer-reviewed report on the trial outcomes has not yet been published, a 2023 press release from Wugen on the first 12 patients with a median follow-up of 107 days reported that WU-CART-007 were well-tolerated in patients with relapsed T-ALL/LBL, with low incidence of high-grade CRS and immune-related adverse events, and achieved a 57% overall response rate dose levels 2 and 3 [86].

Four other types of genetic modification related to γδ T cells are being evaluated in the clinic: Gadeta's optimized Vγ9Vδ2-TCR-T TEG cells, CytoMed's "CTM-N2D" anti-NKG2DL chimeric receptors, In8Bio's chemotherapy-resistance cassettes and αPD-1 mAb-secreting γδ T cells from Changzhou No.2 People's Hospital. "TEG cells" are αβ T cells engineered to express a defined γδ TCR [103]. TEG γδ TCRs are derived from natural Vγ9Vδ2 γδ T cells and may be synthetically matured to increase the affinity of TCR binding to phosphoantigen-driven target butyrophilin conformational changes. TEGs were manufactured by activating leukapheresate αβ T cells using standard CAR-T manufacturing reagents, anti-CD3/CD28 Dynabeads, IL-15, and IL-7, and then transducing with a γ-retroviral vector [103]. The first TEG product, "TEG001," was based on a high-affinity Vγ9Vδ2 TCR derived from the natural repertoire of a healthy donor [103,120]. The autologous TEG001 trial to evaluate product activity against AML, high-risk myelodysplastic syndrome and multiple myeloma (trial ID: NTR6541/NL6357) was initiated in 2017 and conducted at UMC Utrecht in the Netherlands [121]. TEG001 infusion was preceded by 25 mg/m^2 IV fludarabine and 900 mg/m^2 cyclophosphamide preconditioning, and followed by 30 mg IV administration of synthetic butyrophilin-modulating drug, pamidronate on days 0 and 28 of T cell treatment. TEG001 therapy was well-tolerated at the first two dose levels. One of three patients in DL2 achieved a CR 1 month following treatment, and the remaining two exhibited SD. TEG001 persistence was observable for up to 56 days following product infusion. The duration of the complete response was 6 months [102].

TEG002 cells were modified to express a synthetically matured high-affinity γδ TCR, and testing in Gadeta-sponsored trial NCT04688853 was initiated in 2020, evaluating activity in patients with relapsed or refractory myeloma [122]. Patients were preconditioned with unspecified chemotherapy and were treated with three different dose levels of intravenous TEG002. The trial remains open but is not currently recruiting. We were not able to identify published outcomes for this trial.

CytoMed's NKG2DL-targeting CAR consists of an NKG2D ectodomain to bind to NKGD2 ligands on cancer targets, a CD8 hinge and transmembrane, and is linked to a CD3ζ endodomain, with expression driven by a CMV promoter [123]. The constructs are expressed as mRNAs and introduced into γδ T cells via electroporation. Transfected CAR expression is, therefore, transient. We note that the publication laying out the proof-of-concept for the NKG2DL-CAR technology describes optimal functionality in an in vivo model of intraperitoneal SKOV3 ovarian cancer cells when the CARs were delivered locoregionally and across 14 separate weekly injections of 1×10^7 cells each. This suggests that NKG2DL-CAR product performance is highly dependent on high and replenished numbers of effector cells. While trial registration does not include further information on the type of γδ T cell employed, publications on the CytoMed website describe a CAR-T manufacturing process involving a 21-day PBMC culture with zoledronic acid, IFN-γ, IL-2, anti-CD3 mAb, and engineered K562 feeder cells expressing CD64, CD137L, and CD86 [124]. The result is a mixture of Vγ9Vδ2 cells and NK T-like CIK cells, with Vγ9Vδ2 cells constituting only ~20% of the product.

The NKG2DL-CAR product, termed NKG2DL-redirected γδ CAR-T (CTM-N2D), was first tested in the allogeneic 2019 trial NCT04107142 for mixed relapsed or refractory solid tumors. T cells were administered intravenously in a 3+3 dose escalation with four infusions per cycle of the following per-infusion doses: 3×10^8, 1×10^9, 3×10^9. The trial has completed, but trial outcomes publication is still awaited. A second allogeneic product trial, NCT05302037, with the ostensibly same or similar intravenous CTM-N2D product opened in 2022, evaluating performance in mixed solid as well as hematological cancers. The second trial listed unspecified lymphodepleting chemotherapy prior to T cell infusion and will test the following infusion regimen: four infusions per week, at a weekly interval of escalating CTM-N2D doses 1×10^7, 1×10^8, 3×10^8, or 1×10^9 per infusion. The trial is not yet recruiting.

"DeltEx" cells are not engineered with classical efficacy-boosting modalities, but rather with resistance to chemotherapy. DeltEx capitalizes on the observation that, in addition to direct antitumor efficacy, temozolomide chemotherapy upregulates NKG2D ligands on glioblastoma tumors. To enable its use in combination with innately tumor-reactive and NKG2D-expressing Vγ9Vδ2 adoptive immunotherapy, T cells need to be rendered insensitive to temozolomide, as at baseline it is potently lymphotoxic. "DeltEx" cells are manufactured from PBMC, activated using zoledronic acid, and IL-2, and then transduced (with unspecified lentivirus and pseudotype) to express a DNA methyltransferase enzyme, MGMT. MGMT renders T cells resistant to temozolomide via enhanced production of O6-alkylguanine DNA alkyltransferase, thereby enabling γδ T cell function in therapeutic concentrations of chemotherapy [125]. The resulting product is termed drug-resistant immunotherapy, or DRI DeltEx.

Autologous PBMC-derived DRI DeltEx cells were first evaluated in patients with newly diagnosed glioblastoma multiforme in the clinic by the University of Alabama Birmingham and In8Bio in a 2019-registered trial NCT04165941 [126]. Adults with untreated glioblastoma, adequate organ function, and a Karnofsky performance score of ≥70% were enrolled. Subjects underwent subtotal tumor resection and placement of a Rickham reservoir in communication with the resection cavity, followed by apheresis from which γδ T cells were expanded, transduced with MGMT lentivector, harvested, and cryopreserved. Standard-of-care induction temozolomide/radiation therapy was initiated followed by six cycles of maintenance temozolomide. Oral temozolomide ($150\,mg/m^2$) and intracranial dosing of 1×10^7 γδ T cells were carried out on the first day of each maintenance cycle. DL1 subjects received a single fixed dose of γδ T cells, while for DL2 subjects received three doses administered on day one of each of the first three cycles of temozolomide. The study is still recruiting, but interim findings for the study were published in 2021 [104].

Six subjects (four females, two males) were enrolled in DL1. Of these, one generated inadequate γδ T cells and two withdrew consent prior to DRI treatment. For the three patients that received DRI, treatment-related adverse events with maximum CTCAE Grade 3 occurred in one subject. The most common Grade 1/2 events included fever, leukopenia, nausea, and vomiting, which may be attributable to temozolomide or radiotherapy. Circulating T cells remained below normal range throughout maintenance phase in two of three subjects. NK and γδ T cell numbers remained within low normal range for three of three and two of three subjects, respectively. Serum Th1 (IFN-γ, IL-2, TNF-α) and Th2 (IL-4, IL-5, IL-10) cytokines were within range, although TNF-α remained elevated from the γδ T cell infusion through day +30 in two subjects. The interim conclusions were that administration of MGMT-gene modified γδ T cells and temozolomide as DRI is feasible in lymphodepleted subjects during temozolomide maintenance phase, and sufficiently safe to warrant further investigation at additional doses.

Evaluation of DeltEx DRI cells has been registered on a second glioblastoma clinical trial, the 2022 Phase Ib/II trial NCT05664243. As with the first trial the T cell product will be evaluated in combination with surgical resection, standard of care adjuvant induction temozolomide and radiation therapy, followed by cycles of maintenance daily oral temozolomide 150 mg/m^2. The trial, this time sponsored by In8Bio, expects to recruit up to 120 patients, and aims to be autologous for the Phase I and allogeneic for the Phase II component of the study. As with the first study, the effector T cells will be bisphosphonate-expanded Vγ9Vδ2 cells genetically modified to be chemotherapy resistant, though whether the DeltEx DRI product will be identical to that used in the first trial is not entirely clear. A 2024 press release and 2024 ASCO poster from In8Bio [127] described data from the first 13 patients treated. Having progressed through three dose levels, 92% of evaluable patients treated with INB-200 for glioblastoma exceeded the median progression-free survival of 7 months achieved with the standard-of-care (Stupp) regimen. No treatment-related serious adverse events, dose-limiting toxicities, cytokine release syndrome, infusion reactions, or immune effector cell-associated neurotoxicity syndrome were reported in any of the cohorts. Interestingly, γδ T cells were found in a relapsed tumor 148 days after initial locoregional product infusion in one patient with paired biopsies, though it remains unclear whether the found T cells were modified or endogenous patient γδ T cells. The trial remains open and data from the expanded dose level 3 cohort are incoming.

A novel approach to γδ T cell gene modification is being tested by Changzhou No.2 People's Hospital in China. Engineered to secrete an anti-PD-1 mAb, this pan-γδ T cell product is being evaluated in the context of a basket advanced solid tumor clinical trial, NCT06404281 [73]. While overall γδ T cell sensitivity to checkpoint blockade remains poorly understood, recent data indicate that specific tissue-resident γδ T cell subsets, especially expressing a Vδ1 TCR, can exert potent antitumor effector function when de-repressed by the administration of PD-1 checkpoint blockade to patients with a range of solid tumors [13,15,128]. Similar to the BiTe-secreting γδ T cells described above, this product has the potential to boost the efficacy not only of the infused γδ T cells, but also that of bystander PD-1$^+$ cells including endogenous patient αβ T cells and NK cells. No data from this trial have yet been published.

A final modified γδ T cell product type we identified that has reached clinical trials is not gene- but chemically modified. These are the Acepodia 2022- and 2024-registered trials NCT05653271 and NCT06415487, respectively. Both have already been briefly discussed in "Choice of γδ T cell subset for allogeneic trials" section. ACE1831 consists of allogeneic Vδ2 TCR chain-expressing "N"-γδ T cells that have been expanded using an unspecified γδ T and NK T-stimulating cytokine cocktail. These cells have then been covalently linked to anti-CD20 mAb via DNA aptamers. ACE1831 is reported to express high levels of CD56, CD16, NKG2D, and CD69, and low levels of PD-1 and KIRs [69], though the data have not yet been made public. Trial NCT05653271 will be Acepodia's first and will evaluate allogeneic obinutuzumab (anti-CD20 mAb)-linked "N"-γδ T cell product performance in patients with CD20-expressing B cell lymphomas. The trial has dosed its first patient and is currently recruiting, anticipating enrolment of up to 42 patients, without specifying details of the dose escalation. Trial NCT06415487, meanwhile, is evaluating product ACE2016 for $EGFR^+$ locally advanced or metastatic solid tumors. No trial data or detailed characterization of the product have yet been published, though, it was reported that ACE2016 cells are antigen-specific and degranulate in response to $EGFR^+$ targets [70].

Discussion and future perspectives

Our review of the 90 registered γδ T cell oncology clinical trials has enabled us to draw several broad conclusions. The first is that, despite the ever-increasing body of evidence that points to the important role γδ T cells play in situ during natural antitumor immunity [11–13,15,16,34], unmodified γδ T cell adoptive transfer trials—autologous or allogeneic—have yielded tantalizing but ultimately disappointing clinical efficacy outcomes.

This leads us to our second broad conclusion: adoptively transferred, unmodified γδ T cells are susceptible to tumor-mediated suppression. This suppression leads to tumor escape, and ultimately prevents sustained and meaningful γδ T cell therapeutic efficacy in either the hematological or solid tumor setting. Consequently, we draw our third conclusion, which is that genetic modification or combinatorial approaches over and above IL-2 and zoledronic acid are a likely requirement to deliver sustained γδ T cell-mediated remissions in both liquid and solid cancers.

Next, we note the compelling evidence for the importance of lymphodepleting chemotherapy (most often fludarabine and cyclophosphamide) for the therapeutic efficacy of intravenously delivered adoptive γδ T cell immunotherapy. The evidence for this is particularly strong in the hematological cancer context. The significant additive role of lymphodepleting chemotherapy remains an outstanding and difficult question particularly where sponsors envision infusing multiple doses of γδ T cell therapy during each cycle of treatment. The γδ T cell field may have valuable learnings to extract from the parallel and translationally more mature αβ T cell field in terms of approaches to preconditioning and re-infusion [23,129], though it is conceivable that considerations will vary depending on the disease indication, patient population and adjunct treatments. The degree to which systemic chemotherapeutic preconditioning enhances the efficacy of T cell immunotherapy remains unclear when γδ T cells (or other cytotoxic lymphocytes) are delivered locoregionally.

In terms of identifying the most clinically suitable γδ T cell subset as far as efficacy and safety are concerned, ADI001 Adicet Bio trial NCT05001451 has generated a convincing case for TCR-expanded, gene-modified Vδ1 γδ T cell use for the treatment of antigen-positive B cell malignancies. The relative performance of the CD20 CAR construct in different γδ T cell substrates remains unexamined. A potential advantage of the safe but efficacy-wise unproven Vγ9Vδ2 cell subset is its potential for straightforward clinical combination with nontoxic, cheap, and potent TCR-engaging phosphoantigen and bisphosphonate adjuvant drugs. In a broader sense, it remains to be established whether TCR-purified (Vδ1, Vγ9Vδ2, or other) γδ T cell products are to be aspired to at all. It is conceivable that a mixed γδ T cell product, ideally generated using straightforwardly scalable and GMP-compatible methods (as described herein, for example [130]), may therapeutically outperform any one purified cell product. Indeed, it may be advantageous to include a range of nonalloreactive PBMC cytotoxic lymphocytes, including γδ T cells and NK cells. Ultimately, a meaningful value judgment on the optimal choice of γδ T cell subset (and expansion process, phenotype, etc.) is impossible in the absence of empirical clinical evaluation. Even thereafter, conclusions may vary between different cancer indications and treatment strategies.

As has been the case with canonical αβ-CAR-T, translational problem-solving for γδ T cell immunotherapy will be best informed by dissecting case studies of clinically successful products. In this area, γδ T cell immunotherapies have some way to go yet. Transparent reporting from ongoing gene-modified γδ T cell trials, including PA3-17 and TRAC-K/O WU-CART-007 anti-CD7 CAR-T for T cell malignancies (NCT04702841, NCT05377827), further updates (including antigen status) from ADI001 anti-CD20 CAR-T for B cell malignancies (NCT04911478), CTM-N2D NKG2DL-CAR for solid tumors (NCT05302037), TAA06 and QH104 anti-B7H3 CAR-T for neuroblastoma and malignant glioma (NCT05562024, NCT06018363), DeltEx DRI for glioblastoma (NCT05664243), and anti-PSCA CAR-T for prostate cancer (NCT06193486) will be crucial for informing the design of next-generation γδ T cell products. Of particular interest will also be the novel approaches of anti-HLA-G BiTe-secreting CAR-T (NCT06150885) and anti-PD-1 mAb-secreting T cells (NCT06404281) for mixed solid tumors. Product safety, and the depth and duration of the therapeutic effect, along with γδ T cell persistence will be crucial readouts from these trials. The rate of relapse and the antigen status of the relapses will be key in interpreting immunotherapeutic performance.

Valuable learnings will also be made from sophisticated combinatorial trials where γδ T cells form part of a larger therapeutic approach. These include trial NCT05400603, where unmodified, allogeneic V9Vγδ2 cells will be adoptively transferred to neuroblastoma patients, alongside a chemoimmunotherapy regimen of dinutuximab (anti-GD2 mAb), zoledronic acid, and temozolomide-irinotecan chemotherapy.

In addition to efficacy engineering, we anticipate that different γδ T cell cytokine addiction-mitigating strategies will be a product development area of particular translational relevance over the coming decade. Several avenues indicate that secreted IL-15 may be a first γδ T cell armouring strategy to reach the clinic, including preclinical studies from our lab that describe an anti-GD2 scFv-Fc fusion protein (SFP)-secreting Vγ9Vδ2 product armored with stabilized IL-15 for osteosarcoma [66], and a late-stage preclinical Adicet Bio/Regeneron partner program developing soluble IL-15-armored antiglypican 3 (GPC3) CAR-Vδ1 for hepatocellular carcinoma (HCC) [65]. While mostly being developed in a commercial context and, therefore, rarely discussed in peer-reviewed literature, adoptively transferred γδ T cell product stealth engineering too may become a more active area for translational evaluation.

In conclusion, we speculate that the next phase of γδ T cell oncology interventional clinical trials will continue to be dominated by adoptive cell therapy approaches, but in a new, and gene-modified, format. Multicistronic engineering may be required to overcome the current issues with unmodified γδ T cell therapeutic persistence and suppression by tumors. It is likely that immunomodulatory and chemotherapeutic intervention in addition to adoptive cell therapy will yield better outcomes still, whereupon γδ T cell products will become more efficacious but less safe, as has been the case with αβ T cell-based therapeutics.

Appendix

All identified γδ T cell oncology trials; search carried out up to Q2, 2024.

Year	Trial ID (Refs.)	Trial title
2024	NCT06150885 [87]	A safety and efficacy study of allogeneic CAR gamma-delta T cells in subjects with relapsed/refractory solid tumors (CAR001)
2024	NCT06212388	Allogeneic gammadelta T cells combined with interferon-α1b or PD-1 monoclonal antibody in stage III-IV amenable to surgical resection melanoma
2024	NCT06364787	Allogeneic gamma-delta T cells combined with targeted therapy and immunotherapy in a phase 1 clinical trial for first-line treatment of hepatocellular carcinoma
2024	NCT06364800	Allogeneic gamma-delta T cells combined with targeted therapy and immunotherapy in a phase 1 clinical trial of hepatocellular carcinoma resistant to PD-1 monoclonal antibody
2024	NCT06404281 [73]	γδ T-PD-1 Ab cells in the treatment of advanced solid tumors
2024	NCT06193486 [101]	Autologous gamma delta T cells to target prostate stem cell antigen in mCRPC
2024	NCT06415487 [70]	ACE2016 in adult subjects with locally advanced or metastatic solid tumors expressing epidermal growth factor receptor (EGFR)
2023	NCT05377827 [72]	Dose-escalation and dose-expansion study to evaluate the safety and tolerability of anti-CD7 allogeneic CAR T-cells (WU-CART-007) in patients with CD7+ hematologic malignancies
2023	NCT05377827 [85]	A Study of GDX012 in adults with relapsed or refractory acute myeloid leukemia
2023	NCT06018363	Clinical study on the treatment of malignant brain glioma by (CAR-γδ T cell) QH104 cell injection
2023	NCT06069570	Safety study for a gamma delta T cell product used with low dose radiotherapy in patients with stage 4 metastatic NSCLC
2023	NCT05686538	Innate donor effector allogeneic lymphocyte infusion after stem cell transplantation: the IDEAL trial (IDEAL)
2022	NCT05237206 [131]	Study of SUPLEXA in patients with metastatic solid tumors and hematologic malignancies
2022	NCT03183219 [132]	Clinical safety and efficacy of locoregional therapy combined with adoptive transfer of allogeneic γδ t cells for advanced hepatocellular carcinoma and intrahepatic cholangiocarcinoma
2022	NCT05664243	A phase 1b/2 drug resistant immunotherapy with activated, gene modified allogeneic or autologous γδ T Cells (DeltEx) in combination with maintenance temozolomide in subjects with recurrent or newly diagnosed glioblastoma
2022	NCT05562024	TAA06 injection in the treatment of patients with B7-H3-positive relapsed/refractory neuroblastoma
2022	NCT05358808	Efficacy and effectiveness of adoptive cellular therapy with ex vivo expanded allogeneic γδ T-lymphocytes (TCB-008) for patients with refractory or relapsed acute myeloid leukemia (AML) (ACHIEVE)

Year	Trial ID (Refs.)	Trial title
2022	NCT05653271 [69]	ACE1831 in adult subjects with relapsed/refractory CD20-expressing B-cell malignancies
2022	NCT05400603	Allogeneic expanded gamma delta T cells with GD2 chemoimmunotherapy in relapsed or refractory neuroblastoma (Aflac-NBL-2002)
2022	NCT05628545	Adoptive treatment of advanced hepatocellular carcinoma with allogeneic γδ-T cells
2022	NCT05302037	Allogeneic NKG2DL-targeting CAR γδ T cells (CTM-N2D) in advanced cancers (ANGELICA)
2022	NCT05369000	Trial of LAVA-1207 in patients with therapy refractory metastatic castration resistant prostate cancer
2021	NCT02753309 [133]	Rapamycin enhances BCG-specific γδ T cells during intravesical BCG therapy for non-muscle invasive bladder cancer: a randomized, double-blind study
2021	2013-001188-22 [60]	Phase II trial of maintenance treatment with IL2 and zoledronate in multiple myeloma after bone marrow transplantation: biological and clinical results
2021	NCT05001451	Study of GDX012 in patients with MRD positive AML
2021	NCT04911478	Long-term follow-up study of allogeneic gamma Delta (γδ) CAR T cells
2021	NCT05015426 [71]	Gamma delta T-cell Infusion for AML at high risk of relapse after allo HCT
2021	NCT04735471	A safety and efficacy study of ADI-001, an anti-CD20 allogeneic gamma delta CAR-T, in subjects with B cell malignancies (GLEAN-1)
2021	NCT05015426	Gamma delta T-cell infusion for AML at high risk of relapse after allo HCT
2021	NCT04765462	Allogeneic γδ T cell therapy for the treatment of solid tumors
2021	NCT04764513	Safety and efficiency of γδ T cell against hematological malignancies after allo-HSCT
2021	NCT04702841	CAR—γ δ T cells in the treatment of relapsed and refractory CD7 positive T cell-derived malignant tumors
2021	NCT04696705	Allogeneic γδ T cells immunotherapy in r/r non-Hodgkin's lymphoma (NHL) or peripheral T cell lymphomas (PTCL) patients
2021	NCT04887259	Trial of LAVA-051 in patients with relapsed/refractory CLL, MM, or AML
2021	NCT05080790	Treatment with dinutuximab beta, zoledronic acid and low-dose interleukin (IL-2) in patients with leiomyosarcoma (DiTuSarc)
2020	NCT04243499	First-in-human study of ICT01 in patients with advanced cancer (EVICTION)
2020	NCT04688853	A study to investigate the safety and efficacy of TEG002 in relapsed/refractory multiple myeloma patients
2020	NCT04518774	Allogeneic "gammadelta T cells (γδ T cells)" cell immunotherapy in phase 1 hepatocellular carcinoma clinical trial
2020	NCT04165941	Novel gamma-delta (γδ)T cell therapy for treatment of patients with newly diagnosed glioblastoma (DRI)
2019	NCT03885076	Gamma delta T cells in AML

Continued

Year	Trial ID (Refs.)	Trial title
2019	NCT04107142	Haplo/allogeneic NKG2DL-targeting chimeric antigen receptor-grafted γδ T cells for relapsed or refractory solid tumor
2019	NCT04008381	Ex-vivo expanded γδ T lymphocytes in patients with refractory/relapsed acute myeloid leukemia
2019	NCT04032392	Immunotherapy of advanced hepatitis B related hepatocellular carcinoma with γδT cells
2019	NCT03939585	Preemptive infusion of donor lymphocytes depleted of TCR + T cells + CD19+ B cells following ASCT
2019	NCT04028440	γδT cells immunotherapy in patients with relapsed or refractory non-Hodgkin's lymphoma (NHL)
2019	NCT04008381	Ex-vivo expanded γδ T lymphocytes in patients with refractory/relapsed acute myeloid leukemia
2019	NCT02425748	Safety and efficiency of γδ T cell against non small lung cancer (without EGFR mutation)
2019	UMIN000008701 [46]	Combined effects of neoadjuvant letrozole and zoledronic acid on γδT cells in postmenopausal women with early-stage breast cancer
2018	NCT03790072 [62]	Ex-vivo expanded γδ T-lymphocytes (OmnImmune®) in patients with acute myeloid leukemia (AML)
2018	NCT03533816	Expanded/activated gamma delta T-cell infusion following hematopoietic stem cell transplantation and post-transplant cyclophosphamide
2018	000000931 [49]	Adjuvant combination therapy with gemcitabine and autologous γδ T-cell transfer in patients with curatively resected pancreatic cancer
2017	NCT03183232 [53]	Safety and efficiency of γδ T cell against lung cancer
2017	NCT03183206 [53]	Safety and efficiency of γδ T cell against breast cancer
2017	NCT03183219 [53]	Safety and efficiency of γδ T cell against liver cancer
2017	NTR6541 [102]	TEG001 for AML and MM
2017	NCT03180437 [52]	Safety and efficiency of irreversible irradation plus allogeneic γδ T cell against locally advanced pancreatic cancer
2017	NCT02781805	Pilot study of bisphosphonates for breast cancer
2016	NCT01810120 [134]	γδ T-cell reconstitution after HLA-haploidentical hematopoietic transplantation depleted of TCR-αβ+/CD19+ lymphocytes
2016	NCT02656147	Immunotherapy allogeneic With CD19 CAR γδT-cells for B-cell lymphoma, ALL and CLL
2015	NCT02459067	ImmuniCell® in patients with advanced cancers
2015	NCT02425748	Safety and efficiency of γδ T cell against non small lung cancer (without EGFR mutation)
2015	NCT02425735 [53]	Safety and efficiency of γδ T cell against hepatocellular liver cancer
2015	NCT02418481 [53]	Safety and efficiency of γδ T cell against breast cancer(Her-, er-, and pr-)

Year	Trial ID (Refs.)	Trial title
2015	NCT02585908	Safety and efficacy of γδ T cell against gastric cancer
2014	[61]	Successful adoptive transfer and in vivo expansion of haploidentical γδ T cells
2014	UMIN000004130 [48]	Intraperitoneal injection of in vitro expanded Vγ9Vδ2 T cells together with zoledronate for the treatment of malignant ascites due to gastric cancer
2013	UMIN000000854 [135]	Ex vivo characterization of γδ T-cell repertoire in patients after adoptive transfer of autologous Vγ9Vδ2 T cells expressing the interleukin-2 receptor β-chain and the common γ-chain
2012	[136]	Tumor-promoting versus tumor-antagonizing roles of γδ T cells in cancer immunotherapy: results from a prospective phase I/II trial
2011	TRIC-CTR-GU-05-01/ NCT00588913 [137]	Phase I/II study of adoptive transfer of γδ T cells in combination with zoledronic acid and IL-2 to patients with advanced renal cell carcinoma
2011	[138]	Pilot trial of interleukin-2 and zoledronic acid to augment γδ T cells as treatment for patients with refractory renal cell carcinoma
2011	[139]	Clinical evaluation of autologous gamma delta T cell-based immunotherapy for metastatic solid tumors
2011	[140]	Zoledronate-activated Vγ9γδ T cell-based immunotherapy is feasible and restores the impairment of γδ T cells in patients with solid tumors
2011	[141]	Adoptive immunotherapy for advanced non-small cell lung cancer using zoledronate-expanded γδTcells: a phase I clinical study
2011	NCT01404702 [45]	Pilot study of zoledronic acid and interleukin-2 for refractory pediatric neuroblastoma
2011	NCT01367288	Comparative study of neoadjuvant chemotherapy with and without zometa for management of locally advanced breast cancers (NEOZOL)
2011	[142]	Phase I study of bromohydrin pyrophosphate (BrHPP, IPH 1101), a Vgamma9Vdelta2 T lymphocyte agonist in patients with solid tumors
2010	[143]	In vivo manipulation of Vgamma9Vdelta2 T cells with zoledronate and low-dose interleukin-2 for immunotherapy of advanced breast cancer patients
2010	C000000336 [141,144]	A phase I study of adoptive immunotherapy for recurrent non-small-cell lung cancer patients with autologous gammadelta T cells
2010	[145]	In vivo effects of zoledronic acid on peripheral gammadelta T lymphocytes in early breast cancer patients
2009	[146]	Phase-I study of Innacell gammadelta, an autologous cell-therapy product highly enriched in gamma9delta2 T lymphocytes, in combination with IL-2, in patients with metastatic renal cell carcinoma
2009	[147]	Clinical and immunological evaluation of zoledronate-activated Vγ9γδ T-cell-based immunotherapy for patients with multiple myeloma
2008	[148]	Long term disease-free survival in acute leukemia patients recovering with increased gammadelta T cells after partially mismatched related donor bone marrow transplantation

Continued

Year	Trial ID (Refs.)	Trial title
2008	[149]	Phase-I study of Innacell γδ™, an autologous cell-therapy product highly enriched in γ9δ2 T lymphocytes, in combination with IL-2, in patients with metastatic renal cell carcinoma
2007	[150]	Safety profile and anti-tumor effects of adoptive immunotherapy using gamma-delta T cells against advanced renal cell carcinoma: a pilot study
2007	NCT00562666	Immunotherapy of hepatocellular carcinoma with gamma delta T cells (ICAR)
2007	NCT00582790	Study of IL2 in combination with zoledronic acid in patients with kidney cancer
2007	[151]	Targeting human γδ T cells with zoledronate and interleukin-2 for immunotherapy of hormone-refractory prostate cancer
2003	[152]	Gammadelta T cells for immune therapy of patients with lymphoid malignancies

References

[1] Dolgin E. Unconventional γδ T cells 'the new black' in cancer therapy. Nat Biotechnol 2022;40(6):805–8.

[2] Mensurado S, Blanco-Domínguez R, Silva-Santos B. The emerging roles of γδ T cells in cancer immunotherapy. Nat Rev Clin Oncol 2023;20(3):178–91.

[3] Gavriil A, Barisa M, Halliwell E, Anderson J. Engineering solutions for mitigation of chimeric antigen receptor T-cell dysfunction. Cancer 2020;12(8):2326.

[4] Fowler D, Nattress C, Navarrete AS, Barisa M, Fisher J. Payload delivery: engineering immune cells to disrupt the tumour microenvironment. Cancer 2021;13(23):6000.

[5] Harrison RP, Zylberberg E, Ellison S, Levine BL. Chimeric antigen receptor-T cell therapy manufacturing: modelling the effect of offshore production on aggregate cost of goods. Cytotherapy 2019;21(2):224–33.

[6] Tran E, Turcotte S, Gros A, Robbins PF, Lu YC, Dudley ME, et al. Cancer immunotherapy based on mutation-specific CD4+ T cells in a patient with epithelial cancer. Science 2014;344(6184):641–5.

[7] Tran E, Robbins PF, Lu YC, Prickett TD, Gartner JJ, Jia L, et al. T-cell transfer therapy targeting mutant KRAS in Cancer. N Engl J Med 2016;375(23):2255–62.

[8] Robertson J, Salm M, Dangl M. Adoptive cell therapy with tumour-infiltrating lymphocytes: the emerging importance of clonal neoantigen targets for next-generation products in non-small cell lung cancer. Immunooncol Technol 2019;3:1–7.

[9] Kansteiner F. Atara makes history with world-first nod for allogeneic T-cell therapy Ebvallo. Fierce Pharma; 2022. [Cited 2023 May 29]. Available from: https://www.fiercepharma.com/pharma/atara-makes-history-world-first-nod-allogeneic-t-cell-therapy-ebvallo.

[10] Laskowski TJ, Biederstädt A, Rezvani K. Natural killer cells in antitumour adoptive cell immunotherapy. Nat Rev Cancer 2022;22(10):557–75.

[11] Gentles AJ, Newman AM, Liu CL, Bratman SV, Feng W, Kim D, et al. The prognostic landscape of genes and infiltrating immune cells across human cancers. Nat Med 2015;21(8):938–45.

[12] Park JH, Kim HJ, Kim CW, Kim HC, Jung Y, Lee HS, et al. Tumor hypoxia represses γδ T cell-mediated antitumor immunity against brain tumors. Nat Immunol 2021;22(3):336–46.

[13] de Vries NL, van de Haar J, Veninga V, Chalabi M, Ijsselsteijn ME, van der Ploeg M, et al. γδ T cells are effectors of immunotherapy in cancers with HLA class I defects. Nature 2023;613(7945):743–50.

[14] Meraviglia S, Lo Presti E, Tosolini M, La Mendola C, Orlando V, Todaro M, et al. Distinctive features of tumor-infiltrating γδ T lymphocytes in human colorectal cancer. Onco Targets Ther 2017;6(10), e1347742.

[15] Wu Y, Biswas D, Usaite I, Angelova M, Boeing S, Karasaki T, et al. A local human Vδ1 T cell population is associated with survival in nonsmall-cell lung cancer. Nat Cancer 2022;3(6):696–709.

[16] Yin W, Kyle-Cezar F, Woolf TR, Naceur-Lombardelli C, Owen J, Biswas D, et al. An innate-like Vd1 gdT cell compartment in the human breast is associated with remission in triple-negative breast cancer. Sci Transl Med 2019;11(513):eaax9364. https://doi.org/10.1126/scitranslmed.aax9364.

[17] Lamb LS, Henslee-Downey PJ, Parrish RS, Godder K, Thompson J, Lee C, et al. Increased frequency of TCR gamma delta + T cells in disease-free survivors following T cell-depleted, partially mismatched, related donor bone marrow transplantation for leukemia. J Hematother 1996;5(5):503–9.

[18] Perko R, Kang G, Sunkara A, Leung W, Thomas PG, Dallas MH. Gamma delta T cell reconstitution is associated with fewer infections and improved event-free survival after hematopoietic stem cell transplantation for pediatric leukemia. Biol Blood Marrow Transplant 2015;21(1):130–6.

[19] Handgretinger R, Schilbach K. The potential role of γδ T cells after allogeneic HCT for leukemia. Blood 2018;131 (10):1063–72.

[20] Klyuchnikov E, Badbaran A, Massoud R, Fritsche-Friedland U, Janson D, Ayuk F, et al. Enhanced immune reconstitution of γδ T cells after allogeneic stem cell transplantation overcomes the negative impact of pretransplantation minimal residual disease-positive status in patients with acute myelogenous leukemia. Transplant Cell Ther 2021;27(10):841–50.

[21] Arruda LCM, Gaballa A, Uhlin M. Impact of γδ T cells on clinical outcome of hematopoietic stem cell transplantation: systematic review and meta-analysis. Blood Adv 2019;3(21):3436–48.

[22] CAR T cells: engineering immune cells to treat cancer. NCI; 2013. [Cited 2023 May 29]. Available from: https://www.cancer.gov/about-cancer/treatment/research/car-t-cells.

[23] Del Bufalo F, De Angelis B, Caruana I, Del Baldo G, De Ioris MA, Serra A, et al. GD2-CART01 for relapsed or refractory high-risk neuroblastoma. N Engl J Med 2023;388(14):1284–95.

[24] Hunter S, Willcox CR, Davey MS, Kasatskaya SA, Jeffery HC, Chudakov DM, et al. Human liver infiltrating γδ T cells are composed of clonally expanded circulating and tissue-resident populations. J Hepatol 2018;69(3):654–65.

[25] Deseke M, Rampoldi F, Sandrock I, Borst E, Böning H, Ssebyatika GL, et al. A CMV-induced adaptive human Vδ1+ γδ T cell clone recognizes HLA-DR. J Exp Med 2022;219(9), e20212525.

[26] Fichtner AS, Ravens S, Prinz I. Human γδ TCR repertoires in health and disease. Cells 2020;9(4):800.

[27] Willcox BE, Willcox CR. γδ TCR ligands: the quest to solve a 500-million-year-old mystery. Nat Immunol 2019;20 (2):121–8.

[28] Melandri D, Zlatareva I, Chaleil RAG, Dart RJ, Chancellor A, Nussbaumer O, et al. The γδTCR combines innate immunity with adaptive immunity by utilizing spatially distinct regions for agonist selection and antigen responsiveness. Nat Immunol 2018;19(12):1352–65.

[29] Harly C, Guillaume Y, Nedellec S, Peigné CM, Mönkkönen H, Mönkkönen J, et al. Key implication of CD277/butyrophilin-3 (BTN3A) in cellular stress sensing by a major human γδ T-cell subset. Blood 2012;120(11):2269–79.

[30] Barisa M, Kramer AM, Majani Y, Moulding D, Saraiva L, Bajaj-Elliott M, et al. *E. coli* promotes human Vγ9Vδ2 T cell transition from cytokine-producing bactericidal effectors to professional phagocytic killers in a TCR-dependent manner. Nat Sci Rep 2017;7. https://doi.org/10.1038/s41598-017-02886-8.

[31] Davey MS, Willcox CR, Baker AT, Hunter S, Willcox BE. Recasting human Vδ1 lymphocytes in an adaptive role. Trends Immunol 2018;39(6):446–59.

[32] Hayday AC. γδ T cell update: adaptate orchestrators of immune surveillance. J Immunol 2019;203(2):311–20.

[33] Bank I, Silva-Santos B, Kuball J, Kabelitz D, Coffelt S, Born W. γδ T cells in cancer. Frontiers Media SA; 2021. 137 p.

[34] Harmon C, Zaborowski A, Moore H, St. Louis P, Slattery K, Duquette D, et al. γδ T cell dichotomy with opposing cytotoxic and wound healing functions in human solid tumors. Nat Cancer 2023;4(8):1122–37.

[35] Fisher JP, Heuijerjans J, Yan M, Gustafsson K, Anderson J. γδ T cells for cancer immunotherapy: a systematic review of clinical trials. Onco Targets Ther 2014;3(1), e27572.

[36] Fournié JJ, Sicard H, Poupot M, Bezombes C, Blanc A, Romagné F, et al. What lessons can be learned from γδ T cell-based cancer immunotherapy trials? Cell Mol Immunol 2013;10(1):35.

[37] Schaft N. The landscape of CAR-T cell clinical trials against solid tumors—a comprehensive overview. Cancer 2020;12(9):2567.

[38] Springuel L, Lonez C, Alexandre B, Cutsem E, Machiels JP, Van Den Eynde M, et al. Chimeric antigen receptor-T cells for targeting solid tumors: current challenges and existing strategies. BioDrugs 2019;33(5):515–37.

[39] Zlatareva I, Wu Y. Local γδ T cells: translating promise to practice in cancer immunotherapy. Br J Cancer 2023;129:393–405.

[40] First-in-human study of ICT01 in patients with advanced cancer. Full Text View. ClinicalTrials.gov [Internet]. [Cited 2023 July 13]. Available from: https://clinicaltrials.gov/ct2/show/NCT04243499.

[41] Trial of LAVA-1207 in patients with therapy refractory metastatic castration resistant prostate cancer. Full Text View. ClinicalTrials.gov [Internet]. [Cited 2023 July 13]. Available from: https://clinicaltrials.gov/ct2/show/NCT05369000.

[42] Trial of LAVA-051 in patients with relapsed/refractory CLL, MM, or AML. Full Text View. ClinicalTrials.gov [Internet]. [Cited 2023 July 13]. Available from: https://clinicaltrials.gov/ct2/show/NCT04887259.

[43] Allogeneic expanded gamma delta T cells with GD2 chemoimmunotherapy in relapsed or refractory neuroblastoma. Full Text View. ClinicalTrials.gov [Internet]. [Cited 2023 June 30]. Available from: https://clinicaltrials.gov/ct2/show/NCT05400603.

[44] Zakeri N, Hall A, Swadling L, Pallett LJ, Schmidt NM, Diniz MO, et al. Characterisation and induction of tissue-resident gamma delta T-cells to target hepatocellular carcinoma. Nat Commun 2022;13(1):1372.

[45] Pressey JG, Adams J, Harkins L, Kelly D, You Z, Lamb LS. In vivo expansion and activation of γδ T cells as immunotherapy for refractory neuroblastoma: a phase 1 study. Medicine (Baltimore) 2016;95(39), e4909.

[46] Sugie T, Suzuki E, Yamauchi A, Yamagami K, Masuda N, Gondo N, et al. Combined effects of neoadjuvant letrozole and zoledronic acid on γδT cells in postmenopausal women with early-stage breast cancer. Breast 2018;38:114–9.

[47] De Gassart A, Le KS, Brune P, Agaugué S, Sims J, Goubard A, et al. Development of ICT01, a first-in-class, anti-BTN3A antibody for activating Vγ9Vδ2 T cell-mediated antitumor immune response. Sci Transl Med 2021;13 (616), eabj0835.

[48] Wada I, Matsushita H, Noji S, Mori K, Yamashita H, Nomura S, et al. Intraperitoneal injection of in vitro expanded Vγ9Vδ2 T cells together with zoledronate for the treatment of malignant ascites due to gastric cancer. Cancer Med 2014;3(2):362–75.

[49] Aoki T, Matsushita H, Hoshikawa M, Hasegawa K, Kokudo N, Kakimi K. Adjuvant combination therapy with gemcitabine and autologous γδ T-cell transfer in patients with curatively resected pancreatic cancer. Cytotherapy 2017;19(4):473–85.

[50] Alnaggar M, Xu Y, Li J, He J, Chen J, Li M, et al. Allogenic Vγ9Vδ2 T cell as new potential immunotherapy drug for solid tumor: a case study for cholangiocarcinoma. J Immunother Cancer 2019;7(1):36.

[51] Kakimi K, Matsushita H, Masuzawa K, Karasaki T, Kobayashi Y, Nagaoka K, et al. Adoptive transfer of zoledronate-expanded autologous Vγ9Vδ2 T-cells in patients with treatment-refractory non-small-cell lung cancer: a multicenter, open-label, single-arm, phase 2 study. J Immunother Cancer 2020;8(2), e001185.

[52] Lin M, Zhang X, Liang S, Luo H, Alnaggar M, Liu A, et al. Irreversible electroporation plus allogenic Vγ9Vδ2 T cells enhances antitumor effect for locally advanced pancreatic cancer patients. Sig Transduct Target Ther 2020;5 (1):1–9.

[53] Xu Y, Xiang Z, Alnaggar M, Kouakanou L, Li J, He J, et al. Allogeneic Vγ9Vδ2 T-cell immunotherapy exhibits promising clinical safety and prolongs the survival of patients with late-stage lung or liver cancer. Cell Mol Immunol 2021;18(2):427–39.

[54] UMIN Clinical Trials Registry. [Internet]. [Cited 2023 July 17]. Available from: https://center6.umin.ac.jp/cgi-open-bin/ctr_e/ctr_view.cgi?recptno=R000008701.

[55] University of Alabama at Birmingham. Pilot study of zoledronic acid and interleukin-2 for refractory pediatric neuroblastoma: assessment of tolerability and in vivo expansion γδ T-cells, 2014. [Internet]. Report No.: NCT01404702. clinicaltrials.gov; December 2014 [Cited 2023 July 13]. Available from: https://clinicaltrials.gov/study/NCT01404702.

[56] UMIN Clinical Trials Registry. [Internet]. [Cited 2023 July 17]. Available from: https://center6.umin.ac.jp/cgi-open-bin/ctr_e/ctr_view.cgi?recptno=R000004959.

[57] UMIN Clinical Trials Registry. [Internet]. [Cited 2023 July 17]. Available from: https://center6.umin.ac.jp/cgi-open-bin/ctr_e/ctr_view.cgi?recptno=R000001115.

[58] UMIN Clinical Trials Registry. [Internet]. [Cited 2023 July 17]. Available from: https://center6.umin.ac.jp/cgi-open-bin/ctr_e/ctr_view.cgi?recptno=R000007250.

[59] Fuda Cancer Hospital, Guangzhou. Safety and efficiency of IRE plus γδ T cell against locally advanced pancreatic cancer, 2020. [Internet]. Report No.: NCT03180437. clinicaltrials.gov. 2020 October [Cited 2023 July 13]. Available from: https://clinicaltrials.gov/study/NCT03180437.

[60] Fazzi R, Petrini I, Giuliani N, Morganti R, Carulli G, Dalla Palma B, et al. Phase II trial of maintenance treatment with IL2 and zoledronate in multiple myeloma after bone marrow transplantation: biological and clinical results. Front Immunol 2020;11, 573156.

[61] Wilhelm M, Smetak M, Schaefer-Eckart K, Kimmel B, Birkmann J, Einsele H, et al. Successful adoptive transfer and in vivo expansion of haploidentical γδ T cells. J Transl Med 2014;12(1):45.

[62] Vydra J, Cosimo E, Lesný P, Wanless RS, Anderson J, Clark AG, et al. A phase I trial of allogeneic γδ T lymphocytes from haploidentical donors in patients with refractory or relapsed acute myeloid leukemia. Clin Lymphoma Myeloma Leuk 2023. [Internet]. February 11 [Cited 2023 February 21]. Available from: https://www.sciencedirect.com/science/article/pii/S2152265023000381.

[63] Ex-vivo expanded γδ T-LYMPHOCYTES (OmnImmune®) in patients with acute myeloid leukaemia (AML). Full Text View. ClinicalTrials.gov [Internet]. [Cited 2023 June 30]. Available from: https://clinicaltrials.gov/ct2/show/NCT03790072.

[64] Clinical Trials Register. Search for 2013-001188-22. [Internet]. [Cited 2023 July 17]. Available from: https://www.clinicaltrialsregister.eu/ctr-search/search?query=2013-001188-22.

[65] Makkouk A, (Cher) Yang X, Barca T, Lucas A, Turkoz M, Nishimoto K, et al. Allogeneic Vδ1 gamma delta T cells engineered with glypican-3 (GPC3)-specific CAR expressing soluble IL-15 have enhanced antitumor efficacy against hepatocellular carcinoma in preclinical models. J Clin Oncol 2021;39(15_suppl):e14511.

[66] Fowler D, Barisa M, Southern A, Nattress C, Hawkins E, Vassalou E, et al. Payload-delivering engineered γδ T cells display enhanced cytotoxicity, persistence, and efficacy in preclinical models of osteosarcoma. Sci Transl Med. 2024 May 29;16(749):eadg9814.

[67] Almeida AR, Correia DV, Fernandes-Platzgummer A, da Silva CL, da Silva MG, Anjos DR, et al. Delta one T cells for immunotherapy of chronic lymphocytic leukemia: clinical-grade expansion/differentiation and preclinical proof of concept. Clin Cancer Res 2016;22(23):5795–804.

[68] Nishimoto KP, Barca T, Azameera A, Makkouk A, Romero JM, Bai L, et al. Allogeneic CD20-targeted γδ T cells exhibit innate and adaptive antitumor activities in preclinical B-cell lymphoma models. Clin Transl Immunology 2022;11(2), e1373.

[69] Wu TS, Li HK, Hsiao CW, Ou YH, Cheng ZF, Yang HP, et al. Abstract 5573: ACE1831: a novel allogeneic αCD20-conjugated Vδ2 gamma delta T product for non-Hodgkin's lymphoma. Cancer Res 2022;82(12_Supplement):5573.

[70] Li HK, Wu TS, Leng PR, Kuo YC, Cheng ZF, Lee CY, et al. Abstract LB089: ACE2016: an off-the-shelf EGFR-targeting γδ2 T cell therapy against EGFR-expressing solid tumors. Cancer Res 2023;83(8_Supplement): LB089.

[71] Boucher JC, Yu B, Li G, Shrestha B, Sallman D, Landin AM, et al. Large scale ex vivo expansion of γδ T cells using artificial antigen-presenting cells. J Immunother 2023;46(1):5–13.

[72] Leedom T, Hamil AS, Pouyanfard S, Govero J, Langland R, Ballard A, et al. Characterization of WU-CART-007, an allogeneic CD7-targeted CAR-T cell therapy for T-cell malignancies. Blood 2021;138:2772.

[73] Wang Y, Han J, Wang D, Cai M, Xu Y, Hu Y, et al. Anti-PD-1 antibody armored γδ T cells enhance anti-tumor efficacy in ovarian cancer. Sig Transduct Target Ther 2023;8(1):1–11.

[74] Gao X, Mi Y, Guo N, Xu H, Xu L, Gou X, et al. Cytokine-induced killer cells as pharmacological tools for cancer immunotherapy. Front Immunol 2017;8:774.

[75] Sabry M, Lowdell MW. Killers at the crossroads: the use of innate immune cells in adoptive cellular therapy of cancer. Stem Cells Transl Med 2020. Wiley Online Library [Internet]. [Cited 2023 Jun 27]. Available from: https://stemcellsjournals.onlinelibrary.wiley.com/doi/full/10.1002/sctm.19-0423.

[76] Di Lorenzo B, Simões AE, Caiado F, Tieppo P, Correia DV, Carvalho T, et al. Broad cytotoxic targeting of acute myeloid leukemia by polyclonal delta one T cells. Cancer Immunol Res 2019;7(4):552–8.

[77] Study of GDX012 in patients with MRD positive AML. Full Text View. ClinicalTrials.gov [Internet]. [Cited 2023 June 27]. Available from: https://clinicaltrials.gov/ct2/show/NCT05001451.

[78] Neelapu SS, Hamadani M, Miklos DB, Holmes H, Hinkle J, Kennedy-Wilde J, et al. A phase 1 study of ADI-001: anti-CD20 CAR-engineered allogeneic gamma delta (γδ) T cells in adults with B-cell malignancies. J Clin Oncol 2022;40(16_suppl):7509.

[79] Adicet bio reports positive data from ongoing ADI-001 phase 1 trial in patients with relapsed or refractory aggressive b-cell non-Hodgkin's lymphoma (NHL) Adicet Bio [Internet]. [Cited 2023 June 27]. Available from: https://investor.adicetbio.com/news-releases/news-release-details/adicet-bio-reports-positive-data-ongoing-adi-001-phase-1-trial-0/.

[80] Research C for BE and YESCARTA (axicabtagene ciloleucel). FDA; 2022. [Internet]. 2022 April 11 [Cited 2023 June 27]. Available from: https://www.fda.gov/vaccines-blood-biologics/cellular-gene-therapy-products/yescarta-axicabtagene-ciloleucel.

[81] Research C for BE and. KYMRIAH (tisagenlecleucel). FDA; 2022. [Internet]. 2022 July 7 [Cited 2023 June 27]. Available from: https://www.fda.gov/vaccines-blood-biologics/cellular-gene-therapy-products/kymriah-tisagenlecleucel.

[82] Acepodia Inc. Acepodia announces first patient dosed in phase 1 clinical trial of ACE1831, an anti-CD20 armed allogeneic gamma delta T-cell therapy in development to treat patients with non-Hodgkin's lymphoma. [Internet]. [Cited 2023 June 27]. Available from: https://www.prnewswire.com/news-releases/acepodia-announces-first-patient-dosed-in-phase-1-clinical-trial-of-ace1831-an-anti-cd20-armed-allogeneic-gamma-delta-t-cell-therapy-in-development-to-treat-patients-with-non-hodgkins-lymphoma-301830957.html.

[83] Gamma delta T-cell infusion for AML at high risk of relapse after allo HCT. Full Text View. ClinicalTrials.gov [Internet]. [Cited 2023 June 28]. Available from: https://clinicaltrials.gov/ct2/show/NCT05015426.

[84] Kiromic BioPharma reports six-month results from first patient enrolled in Deltacel-01 Clinical Trial—company announcement., 2024. FT.com [Internet]. [Cited 2024 June 13]. Available from: https://markets.ft.com/data/announce/full?dockey=600-202406060800BIZWIRE_USPRX____20240606_BW202467-1.

[85] Knowles LJ, Malik M, Nussebaumer O, Brown A, van Wetering S, Koslowski M. Abstract CT525: GDX012U-001 a phase 1, open-label, dose escalation, and dose expansion study to assess the safety, tolerability, and preliminary antileukemic activity of GDX012 in patients with MRD positive AML. Cancer Res 2022;82(12_Supplement):CT525.

[86] Wugen SE. Wugen presents initial data from first-in-human phase 1/2 trial of WU-CART-007 at the European Hematology Association (EHA) 2023 Congress, 2023. [cited 2024 Jun 13]. Available from: https://wugen.com/wugen-presents-initial-data-from-first-in-human-phase-1-2-trial-of-wu-cart-007-at-the-european-hematology-association-eha-2023-congress/.

[87] Huang S, Pan C, Lin Y, Chen M, Chen Y, Jan C, et al. BiTE-Secreting CAR-γδT as a Dual Targeting Strategy for the Treatment of Solid Tumors. Adv Sci (Weinh) 2023;10(17):2206856.

[88] Allogeneic γδ T cell therapy for the treatment of solid tumors. Full Text View. ClinicalTrials.gov [Internet]. [Cited 2023 June 30]. Available from: https://clinicaltrials.gov/ct2/show/NCT04765462.

[89] ACE1831 in adult subjects with relapsed/refractory CD20-expressing B-cell malignancies. Full Text View. ClinicalTrials.gov [Internet]. [Cited 2023 June 30]. Available from: https://clinicaltrials.gov/ct2/show/NCT05653271.

[90] ACHIEVE—Efficacy and effectiveness of adoptive cellular therapy with ex-vivo expanded allogeneic γδ T-lymphocytes (TCB-008) for patients with refractory or relapsed acute myeloid leukaemia (AML). Full Text View. ClinicalTrials.gov [Internet]. [Cited 2023 June 30]. Available from: https://clinicaltrials.gov/ct2/show/NCT05358808.

[91] A safety and efficacy study of ADI-001, an anti-CD20 allogeneic gamma delta CAR-T, in subjects with B cell malignancies. Full Text View. ClinicalTrials.gov [Internet]. [Cited 2023 June 30]. Available from: https://clinicaltrials.gov/ct2/show/NCT04735471.

[92] Acepodia Biotech, Inc. A phase 1 multicenter study evaluating the safety and efficacy of ACE2016, an allogeneic anti-EGFR conjugated gamma delta T cell (gdT) therapy in adult subjects with locally advanced or metastatic solid tumors expressing epidermal growth factor receptor (EGFR); 2024. [Internet]. Report No.: NCT06415487. clinicaltrials.gov; 2024 May [Cited 2024 Jan 1]. Available from: https://clinicaltrials.gov/study/NCT06415487.

[93] Allogeneic NKG2DL-targeting CAR γδ T Cells (CTM-N2D) in advanced cancers. Full Text View. ClinicalTrials.gov [Internet]. [Cited 2023 June 30]. Available from: https://clinicaltrials.gov/ct2/show/NCT05302037.

[94] Allogeneic γδ T cells immunotherapy in r/r non-Hodgkin's lymphoma (NHL) or peripheral T cell lymphomas (PTCL) patients. Full Text View. ClinicalTrials.gov [Internet]. [Cited 2023 June 30]. Available from: https://clinicaltrials.gov/ct2/show/NCT04696705.

[95] Minculescu L. Innate donor effector allogeneic lymphocyte infusion after stem cell transplantation: the IDEAL trial; 2024. [Internet]. Report No.: NCT05686538. clinicaltrials.gov; 2024 April [Cited 2024 January 1]. Available from: https://clinicaltrials.gov/study/NCT05686538.

[96] Takeda. A phase 1/2a, open-label, dose escalation, and dose expansion study to assess the safety and efficacy of GDX012 in patients with relapsed or refractory acute myeloid leukemia; 2024. [Internet]. Report No.: NCT05886491. clinicaltrials.gov; 2024 May [Cited 2024 January 1]. Available from: https://clinicaltrials.gov/study/NCT05886491.

[97] Washington University School of Medicine. Phase 1 dose-escalation and dose-expansion study to evaluate the safety and tolerability of anti-CD7 allogeneic CAR T-cells (WU-CART-007) in patients with CD7+ hematologic malignancies; 2023. [Internet]. Report No.: NCT05377827. clinicaltrials.gov; 2023 October [Cited 2024 January 1]. Available from: https://clinicaltrials.gov/study/NCT05377827.

[98] Immunotherapy with CD19 CAR γδT-cells for B-cell lymphoma, ALL and CLL., 2023. Full Text View. ClinicalTrials.gov [Internet]. [Cited 2023 June 30]. Available from: https://clinicaltrials.gov/ct2/show/NCT02656147.

[99] Safety and efficiency of γδ T cell against hematological malignancies after allo-HSCT., 2023. Full Text View. ClinicalTrials.gov [Internet]. [Cited 2023 June 30]. Available from: https://clinicaltrials.gov/ct2/show/NCT04764513.

[100] Expanded/activated gamma delta T-cell infusion following hematopoietic stem cell transplantation and post-transplant cyclophosphamide., 2023. Full Text View. ClinicalTrials.gov [Internet]. [cited 2023 June 30]. Available from: https://clinicaltrials.gov/ct2/show/NCT03533816.

[101] Frieling JS, Tordesillas L, Bustos XE, Ramello MC, Bishop RT, Cianne JE, et al. γδ-Enriched CAR-T cell therapy for bone metastatic castrate-resistant prostate cancer. Sci Adv 2023;9(18):eadf0108.

[102] de Witte M, Scheepstra J, Weertman N, Daudeij A, van der Wagen L, Oostvogels R, et al. First in human clinical responses and persistence data on TEG001: a next generation of engineered Aβ T cells targeting AML and MM with a high affinity γ9δ2TCR. Blood 2022;140(Supplement 1):12737–9.

[103] Straetemans T, Kierkels GJJ, Doorn R, Jansen K, Heijhuurs S, Dos Santos JM, et al. GMP-grade manufacturing of T cells engineered to express a defined γδTCR. Front Immunol 2018;9:1062.

[104] Nabors LB, Lamb LS, Beelen MJ, Pillay T, ter Haak M, Youngblood S, et al. Phase 1 trial of drug resistant immunotherapy: a first-in-class combination of MGMT-modified γδ t cells and temozolomide chemotherapy in newly diagnosed glioblastoma. J Clin Oncol 2021;39(15_suppl):2057.

[105] Lamb Jr LS, Bowersock J, Dasgupta A, Gillespie GY, Su Y, Johnson A, et al. Engineered drug resistant γδ T cells kill glioblastoma cell lines during a chemotherapy challenge: a strategy for combining chemo- and immunotherapy. PLoS One 2013;8(1), e51805.

[106] Dushu Lake Hospital Affiliated to Soochow University. Allogenic B7H3-targeting CAR-γδT Cell therapy in r/r GBM; 2023. [Internet]. Report No.: NCT06018363. clinicaltrials.gov; 2023 August [Cited 2024 January 1]. Available from: https://clinicaltrials.gov/study/NCT06018363.

[107] CAR—γ δ T cells in the treatment of relapsed and refractory CD7 positive T cell-derived malignant tumors., 2023. Full Text View. ClinicalTrials.gov [Internet]. [Cited 2023 July 3]. Available from: https://clinicaltrials.gov/ct2/show/NCT04702841.

[108] Long-term follow-up study of allogeneic gamma delta (γδ) CAR T cells. Full Text View. ClinicalTrials.gov [Internet]. [Cited 2023 July 3]. Available from: https://clinicaltrials.gov/ct2/show/NCT04911478.

[109] TAA06 Injection in the treatment of patients with B7-H3-positive relapsed/refractory neuroblastoma; 2023. Full Text View. ClinicalTrials.gov [Internet]. [Cited 2023 July 3]. Available from: https://clinicaltrials.gov/ct2/show/NCT05562024.

[110] Lee H, Moffitt Cancer Center and Research Institute. A phase I clinical trial of an infusion of autologous gamma delta T cells genetically engineered with a chimeric receptor to target the prostate stem cell antigen in patients with metastatic castration resistant prostate cancer; 2024. [Internet]. Report No.: NCT06193486. clinicaltrials.gov; 2024 March [Cited 2024 January 1]. Available from: https://clinicaltrials.gov/study/NCT06193486.

[111] Ever Supreme Bio Technology Co., Ltd. A single arm, open label, dose-escalation phase I and dose-expansion phase IIa clinical study to evaluate the feasibility, safety, and efficacy of allogeneic chimeric antigen receptor (CAR) gamma-delta T cells CAR001 in subjects with relapsed/refractory solid tumors; 2023. [Internet]. Report No.: NCT06150885. clinicaltrials.gov; 2023 November [Cited 2024 January 1]. Available from: https://clinicaltrials.gov/study/NCT06150885.

[112] Zhang M. First-in-class autologous CD7-CAR-T cells exhibited promising clinical efficacy for relapsed and refractory T-lymphoblastic leukemia/lymphoma. ASH; 2020. [Cited 2023 July 13]. Available from: https://ash.confex.com/ash/2020/webprogram/Paper136508.html.

[113] Technology platform. PersonGen BioTherapeutics (Suzhou)Co., Ltd. [Internet]. [Cited 2023 July 3]. Available from: https://en.persongen.com/technology-platform.html.

[114] Annesley C. Phase 1 study of B7-H3-specific CAR T cell locoregional immunotherapy for diffuse intrinsic pontine glioma/diffuse midline glioma and recurrent or refractory pediatric central nervous system tumors; 2023. [Internet]. Report No.: NCT04185038. clinicaltrials.gov; 2023 December [Cited 2024 January 1]. Available from: https://clinicaltrials.gov/study/NCT04185038.

[115] Mackall C. Phase I Clinical Trial of Locoregionally (LR) Delivered Autologous B7-H3 Chimeric Antigen Receptor T Cells (B7-H3CART) in Adults With Recurrent Glioblastoma Multiforme (GBM); 2022. [Internet]. Report No.: NCT05474378 clinicaltrials.gov; 2022 July [Cited 2023 January 1]. Available from: https://clinicaltrials.gov/study/NCT05474378.

[116] Majzner RG, Theruvath JL, Nellan A, Heitzeneder S, Cui Y, Mount CW, et al. CAR T cells targeting B7-H3, a pan-cancer antigen, demonstrate potent preclinical activity against pediatric solid tumors and brain tumors. Clin Cancer Res 2019;25(8):2560–74.

[117] Vitanza NA, Wilson AL, Huang W, Seidel K, Brown C, Gustafson JA, et al. Intraventricular B7-H3 CAR T cells for diffuse intrinsic pontine glioma: preliminary first-in-human bioactivity and safety. Cancer Discov 2023;13 (1):114–31.

[118] Barisa M, Zappa E, Muller H, Shah R, Buhl J, Draper B, et al. Functional avidity of anti-B7H3 CAR-T constructs predicts antigen density thresholds for triggering effector function; 2024. [Internet]. bioRxiv [Cited 2024 February 25]. p. 2024.02.19.580939. Available from: https://www.biorxiv.org/content/10.1101/2024.02.19.580939v1.

[119] Abate-Daga D, Lagisetty KH, Tran E, Zheng Z, Gattinoni L, Yu Z, et al. A novel chimeric antigen receptor against prostate stem cell antigen mediates tumor destruction in a humanized mouse model of pancreatic cancer. Hum Gene Ther 2014;25(12):1003–12.

[120] Gründer C, van Dorp S, Hol S, Drent E, Straetemans T, Heijhuurs S, et al. γ9 and δ2CDR3 domains regulate functional avidity of T cells harboring γ9δ2TCRs. Blood 2012;120(26):5153–62.

[121] Fuda Cancer Hospital, Guangzhou. γδ T cell immunotherapy for treatment of liver cancer; 2020. [Internet]. Report No.: NCT03183219. clinicaltrials.gov; 2020 July [Cited 2023 July 13]. Available from: https://clinicaltrials.gov/study/NCT03183219.

[122] A study to investigate the safety and efficacy of TEG002 in relapsed/refractory multiple myeloma patients. Full Text View. ClinicalTrials.gov [Internet]. [Cited 2023 July 4]. Available from: https://clinicaltrials.gov/ct2/show/NCT04688853.

[123] Ang WX, Ng YY, Xiao L, Chen C, Li Z, Chi Z, et al. Electroporation of NKG2D RNA CAR improves Vγ9Vδ2 T cell responses against human solid tumor xenografts. Mol Ther Oncolytics 2020;17:421–30.

[124] Du SH, Li Z, Chen C, Tan WK, Chi Z, Kwang TW, et al. Co-expansion of cytokine-induced killer cells and Vγ9Vδ2 T cells for CAR T-cell therapy. PLoS One 2016;11(9), e0161820.

[125] Lamb LS, Pereboeva L, Youngblood S, Gillespie GY, Nabors LB, Markert JM, et al. A combined treatment regimen of MGMT-modified γδ T cells and temozolomide chemotherapy is effective against primary high grade gliomas. Sci Rep 2021;11(1):21133.

[126] A phase I study of drug resistant immunotherapy (DRI) with activated, gene modified γδ T cells in patients with newly diagnosed glioblastoma multiforme receiving maintenance temozolomide chemotherapy; 2021. [Internet]. Report No.: NCT04165941. clinicaltrials.gov; 2021 April [Cited 2021 November 9]. Available from: https://clinicaltrials.gov/ct2/show/NCT04165941.

[127] IN8bio presents progression-free survival update from phase 1 study of INB-200 at 2024 American Society of Clinical Oncology Annual Meeting. IN8bio, Inc. [Internet]. [Cited 2024 June 15]. Available from: https://investors.in8bio.com/news-releases/news-release-details/in8bio-presents-progression-free-survival-update-phase-1-study/.

[128] Davies D, Kamdar S, Woolf R, Zlatareva I, Iannitto ML, Morton C, et al. PD-1 defines a distinct, functional, tissue-adapted state in Vδ1+ T cells with implications for cancer immunotherapy. Nat Cancer 2024;5:420–32.

[129] Hirayama AV, Gauthier J, Hay KA, Voutsinas JM, Wu Q, Gooley T, et al. The response to lymphodepletion impacts PFS in patients with aggressive non-Hodgkin lymphoma treated with CD19 CAR T cells. Blood 2019;133(17):1876–87.

[130] Ferry GM, Agbuduwe C, Forrester M, Dunlop S, Chester K, Fisher J, et al. A simple and robust single-step method for CAR-Vδ1 γδT cell expansion and transduction for cancer immunotherapy. Front Immunol 2022;13. [Cited 2022 May 31]. Available from: https://www.frontiersin.org/article/10.3389/fimmu.2022.863155.

[131] Joshi R, Goh J, Gargosky S, Kitchenadasse G, Atkinson V, Joubert W, et al. 727 Topline safety and efficacy update of SUPLEXA-101, a first-in-human, single agent study of SUPLEXA therapeutic cells in 28 patients with metastatic solid tumours. J Immunother Cancer 2023;11(Suppl. 1). [Cited 2024 June 16]. Available from: https://jitc.bmj.com/content/11/Suppl_1/A822.

[132] Zhang T, Chen J, Niu L, Liu Y, Ye G, Jiang M, et al. Clinical safety and efficacy of locoregional therapy combined with adoptive transfer of allogeneic γδ T cells for advanced hepatocellular carcinoma and intrahepatic cholangiocarcinoma. J Vasc Interv Radiol 2022;33(1):19–27.e3.

[133] Ji N, Mukherjee N, Reyes RM, Gelfond J, Javors M, Meeks JJ, et al. Rapamycin enhances BCG-specific γδ T cells during intravesical BCG therapy for non-muscle invasive bladder cancer: a randomized, double-blind study. J Immunother Cancer 2021;9(3), e001941.
[134] Airoldi I, Bertaina A, Prigione I, Zorzoli A, Pagliara D, Cocco C, et al. γδ T-cell reconstitution after HLA-haploidentical hematopoietic transplantation depleted of TCR-αβ+/CD19+ lymphocytes. Blood 2015;125 (15):2349–58.
[135] Izumi T, Kondo M, Takahashi T, Fujieda N, Kondo A, Tamura N, et al. Ex vivo characterization of γδ T-cell repertoire in patients after adoptive transfer of Vγ9Vδ2 T cells expressing the interleukin-2 receptor β-chain and the common γ-chain. Cytotherapy 2013;15(4):481–91.
[136] Kunzmann V, Smetak M, Kimmel B, Weigang-Koehler K, Goebeler M, Birkmann J, et al. Tumor-promoting versus tumor-antagonizing roles of γδ T cells in cancer immunotherapy: results from a prospective phase I/II trial. J Immunother 2012;35(2):205–13.
[137] Kobayashi H, Tanaka Y, Yagi J, Minato N, Tanabe K. Phase I/II study of adoptive transfer of γδ T cells in combination with zoledronic acid and IL-2 to patients with advanced renal cell carcinoma. Cancer Immunol Immunother 2011;60(8):1075–84.
[138] Lang JM, Kaikobad MR, Wallace M, Staab MJ, Horvath DL, Wilding G, et al. Pilot trial of interleukin-2 and zoledronic acid to augment γδ T cells as treatment for patients with refractory renal cell carcinoma. Cancer Immunol Immunother 2011;60(10):1447–60.
[139] Nicol AJ, Tokuyama H, Mattarollo SR, Hagi T, Suzuki K, Yokokawa K, et al. Clinical evaluation of autologous gamma delta T cell-based immunotherapy for metastatic solid tumours. Br J Cancer. 2011 Sep;105(6):778–86.
[140] Noguchi A, Kaneko T, Kamigaki T, Fujimoto K, Ozawa M, Saito M, et al. Zoledronate-activated Vγ9γδ T cell-based immunotherapy is feasible and restores the impairment of γδ T cells in patients with solid tumors. Cytotherapy 2011;13(1):92–7.
[141] Sakamoto M, Nakajima J, Murakawa T, Fukami T, Yoshida Y, Murayama T, et al. Adoptive immunotherapy for advanced non-small cell lung cancer using zoledronate-expanded γδTcells: a phase I clinical study. J Immunother 2011;34(2):202–11.
[142] Bennouna J, Levy V, Sicard H, Senellart H, Audrain M, Hiret S, et al. Phase I study of bromohydrin pyrophosphate (BrHPP, IPH 1101), a Vgamma9Vdelta2 T lymphocyte agonist in patients with solid tumors. Cancer Immunol Immunother 2010;59(10):1521–30.
[143] Meraviglia S, Eberl M, Vermijlen D, Todaro M, Buccheri S, Cicero G, et al. In vivo manipulation of Vgamma9Vdelta2 T cells with zoledronate and low-dose interleukin-2 for immunotherapy of advanced breast cancer patients. Clin Exp Immunol 2010;161(2):290–7.
[144] Nakajima J, Murakawa T, Fukami T, Goto S, Kaneko T, Yoshida Y, et al. A phase I study of adoptive immunotherapy for recurrent non-small-cell lung cancer patients with autologous γδ T cells. European Journal of Cardio-Thoracic Surgery. 2010 May 1;37(5):1191–7.
[145] Santini D, Martini F, Fratto ME, Galluzzo S, Vincenzi B, Agrati C, et al. In vivo effects of zoledronic acid on peripheral γδ T lymphocytes in early breast cancer patients. Cancer Immunol Immunother 2009;58(1):31–8.
[146] Bennouna J, Bompas E, Neidhardt EM, Rolland F, Philip I, Galéa C, et al. Phase-I study of Innacell gammadelta, an autologous cell-therapy product highly enriched in gamma9delta2 T lymphocytes, in combination with IL-2, in patients with metastatic renal cell carcinoma. Cancer Immunol Immunother 2008;57(11):1599–609.
[147] Abe Y, Muto M, Nieda M, Nakagawa Y, Nicol A, Kaneko T, et al. Clinical and immunological evaluation of zoledronate-activated Vγ9γδ T-cell-based immunotherapy for patients with multiple myeloma. Exp Hematol 2009;37(8):956–68.
[148] Godder KT, Henslee-Downey PJ, Mehta J, Park BS, Chiang KY, Abhyankar S, et al. Long term disease-free survival in acute leukemia patients recovering with increased γδ T cells after partially mismatched related donor bone marrow transplantation. Bone Marrow Transplant 2007;39(12):751–7.
[149] Bennouna J, Bompas E, Neidhardt EM, Rolland F, Philip I, Galéa C, et al. Phase-I study of Innacell gammadelta, an autologous cell-therapy product highly enriched in gamma9delta2 T lymphocytes, in combination with IL-2, in patients with metastatic renal cell carcinoma. Cancer Immunol Immunother. 2008 Nov;57(11):1599–609.
[150] Kobayashi H, Tanaka Y, Yagi J, Osaka Y, Nakazawa H, Uchiyama T, et al. Safety profile and anti-tumor effects of adoptive immunotherapy using gamma-delta T cells against advanced renal cell carcinoma: a pilot study. Cancer Immunol Immunother 2007;56(4):469–76.
[151] Dieli F, Vermijlen D, Fulfaro F, Caccamo N, Meraviglia S, Cicero G, et al. Targeting human γδ T cells with zoledronate and INTERLEUKIN-2 for immunotherapy of hormone-refractory prostate cancer. Cancer Res 2007;67(15):7450–7.
[152] Wilhelm M, Kunzmann V, Eckstein S, Reimer P, Weissinger F, Ruediger T, et al. Gammadelta T cells for immune therapy of patients with lymphoid malignancies. Blood 2003;102(1):200–6.

CHAPTER

6

Allograft persistence: The next frontier for allogeneic γδ T cell therapy

Daniel Fowler and Jonathan Fisher

UCL Great Ormond Street Institute of Child Health, University College London, London, United Kingdom

Abstract

Although γδ T cells are a promising allo-friendly alternative to αβ T cells for cellular therapies against hematological and solid malignancies, there are still important considerations to be made regarding their capacity to persist long enough in an human leucocyte antigen (HLA) disparate patient in order for them to exert durable clinical responses. Specifically, the phenomenon of host versus graft rejection poses a significant barrier to their long-term survival in vivo, necessitating some form of intervention to prevent the patient's immune system from destroying the cells before they have had the chance to exert their program of antitumor effector responses in full. Despite only a limited number of peer-reviewed clinical trials on the persistence and efficacy of donor-derived γδ T cells in vivo, we can glean valuable insight from parallel studies using αβ T cells and natural killer cells, which will help guide the clinical application of future cellular therapies aimed at exploiting the universal cell therapy potential of γδ T cells. Herein, we discuss the steps that can be taken to prevent graft rejection so as to prolong the persistence and survival of allogeneic γδ T cell therapies within a patient. Two fundamental themes will be explored: the first involves lymphodepleting the patient in order to temporarily incapacitate their ability to mount immune responses against mismatched HLA, and the second involves genetic modifications to the cell therapy itself in order to render it resistant to attack from allogeneic immune cells. Within these themes, we will examine the various drugs and technologies that have been used to achieve the common goal of overcoming host versus graft rejection, and discuss the utility of these approaches in building the next generation of universal γδ T cell therapies.

Abbreviations

ADCC antibody-dependent cell cytotoxicity
ADCP antibody-dependent cell phagocytosis
AGT alkylguanine DNA alkyltransferase
ALL acute lymphoblastic leukemia
AML acute myeloid leukemia
ATG antithyroglobulin
BCMA B-cell maturation antigen
CAR chimeric antigen receptor
CLL chronic lymphoblastic leukemia

https://doi.org/10.1016/B978-0-443-21766-1.00001-1

CMV cytomegalovirus
CNi calcineurin inhibitors
CRS cytokine release syndrome
CTIIA class II major histocompatibility complex transactivator
CTL cytotoxic T lymphocyte
DC dendritic cell
EBV Epstein Barr virus
ER endoplasmic reticulum
Flu/Cy Fludarabine/Cyclophosphamide
GvHD graft versus host disease
hiPSC human induced pluripotent stem cell
HLA human leucocyte antigen
HSCT hematopoietic stem cell transplantation
ILC innate lymphoid cells
IMPDH inosine monophosphate dehydrogenase enzyme
LAG-3 lymphocyte activation gene-3
MAIT mucosal-associated invariant T
MDSC myeloid-derived suppressor cells
MPA mycophenolic acid
mTORi mTOR inhibitor
NFATc nuclear factor of activated T cell cytoplasmic
NHL non-Hodgkin's lymphoma
NK natural killer
PD-1 programmed cell death protein-1
PTLD posttransplantation lymphoproliferative disease
RFX5 regulatory factor X5
SLE systemic lupus erythematosus
TCR T cell receptor
TIGIT T cell immunoreceptor with Ig and immunoreceptor tyrosine-based inhibitory motif domain
TIL tumor-infiltrating lymphocyte
TIM-3 T cell immunoglobulin and mucin domain-containing protein-3
TME tumor microenvironment
TRAC T cell receptor constant alpha chain
Treg regulatory T cell
VISTA V-domain Ig suppressor of T cell activation

Conflict of interest

J.F. is an inventor of patents pertaining to the engineering of gamma delta T cells (WO2021148788A1 and WO2016174461A1). J.F and D.F are inventors of a patent pertaining to methods of engineering innate lymphocytes (WO2023180759A1). D.F. is a listed author on patents pertaining to the use of Vdelta1+ T cells in myeloid malignancies (WO2021186137A1).

Introduction

Recent advances in gene engineering have set a new and exciting precedent for how cell-based immunotherapies can be used in the treatment of diseases with highly complex pathologies such as cancer. No longer are we solely dependent on an immune cell's inherent biology to deliver the potent and durable antitumor effector responses required to overcome a rapidly growing, highly immunosuppressive tumor. Bespoke programs of performance-enhancing gene modifications can now be used to sculpt the protein repertoire of an immune cell, and thus create highly customized cellular therapies that are functionally tailored to overcome the unique set of challenges identified for a given tumor. Furthermore, we now have the expertise to engineer a variety of different cell types, including αβ T cells, natural killer (NK) cells [1], NKT cells [2], γδ T cells [3], macrophages [4], neutrophils [5], mucosal-associated invariant T (MAIT) cells [6] and innate lymphoid cells (ILCs) [7], allowing us the freedom to pick and choose the best cell chassis for a given clinical application, and thus exploit fully the intrinsic diversity of our immune system and the unique functional properties of its different immune cell subsets. These technological advancements—together with our ever-expanding knowledge of the etiology of different tumors and their complex mechanisms of immunosuppression—have led to a boom in the development of highly sophisticated cellular therapies that are better equipped to generate durable clinical responses across different tumors.

Sculpting the transcriptome of cellular therapies has seemingly limitless potential for enhancing their efficacy in the clinic. Gene editing tools, for example, can be used to remove undesirable proteins that potentially interfere with their performance in vivo, including immune checkpoints such as programmed cell death protein 1 (PD-1), T cell immunoglobulin and mucin domain-containing protein 3 (TIM-3), lymphocyte activation gene-3 (LAG-3), T cell immunoreceptor with Ig and immunoreceptor tyrosine-based inhibitory motif domain (TIGIT) and V-domain Ig suppressor of T cell activation (VISTA) [8]. Conversely, viral vectors can be used to deliver polycistronic gene constructs into the cell's genome that encode proteins associated with key antitumor effector functions, such as homing, cytotoxicity, immune orchestration, persistence, and survival. Conventional chimeric antigen receptor (CAR) T cells are a good example of just how effective this approach can be. These cells have shown remarkable efficacy against hematological malignancies of B cell origin, simply by inserting a single transgene that encodes a synthetic surface-bound cytotoxicity receptor into a polyclonally expanded pool of αβ T cells, subsequently redirecting their reactivity against a single defined tumor-associated antigen. Genetic modifications that further complement the CAR and broaden their clinical application to effectively target other types of cancer and overcome the challenges associated with exhaustion, immunosuppression, persistence, antigen loss, and antigen heterogeneity are currently being tested [9,10].

A particularly important limitation of conventional CAR T cell therapies is that the αβ T cells used to create them are inherently alloreactive, thus restricting their clinical application to an autologous cell therapy. CAR T cell-based therapies, therefore, can only be manufactured on a per-patient basis, which is fraught with complications, including: (1) relatively high production costs; (2) a potentially suboptimal cell therapy due to inherent flaws in a patient's immune cells; (3) inconsistency in efficacy between different patients due to potential variation between the different cell therapy preparations; (4) poor T cell expansion ex vivo due to the potential immunosuppressive effects of previous treatments such as chemotherapy and radiotherapy, and/or the tumor itself; (5) additional stress to the patient having to undergo large volume blood collection by apheresis; (6) delays in treatment while cells are manufactured and validated; and (7) potential loss of treatment if the cell manufacture fails unexpectedly. Using healthy donor material to manufacture CAR T cells would circumvent all of these complications; however, healthy donor allogeneic αβ T cells bearing mismatched human leucocyte antigen (HLA) can cause the potentially life-threatening complications associated with graft versus host disease (GvHD) and display limited longevity in vivo due to host versus graft rejection. Using partially or fully HLA-matched donors may reduce GvHD and graft rejection; indeed, haploidentical and HLA-matched donor-derived CD19 CAR T cells did not induce GvHD in a small cohort of B-cell acute lymphoblastic leukemia (ALL) patients prior to receiving hematopoietic stem cell transplantation (HSCT); however, only the HLA-matched CAR T cells displayed meaningful expansion and efficacy in vivo, suggesting that although partial HLA matching can overcome GvHD, it does not address the issue of host versus graft rejection [11]. Fully HLA-matched donors though are rare and often unavailable, and so a readily-available, "off-the-shelf," universal cell therapy would be highly desirable.

Additional gene editing to protect αβ T cells from allo-rejection and concomitantly ablate their own reactivity against mismatched HLA is currently under investigation and has the potential to convert conventional autologous-restricted CAR T cells into the highly coveted donor-derived cell therapy; for example, gene editing of the T cell receptor constant alpha chain (TRAC) gene locus to knock-out T cell receptor (TCR) α has been used to reduce the alloreactivity of HLA mismatched CAR T cells and thus prevent GvHD [12]. In parallel, there has been considerable interest in the use of nonconventional "allo-friendly" immune cells as an alternative immune cell chassis. γδ T cells, in particular, have recently captured a significant portion of that limelight. Unlike their αβ T cell counterparts, γδ T cells possess a TCR that does not recognize nonself peptides presented in the context of self MHC, and thus pose a lower risk in terms of GvHD when used as an allogeneic cell therapy [13]. Multivariate analysis of the cellular infiltrate of a broad range of solid tumors has revealed that γδ T cells are the best correlate of a positive clinical outcome, strongly suggesting that these immune cells in particular play a fundamental role in tumor surveillance [14]. Indeed, they display innate-like recognition and killing of a broad range of tumors via multiple stress-sensing cytotoxicity receptors, including the γδTCR itself, NKG2D, DNAM-1, NKp30, NKp44, and NKp46 [15]. Moreover, from a cell therapy manufacturing perspective, these cells are: (a) easily expanded; (b) readily transduced; (c) easily detected and specifically activated via their unique TCR; and (d) capable of expressing relatively large polycistronic transgene constructs without undergoing endoplasmic reticulum (ER) stress. Although the clinical responses observed during early-phase trials using adoptive transfer of unmodified ex vivo expanded autologous $V\delta2^+$ γδ T cells were somewhat disappointing [16–26], advances in our understanding of these unique cells and how we can amplify their function through genetic modification have resulted in a resurgent interest in their use as an adoptive cell therapy. Furthermore, good manufacturing process (GMP)-compatible and scalable expansion protocols for the non-Vδ2 subset of γδ T cells, the $V\delta1^+$ subset, which represents a lower proportion of circulating γδ T cells in humans, have recently been developed and early-phase clinical trials using this cell type are underway [27–29].

Although γδ T cells are relatively safe for use as a universal therapy due to their lack of alloreactivity against mismatched HLA, they themselves express high levels of HLA class I and can upregulate—in response to certain antigenic and cytokine stimulation—marked levels of HLA class II [30]. In patients receiving allogeneic γδ T cells from an HLA-unmatched donor, professional antigen presenting cells will process and present the foreign HLA/peptide complexes and stimulate clonal expansion of alloreactive T cells from the polyclonal pool of naïve lymphocytes. Once fully differentiated into effector cells, these alloreactive T cells will be capable of directly killing the γδ T cell graft (see Fig. 1). Alloreactive B cells will also be activated to produce graft-specific antibodies that are potentially capable of opsonizing the γδ T cell therapy, thus rendering it susceptible to antibody-dependent cell cytotoxicity (ADCC) and antibody-dependent cell phagocytosis (ADCP) by

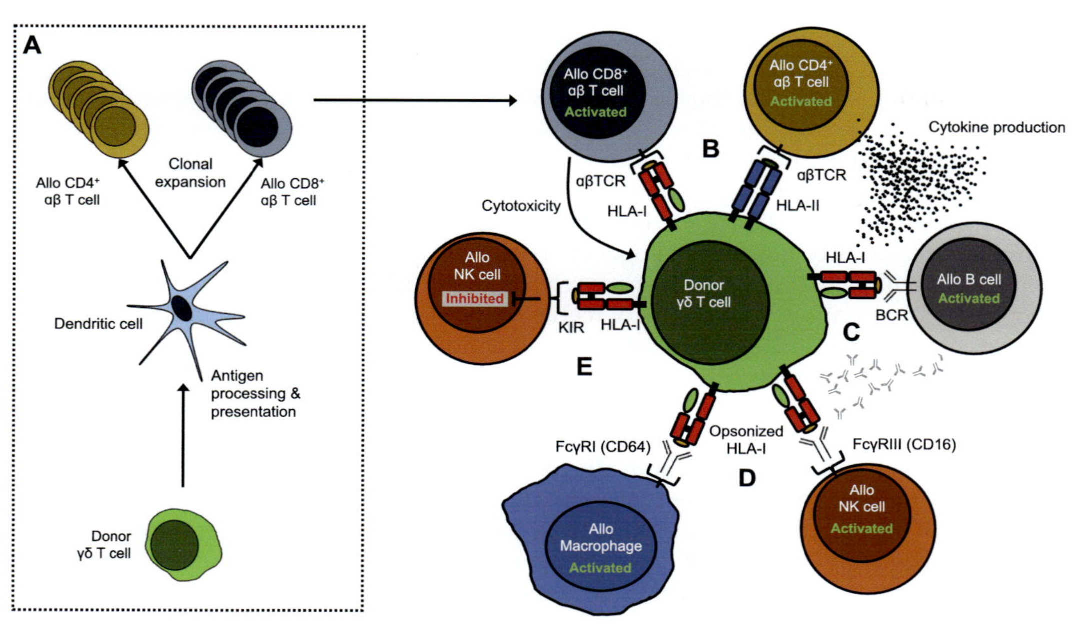

FIG. 1 Schematic depicting the potential mechanisms of rejection of an allogeneic γδ T cell therapy by the immune system of an HLA-mismatched patient. (A) Professional antigen presenting cells, such as dendritic cells, will capture nonself proteins (typically HLA/peptide complexes) from the surface of allogeneic γδ T cells, then process and present these antigens to naïve $CD4^+$ and $CD8^+$ αβ T cells, causing clonal expansion of the reactive clones. (B) Clonally expanded effector $CD4^+$ and $CD8^+$ αβ T cells will recognize the allogeneic γδ T cells and respond with granzyme B/perforin-mediated cytotoxic responses and the production of a myriad of cytokines, including IFN-γ, IL-4, TNF-α and IL-2. (C) With the aid of cytokine support from activated $CD4^+$ αβ T cells, B cells bearing B cell receptors (BCRs) complimentary to nonself proteins on the allogeneic γδ T cell will be stimulated to undergo differentiation and clonal expansion into antibody-secreting plasma cells. These cells will secrete antibody specific for the nonself proteins, subsequently opsonizing the allogeneic γδ T cell. (D) NK cells bearing the low-affinity Fc receptor FcγRIII (CD16) will mediate ADCC against the opsonized allogeneic γδ T cell, ultimately resulting in cell death. Macrophages bearing the high-affinity Fc receptor FcγRI (CD64) will also engage with the opsonized allogeneic γδ T cell, resulting in ADCP. (E) NK cells will also receive inhibitory signals via HLA-specific killer-immunoglobulin-like receptor (KIR) recognition of allogeneic HLA.

fratricide, endogenous γδ T cells, NK cells, neutrophils and/or macrophages [31,32]. Although these alloreactive adaptive immune responses will ultimately result in the rejection of a donor-derived γδ cell therapy, it will take time for the patient to generate them from low numbers of naïve T and B cell clones. This process is relatively slow, thus creating a potential therapeutic window for donor-derived γδ T cells to carry out their antitumor effector response. Whether this is sufficient to generate a durable and long-lasting clinical response, however, remains somewhat unclear. We know from autologous CAR T cell trials that in vivo persistence correlates well with a positive clinical outcome in hematological malignancies [33], highlighting the importance of longevity and the potential implications that graft rejection will have on the efficacy of an allogeneic cell therapy. This problem cannot be overcome by simply repeatedly infusing the patient with more cells; in contrast to autologous cells, repeated infusions of allogeneic cells can potentially produce a vaccination effect whereby the patient's adaptive immune system becomes primed to mount faster and more effective antigraft immune responses, thus enhancing the rate of rejection with each infusion. Some form of intervention may, therefore, be necessary in order to protect allogeneic cell therapies from being rejected by the patient's immune system so as to prolong persistence in vivo and maximize efficacy. Quite surprisingly though, Brudno et al. reported that donor-derived CD19-targeting CAR T cells were efficacious against adult B-cell malignancies without any additional measures to suppress allogeneic rejection [34]. It remains to be seen, however, whether this will be the case for the more resilient and difficult-to-treat solid tumors, where homing and long-term persistence are perceived to be of greater importance in achieving clinical responses. As we will explore herein, interventions designed specifically to block graft rejection are more common than not in clinical trials for allogeneic cell therapies, and this certainly seems to be the direction in which the field is moving.

Managing host versus graft rejection in some form or other will no doubt be an important consideration in the design of all future allogeneic γδ T cell-based therapies in order to ensure prolonged persistence of the cells in vivo and thus maximize their overall efficacy. In this review, we will explore the potential strategies that can be employed to prevent host versus graft rejection and maximize in vivo persistence of an allogeneic γδ T cell therapy. These strategies have been categorized into either: (a) preconditioning, which essentially involves temporary ablation or suppression of the patient's immune system via chemotherapeutics, antibody therapies, and immunosuppressants; or (b) stealth engineering, which involves genetic modifications that enable the cell therapy to evade allogeneic recognition and subsequent rejection.

Preconditioning

To prevent host versus graft rejection of an allogeneic cell therapy, a patient's immune system can be dampened temporarily using immune ablating or suppressing drugs, including various chemotherapeutics, antibodies, antibody-drug conjugates, and immunosuppressants. Their effects, however, are often far-reaching, hitting not just the alloreactive immune cells responsible for host versus graft immunity, but also nonredundant immune effector cells, including the patient's endogenous antitumor $CD8^+$ T cells and NK cells, and more importantly, the allogeneic cell therapy itself. Collateral targeting of the cell therapy can often be circumvented—at least in part—by episodic and sequential dosing. Typically, this involves pretreating the patient with a relatively short dosing regimen that causes an acute but reversible state of lymphopenia predominantly affecting T cells, B cells, and NK cells, which is then followed by a gradual reconstitution of the immune system, often taking many months to fully recover. Rapid clearance of the drug(s) combined with a gradual period of immune recovery provides a therapeutic window in which an allogeneic cell therapy can be infused and allowed to deliver its antitumor response with a much lower risk of being rejected. Given the empirical evidence that cell therapy persistence is a fundamental requirement for optimal clinical responses [33], the rejection-free therapeutic window that is created by this preconditioning strategy is critical, and largely dependent on the unique pharmacokinetics and immune reconstitution profile associated with the chosen drug regimen.

Patient preconditioning is not only used for delaying rejection of an allogeneic cell therapy, it is also used frequently in autologous adoptive cell transfer. In this context, nonmyeloablative regimens of lymphodepleting chemotherapeutic agents, which typically do not require stem cell support for immune recovery, are administered to the patient to temporarily ablate their immune system prior to autologous cell therapy infusion. This has shown demonstrable capacity to enhance greatly the overall efficacy of different autologous cell-based therapies, including tumor-infiltrating lymphocytes (TILs), TCR-engineered T cells, and CAR T cells [35]. Suryadevara et al., for example, showed that in a syngeneic murine model of glioblastoma, preconditioning the mice with Temozolomide—an alkylating chemotherapy that induces lymphopenia—improved the proliferation, persistence and efficacy of EGFRvIII-targeting CAR T cells [36]. Potentially, there are multiple benefits of lymphodepletion prior to cell therapy infusion, including: (a) reduction in overall tumor burden to create a higher ratio of effector to target cells; (b) removal of the patient's endogenous pool of immune cells, which can act as a sink for cytokines and nutrients, thus increasing the availability of important factors such as IL-7, IL-15, IL-21 and glucose, which are known to be essential for optimal proliferation and cytotoxicity in immune effector cells; (c) recontouring the immune landscape of the tumor microenvironment (TME) by depleting immunosuppressive cells such as regulatory T cells (Tregs) and myeloid-derived suppressor cells (MDSCs) to create more favorable conditions for generating antitumor immune responses and revert T cell exclusion within the tumor mass; (d) sensitization of tumors to immune attack by upregulating costimulatory molecules and stress ligands such as MICA and MICB; and (e) induction of apoptotic and necrotic immunogenic cell death for efficient stimulation of antigen processing and presentation by dendritic cells (DCs) and other professional antigen presenting cells [37,38].

In the following subsections entitled *Chemotherapy*, *Antibody therapy*, and *Immunosuppressants*, we will discuss some of the main examples of preconditioning that may have utility in preventing host versus graft rejection of an allogeneic γδ T cell therapy.

Chemotherapy

Chemotherapy is a means of lymphodepletion that is widely used as a preconditioning strategy for both autologous and allogeneic cellular therapies. Cyclophosphamide, for example, like the previously mentioned Temozolomide, is a DNA alkylating agent that, although typically used to treat a range of different malignancies, including lymphomas, myelomas, leukemia, mycosis fungoides, neuroblastoma, ovarian adenocarcinoma, retinoblastoma, and breast carcinoma, is used widely in adoptive cell transfer [39]. It is toxic to both cancer cells and proliferating lymphocytes via three known mechanisms of action: (1) attaching alkyl groups to DNA bases, which causes DNA fragmentation and thus prevents DNA synthesis and RNA transcription; (2) DNA crosslinking, which prevents DNA from unzipping during synthesis or transcription; and (3) DNA mutagenesis by causing nucleotide mispairing [40]. Interestingly, Tregs seem to be more susceptible to the toxic effects of Cyclophosphamide compared with other T cells such as cytotoxic $CD8^+$ T cells due to their low expression of glutathione, an antioxidant that can counteract the DNA-damaging effects of Cyclophosphamide [41]. When administered at the right dose, Cyclophosphamide can preferentially eliminate Tregs and concomitantly support antitumor effector immune cells; indeed, in a mouse model of Lewis lung cancer, continuous administration of low-dose Cyclophosphamide was reported to deplete Tregs, and restore $CD4^+/CD8^+$ T cells, not just in the circulatory and splenic compartments, but also within the tumor itself [42]. Similarly, a low-dose regimen of Cyclophosphamide in a murine model of neuroblastoma has been shown to cause marked depletion of Tregs in the tumor and enhance the efficacy of anti-PD-1 therapy [43]. Rebalancing the T effector to Treg ratio in this way contributes to the overall antitumor efficacy of Cyclophosphamide; in a cohort of patients with metastatic colorectal cancer, for example, depletion of Tregs and upregulation of cytotoxic $CD8^+$ T cells following low dose Cyclophosphamide treatment was reported to correlate well with delayed disease progression [44].

Although Cyclophosphamide can be used as a stand-alone preconditioning agent for autologous cell therapies, it has been reported to be more effective in this role when given in combination with a second chemotherapeutic agent called Fludarabine. Specifically, Turtle et al. found that the addition of Fludarabine to Cyclophosphamide-based lymphodepletion improved in vivo persistence and disease-free survival for autologous CD19-targeting CAR T cells in 32 adults with relapsed and/or refractory B cell non-Hodgkin lymphoma [45,46]. In contrast to Cyclophosphamide, Fludarabine is an antimetabolite—more precisely a purine analog—that inhibits DNA synthesis by interfering with ribonucleotide reductase and DNA polymerase α [47]. It is typically used as a chemotherapeutic against chronic lymphoblastic leukemia (CLL) [48], but has demonstrable capacity to function effectively as a preconditioning agent in and of itself. Wallen et al., for instance, published a clinical trial in which metastatic melanoma patients were treated with two infusions of a single autologous tumor-reactive cytotoxic T lymphocyte (CTL) clone; the first infusion had no prior Fludarabine preconditioning, whereas the second infusion involved a course of Fludarabine lymphodepletion. In this study, Fludarabine was reported to increase plasma levels of IL-7 and IL-15, and subsequently improve in vivo persistence of the CTLs. Interestingly, Fludarabine—which was used as a single agent preconditioning in this study—caused an unexpected increase in Tregs, suggesting that it may preferentially target T effector cells somehow, which further supports a combination approach with cyclophosphamide [49].

Fludarabine in combination with Cyclophosphamide (referred to here as Flu/Cy) is frequently used as a preconditioning regimen in clinical trials of different autologous cell-based therapies. In 2022, for example, in a phase I clinical trial conducted by Yu et al. (NCT02765243), autologous GD2-targeting CAR T cells were administered to pediatric patients with refractory and/or recurrent neuroblastoma following a course of Fludarabine at $25\,mg/m^2$ and Cyclophosphamide at $300\,mg/m^2$ on days −4, −3, and −2 [50]. In the same year, Kekre et al. reported a Canadian trial (NCT03765177) that tested autologous CD19-targeting CAR T cells against adult $CD19^+$ non-Hodgkin's lymphoma (NHL) and ALL in which slightly higher doses of Fludarabine at $40\,mg/m^2$ on days −4, −3, and −2 and Cyclophosphamide at $500\,mg/m^2$ on days −4 and −3 were administered prior to cell infusion [51]. Mackensen et al. reported daily dosing of Fludarabine at $25\,mg/m^2$ from days −5 to day −3 and Cyclophosphamide at $1000\,mg/m^2$ on day −3 before infusion of autologous CD19-targeting CAR T cells in a clinical trial for treating systemic lupus erythematosus (SLE) [52]. Interim results reported by Heczey et al. in 2020 and 2023 from the first-in-human phase I trial of autologous iNKT cells engineered with a GD2-targeting CAR and IL-15 in pediatric patients with relapsed or resistant neuroblastoma (NCT03294954), used a course of

TABLE 1 Summary of the autologous cell therapies cited herein that used Flu/Cy as a preconditioning regimen.

Year	Authors	Cell therapy	Patients	Flu/Cy (mg/m^2)	Persistence
2018	Ramos et al. [55]	CD19 CAR T	Adult NHL (16)	30/500	Detectable at 6 months
2022	Yu et al. [50]	GD2 CAR T	Pediatric NB (32)	25/300	▪ Monitored in 10 patients only ▪ Peaked at day 20 ▪ Detectable after 1 year in 1 patient
2022	Kekre et al. [51]	CD19 CAR T	Adult NHL (25) Adult ALL (5)	40/500	▪ Peaked at day 7 or 14 ▪ Detectable at 3 months in some patients
2022	Mackensen et al. [52]	CD19 CAR T	SLE (5)	25/1000	Peaked at day 9
2023 (interim)	Heczey et al. [53,54]	GD2 CAR iNKT	Pediatric NB (12)	30/500	Detectable at 3 weeks

preconditioning consisting of Fludarabine at 30 mg/m^2 and Cyclophosphamide at 500 mg/m^2 on days −4, −3, and −2 [53,54]. Although in vivo expansion and persistence were reported in this selection of autologous clinical trials (summarized in Table 1), it is difficult to discern to what extent this can attributed to the use of Flu/Cy; indeed, side-by-side comparisons of preconditioning versus no preconditioning were not carried out. In a study by Ramos et al., however, 2 doses of autologous CD19-targeting CAR T cells were administered to a cohort of patients with NHL, the first dose was preceded by daily dosing for 3 days of Fludarabine (30 mg/m^2) and Cyclophosphamide (500 mg/m^2), which finished 24–72 h before administration of the CAR T cells, and resulted in marked in vivo expansion, whereas the second dose was given without preconditioning and resulted in a relatively poor expansion [55]. Interestingly, published clinical trials using autologous γδ T cells tended not to use Flu/Cy or any other form of preconditioning [16–26]. Parenthetically, two clinical trials infused autologous γδ T cells into patients receiving chemotherapy; however, the remit of these studies was to test primarily the efficacy of a chemotherapy/immunotherapy combination approach, in which chemotherapy was not used as preconditioning [56,57].

Although Flu/Cy is considered a standard preconditioning regimen for adoptive cell transfer of autologous cellular immunotherapies, in particular those that are CAR T cell-based, there has been considerable effort to find alternative agents, which has been driven in part by the need for lower toxicities and alternative drugs during Fludarabine shortages [58,59]. Bendamustine, for example, is a nitrogen mustard drug that functions as both an alkylating agent and antimetabolite. It is typically used as an antineoplastic chemotherapy for CLL and NHL, but may also have utility as a preconditioning agent prior to infusion of cellular immunotherapies. In the clinical trial published by Schuster et al. for autologous CD19-targeting CAR T cells in adult relapsed or refractory diffuse large B-cell lymphoma (NCT02445248), 103 patients were preconditioned with either Flu/Cy (73%) or Bendamustine (20%) prior to CAR infusion, and the authors did not report any differences in safety or efficacy between the two different preconditioning strategies [60]. Also, Ghilardi et al.

compared autologous CD19-targeting CAR T cells in two cohorts of patients with relapsed or refractory large B-cell lymphomas: one preconditioned with Flu/Cy (42 patients) and one with Bendamustine (90 patients) [61]. CAR efficacy was comparable between the two preconditioning regimens; however, Flu/Cy preconditioning led to greater lymphocytopenia and more adverse events, including CRS, neurotoxicity, hematological toxicities, infections, and neutropenic fevers. Ramos et al. have also published results from two parallel clinical trials (NCT02690545 and NCT02917083) comparing preconditioning with Flu/Cy (17 patients), Fludarabine and Bendamustine (15 patients) or Bendamustine alone (5 patients) prior to infusion of anti-CD30 CAR-T cells in relapsed/refractory Hodgkin's lymphoma [62]. Efficacy in terms of overall response rates was 65% for Flu/Cy, 80% for Fludarabine and Bendamustine, and strikingly, 0% for Bendamustine alone, which was consistent with the Bendamustine only cohort displaying lower IL-7 and IL-15 availability and reduced CAR T cell persistence. These contrasting results suggest that Bendamustine alone may not be sufficient for lymphodepletion and highlight Fludarabine as a particularly important component of the preconditioning regimen.

In the context of allogeneic cell therapy, Flu/Cy can be used as a preconditioning regimen to support in vivo persistence in the same way that it does for autologous cell therapy, but with the added benefit of prolonging graft survival. To date, there have been a number of clinical trials demonstrating the utility of Flu/Cy in supporting the persistence of allogeneic cell therapies. In a clinical trial published by Otegbeye et al. in 2022, for example, nine patients with mixed malignancies of hematological and colorectal origin were infused with ex vivo expanded NK cells derived from unrelated, HLA-disparate donors [63]. Fludarabine at 25 mg/m^2 per day for 5 days (on day −6 to −2) and Cyclophosphamide at 60 mg/kg as a single dose on day −7 were used to precondition the patients. No GvHD was observed and donor NK cells persisted—albeit at low frequencies—for up to 4 weeks at all dose levels tested, with clinical responses reported in some patients. Similarly, Dok Hyun Yoon et al. in 2023 reported a phase I trial of ex vivo expanded allogeneic NK cells in relapsed/refractory B cell NHL consisting of multiple treatment cycles of Fludarabine at 20 mg/m^2 and Cyclophosphamide at 250 mg/m^2 preconditioning followed by NK cell infusion [64]. In this trial, NK cells persisted for at least 7 days postinfusion in 7/9 patients, and a 55.6% response rate was observed. In a phase I clinical trial reported by Hu et al. in 2022 (NCT04538599), donor-derived allogeneic CD7-targeting CAR T cells were administered to patients with CD7$^+$ hematological malignancies [65]. Additional genetic modifications were incorporated into this cell therapy, including CD7 knock-out to reduce fratricide, TCR knock-out to prevent GvHD, and HLA class II knock-out plus overexpression of an E-cadherin-based NK inhibitor to evade allogeneic rejection. The trial consisted of 12 patients with different hematological tumors: 7 with T cell ALL, four with T cell lymphoma, and 1 with acute myeloid leukemia (AML). Lymphodepletion consisted of Fludarabine at 30 mg/m^2, Cyclophosphamide at 300 mg/m^2, and Etoposide at 100 mg/day, all over 4 days prior to infusion. Note that Etoposide is an inhibitor of Topoisomerase II—an enzyme responsible for the winding and unwinding of DNA during synthesis and replication—that causes apoptosis in cells undergoing division. The reasons for including Etoposide in the preconditioning regimen are not explicitly stated by the authors. No cases of dose-limiting toxicity, GvHD, or immune effector cell-associated neurotoxicity syndrome were reported; however, cytokine release syndrome (CRS) was observed in some patients. In terms of preconditioning-related toxicities, neutropenia was observed in all patients, and a fatal incident of sepsis in one of the patients. Expansion of the CAR T cells—as monitored using flow cytometry—peaked at day 10–14 postinfusion and persisted for a median of 28 days within a broad range of 10–120 days. Profiling of endogenous immune cells showed outgrowth of a subpopulation of $CD8^{+}CD7^{-}$ T cells 14–21 days after CAR T cell infusion, which negatively correlated with the number of CAR T cells, suggesting that CAR T cells were being rejected by this specific immune cell subset; indeed, this was confirmed by isolating $CD8^{+}CD7^{-}$ T cells from one of the treated patients and demonstrating alloreactivity against the donor-derived CAR T cells in vitro. With specific regards to γδ T cell-based therapies, Wilhelm et al. in 2014 treated a cohort of patients with advanced hematological malignancies using a combination of zoledronic acid, IL-2, and CD4-/CD8-depleted PBMCs from HLA half-matched family members [66]. These patients received Fludarabine at $25\,mg/m^2$ on day −6 until day −2 and Cyclophosphamide at 60 mg/kg on days −6 and −5 prior to cell infusion and showed marked expansion of donor γδ T cells that persisted for 1 month. More recently, Vydra et al. documented a small-scale early-phase clinical trial of haploidentical donor-derived γδ T cells in combination with zoledronic acid and IL-2 in 7 adult patients with AML (NCT03790072) [67]. Preconditioning involved a course of Fludarabine at $25\,mg/m^2$ (maximum dose 50 mg) from day −6 until day −2 and Cyclophosphamide at $500\,mg/m^2$ on days −6 and −5. Persistence of the adoptively transferred γδ T cells of at least 100 days was observed in two of the patients. See Table 2 for a summary of these examples of allogeneic trials.

In addition to relying on episodic and sequential administration as a means of safeguarding the cell therapy from the harmful effects of a chemotherapy preconditioning regimen, the allogeneic cells can be given further protection by genetically engineering them with transgenes that confer specific resistance. This permits the concomitant delivery of the preconditioning chemotherapy and cell therapy, which can allow patients to start their treatment sooner, without having to wait for complete drug clearance, as well as potentially receive maintenance doses to delay graft rejection throughout the treatment phase. Although potential transgenes that confer specific resistance against the frequently used Fludarabine and Cyclophosphamide have yet to be tested—such as Glutathione synthetase for enhanced resistance to Cyclophosphamide—the proof-of-concept that γδ T cells can be engineered with some form of drug resistance has already been demonstrated; indeed, Lamb et al. have shown that a transgene encoding the DNA repair enzyme O^6-alkylguanine DNA alkyltransferase (AGT) can provide demonstrable protection against the DNA alkylating effects of Temozolomide [71]. Early-phase clinical trials testing the feasibility of these drug-resistant γδ T cells (both autologous and allogeneic) during maintenance dosing of Temozolomide in glioblastoma are ongoing (NCT04165941 and NCT05664243).

Antibody therapy

Flu/Cy preconditioning has also been combined with immune targeting antibodies against defined immune-associated antigens for additional lymphodepleting specificity and potency. Antibodies can mediate the elimination of target cells via multiple mechanisms, including direct induction of apoptosis, complement-dependent cytotoxicity, cell-mediated cytotoxicity via NK cells, neutrophils, and γδ T cells, as well as phagocytosis via macrophages [72].

TABLE 2 Summary of the allogeneic cell therapies cited herein that used Flu/Cy ± Alemtuzumab as a preconditioning regimen.

Year	Authors	Cell therapy	Patients	Preconditioning			Persistence
				Flu	Cy	Other	
2014	Wilhelm et al. [66]	γδ T	Advanced hematological malignancies (4)	$25 mg/m^2$	60 mg/kg	–	Expansion observed that persisted for 1 month
2022	Benjamin et al. [68]	CD19 CAR T (TCR & CD52 KO)	Adult ALL (25)	$30 mg/m^2$	$500 mg/m^2$	–	▪ Peak at day 14 ▪ Median persistence of 28 days
2022	Hu et al. [65]	CD7 CAR T	Adult ALL (7) Lymphoma (4) AML (1)	$30 mg/m^2$	$300 mg/m^2$	Etoposide 100 mg/day for 4 days	▪ Peak expansion 10–14 ▪ Median persistence 28 days
2022	Otegbeye et al. [63]	NK	Mixed haem and colorectal (9)	$25 mg/m^2$	60 mg/kg	–	Low levels detected at 4 weeks
2022	Ottaviano et al. [69]	CD19 CAR T (TCR & CD52 KO)	Pediatric ALL (6)	$150 mg/m^2$	120 mg/kg	Alemtuzumab at 1 mg/kg	▪ Peak at 7–14 ▪ Undetectable at day 28
2023	Yoon et al. [64]	NK	NHL (9)	$20 mg/m^2$	$250 mg/m^2$	–	Cells persisted for at least 7 days
2023	Mailankody et al. [70]	BCMA CAR T (TCR & CD52 KO)	Multiple myeloma (43)	$90 mg/m^2$	$900 mg/m^2$	ALLO-647 (anti-CD52) at 39, 60, or $90 mg/m^2$	▪ Peak at day 10 ▪ Variable persistence ▪ Gone by day 28
2023	Vydra et al. [67]	γδ T (haploidentical)	AML (7)	$25 mg/m^2$	$500 mg/m^2$	–	At least 100 days in 2/7 patients

Flu/Cy is frequently combined with an antibody called Alemtuzumab, which is a recombinant DNA-derived IgG1 kappa monoclonal antibody that binds to the glycoprotein CD52. Alemtuzumab has been manufactured under the brand names Campath and Lemtrada, and has a diverse range of clinical applications, from targeting malignancies such as skin lymphoma, B cell CLL, and T cell prolymphocytic leukemia, to dampening autoimmune responses in diseases such as multiple sclerosis, to suppressing allogeneic tissue rejection during stem cell and organ transplantation. The precise role of its target protein, CD52, is poorly understood; however, it has been implicated in various mechanisms of general immune function, including costimulation, migration, and adhesion [73], and is highly expressed on most immune cells, including mature T and B cells, NK cells, monocytes, macrophages, DCs, and granulocytes, yet absent on erythrocytes, platelets, and hematopoietic progenitor cells. Alemtuzumab treatment thus causes rapid and long-lasting depletion of T cells, B cells, NK cells, DCs, granulocytes, and monocytes; for example, immune profiling of patients receiving Alemtuzumab treatment during renal transplantation shows that lymphocytes are rapidly depleted from the peripheral blood within 1 h and from the secondary lymphoid tissues over 3–5 days, with T cell recovery taking up to 36 months to return to 50% of baseline [74]. Innate cells were notably faster to recover, with NK cells constituting over half of peripheral lymphocytes after 1 month, recovering to 60%–80% of baseline after 6 months. Monocytes and DCs were also depleted, with recovery back to baseline levels taking 3 and 6 months, respectively.

Alemtuzumab has been added to the Flu/Cy preconditioning regimens of a number of allogeneic cell therapies. Concomitantly, CRISPR-Cas9 or TALEN gene editing tools have been used to knock-out CD52 protein expression from the cell therapy, thus providing extra protection against any residual Alemtuzumab left after the preconditioning phase. Mailankody et al., for example, in 2023 published interim results from an ongoing phase I trial of allogeneic donor-derived B-cell maturation antigen (BCMA)-targeting second-generation CAR T cells with a Rituximab-based off-switch (designated ALLO-715) in 43 patients with relapsed/refractory multiple myeloma (NCT04093596) [70]. GvHD was minimized by knocking out the TCR and protection against Alemtuzumab-based preconditioning by knocking out CD52. Patients were preconditioned using a CD52-targeting antibody designated ALLO-647 at either 39, 60, or 90 mg/m^2, Fludarabine at 90 mg/m^2, and Cyclophosphamide at 900 mg/m^2 administered over the course of 3 days prior to infusion. A dose escalation of 40, 160, 320, and 480 million CAR T cells was tested; however, most patients (24/43) received the 320 million dose and within this cohort, a complete dose escalation of ALLO-647 combined with Flu/Cy was carried out. Assessment of ALLO-647 pharmacokinetics showed that serum concentrations were dose-dependent and clearance started within hours of administering ALLO-647, with a steady decline thereafter. For the 39 and 60 mg doses of ALLO-647, serum concentrations were almost at baseline after 28 days, whereas for the 90 mg dose, they were still detectable. At 3 months, however, ALLO-647 was undetectable for all doses tested. Immune profiling showed an ALLO-647 dose-dependent reduction in T cell, B cell, and NK cell counts, with NK cells taking 1–3 months to recover back to predepletion levels, T cells 3 months, and B cells 6 months. Responders displayed lower T cells postlymphodepletion, suggesting that lymphodepletion of T cells was important for efficacy. Importantly, out of the 24 patients that received the 320 million dose of CAR T cells, expansion was observed in 20 patients, which peaked at day 10. Variable CAR T cell persistence was reported with

16/24 patients having no detectable CAR T levels by day 28. CRS and neurotoxicity were observed, and of the 24 patients that received the 320 million dose with Flu/Cy/ALLO-647 preconditioning, 70.8% displayed a clinical response, with responders having higher levels of CAR T cells than nonresponders. In terms of ALLO-647 safety and tolerability, neutropenia (69.8%), anemia (55.8%), and thrombocytopenia (51.2%) were observed. 53.5% of the patients displayed infections—mostly cytomegalovirus reactivation—of which 23.3% were equal to or greater than grade 3. In total, 3 deaths were related to grade 5 infections, specifically fungal pneumonia, adenoviral hepatitis, and sepsis. In a separate phase I trial published by Ottaviano et al. (NCT04557436), six children with B cell ALL were treated with TCR and CD52 knock-out donor-derived allogeneic CD19-targeting CAR T cells after a 7-day preconditioning regimen of Fludarabine (150 mg/m^2), Cyclophosphamide (120 mg/kg) and Alemtuzumab (1 mg/kg) [69]. Cell expansion and clinical responses were observed in 4/6 patients; specifically, CAR T cell numbers peaked at day 7–14 postinfusion and were undetectable by day 28. In terms of CAR T cell-dependent adverse events, two patients developed grade II CRS, 1 had transient grade IV neurotoxicity, and 1 presented with skin GvHD. In addition, Benjamin et al. published a similar phase I multicentre clinical trial of TRAC and CD52 knock-out donor-derived allogeneic CD19-targeting CAR T cells in 25 adult patients with B cell ALL (NCT02746952) [68]. Preconditioning prior to cell infusion consisted of Fludarabine at 30 mg/m^2 and Cyclophosphamide at 500 mg/m^2 for 3 days, with 22/25 patients receiving additional Alemtuzumab preconditioning over 5 days at various doses. A 48% response rate was reported, with 14 of the patients having discernible CAR T cell expansion and a peak and median persistence in the blood of 14 and 28 days, respectively. In terms of adverse events, various incidences of CRS, neurotoxicity, prolonged cytopenia, GvHD, and infections were recorded. Of particular note, all of the patients that displayed CAR T cell expansion received Alemtuzumab, whereas the three patients that did not receive Alemtuzumab showed no CAR T cell expansion, tentatively supporting the notion that the addition of Alemtuzumab to Flu/Cy may provide better persistence for an allogeneic cell therapy. See Table 2 for a summary of these examples of allogeneic trials.

Although the abovementioned clinical trials using either Flu/Cy alone or in combination with Alemtuzumab suggest that these are feasible preconditioning strategies for supporting allogeneic cell therapy persistence in vivo, it is important to highlight that—as with the autologous trials—these studies have not directly compared persistence in patients that received preconditioning with those that did not, thus making it difficult to discern precisely to what extent antigraft immunity affects persistence and moreover to what extent preconditioning can suppress it. Interpretations are confounded further by the enhanced persistence associated with the depletion of suppressor cells and cytokine sinks, which would benefit allogeneic cell therapies in the same way as autologous. There is clearly an unmet need for more empirical evidence on the direct impact of different preconditioning regimens on the persistence of allogeneic cells—specifically γδ T cells—that requires more thorough longitudinal cell therapy tracking postpreconditioning, and deeper interrogation of different preconditioning drug combinations and doses, not only in hematological tumors, but also solid malignancies. The importance of identifying an optimal dosing regimen is likely to be critical; too severe and immune reconstitution may take too long, leaving the patient vulnerable and at risk from infections; too little and the cell therapy will be rejected before it has the chance to exert a meaningful antitumor effect. Moreover, there are other collateral effects of preconditioning that are not yet well understood; for example, although the barren immune landscape left behind after preconditioning may be rich in cytokines and less immunosuppressive, the adoptively transferred cells will lack any kind of support from endogenous bystander immune cells. Although this seems to have little if any effect on the overall persistence and efficacy of autologous CAR T cells in the treatment of hematological tumors, which have been consistently reported to benefit markedly from lymphodepletion [75], it may have a more significant impact in the treatment of other types of cancer and/or in the performance of other types of cell therapies; for instance, in targeting the highly heterogenous and notoriously difficult-to-treat solid tumors where orchestrated immunity may be paramount and/or in allogeneic cell therapies designed to have inherent and/or synthetic immune orchestrating capabilities. In this case, the fine balance between rejection and immune reconstitution becomes critical. A further consideration is that the recovering immune system postpreconditioning may be somewhat skewed toward an unfavorable antiinflammatory profile, which could potentially render the patient more vulnerable to outgrowth and dissemination of any residual tumor. In certain disease settings, for example, immune reconstitution post-Alemtuzumab treatment seems to be a skewed toward an antiinflammatory immune profile; indeed, in Alemtuzumab-treated multiple sclerosis patients, elevated IL-10- and TGF-β-producing $CD4^+$ and $CD8^+$ T cells has been reported, along with increased IL-4-producing Th2 cells and reduced IL-17- and IFN-γ-producing cells [76]. Further to this, there are underlying safety concerns associated with lymphodepletion in general, including the increased risk of infections that are associated with neutropenia and systemic immunosuppression, as well as other complications such as anemia and thrombocytopenia. Fludarabine treatment has also been associated with fevers and neurotoxicity, and Cyclophosphamide has been reported to cause hemorrhagic cystitis, pericarditis, and neurotoxicity.

There are various other antibody-based drugs that could prove to be useful in preventing rejection of allogeneic cell therapies that, although not typically used to precondition patients for adoptive cell therapy, are used routinely in delaying graft rejection and GvHD during transplantation, and thus worthy of a mention here. Antithyroglobulins (ATGs), for example, are polyclonal cocktails of either horse (hATG) or rabbit (rATG) derived antibodies that are routinely used as an immunosuppressant during allogeneic solid organ, stem cell, or bone marrow transplantation [77,78]. ATGs are generated by purifying the peripheral blood IgG component from horses or rabbits that have been immunized with human thymocytes. This pool of IgG contains clones specific for a myriad of human cell surface proteins, including CD2, CD3, CD4, CD5, CD7, CD8, CD11a, CD19, CD20, CD25, CD28, HLA II and the αβTCR, and thus binds to and kills multiple immune cell subsets, including T cells, B cells, plasma cells, NK cells, monocytes, DCs, and macrophages. Their mechanism of action involves complement-mediated apoptosis [79], and for rATG specifically, Fc-mediated ADCC/ADCP due to the intrinsic cross-reactivity of rabbit IgGs with human Fc receptors; note that horse IgGs do not bind to human Fc receptors [80]. Clearance is seemingly quick; for example, administration of rATG at a dose of 6 mg/kg in patients due to receive partially HLA-matched blood hematopoietic progenitor cell transplants resulted in a median clearance of 6 days, reaching subtherapeutic levels by a median of 15 days [81]. All lymphocyte subsets were undetectable by day seven, with T and B cell depletion lasting for up to 12 months. As with Alemtuzumab, innate cells were the fastest to recover, followed by B cells and $CD8^+$ T cells, and a very slow recovery of CD4$^+$ T cells and iNKT cells. Differences between rATG and hATG have also been reported; most notably, rATG seems to remain in the blood longer and results in more durable reduction in neutrophils, CD4$^+$ T cells and Tregs compared with hATG [82]. Interestingly, nonhuman primate models have suggested that depletion of T and B cells by ATG is less complete compared with the monoclonal antibody Alemtuzumab [83].

Another example is T-Guard, which is essentially a mixture of two antibody-drug conjugates: 1 targeting the T cell coreceptor CD3 and the other targeting CD7, which is expressed on thymocytes, NK cells, and mature T cells. The antibodies are conjugated to ricin A toxin, which becomes internalized by the target cell and induces apoptotic cell death by inhibiting protein synthesis. In addition to delivering the cytotoxic drug, the CD3 antibodies themselves physically block and interfere with TCR ligand binding. Unlike ATG, this reagent is relatively new and currently undergoing development as a therapeutic for steroid-refractory GvHD [84]. In terms of its pharmacokinetics, when used in this particular disease setting, a 1-week course of treatment consisting of four intravenous infusions of 4 mg/m^2 administered at 48-h intervals had a mean serum half-life and maximum concentration of 8.59±3.04 h and 1231 ± 671 μg/mL, respectively. Rapid depletion of NK cells and T cells were observed posttreatment, which started to recover in the second week after treatment and were back to normal after 1 and 3 months, respectively. The immune profile pre and posttreatment showed comparable memory distribution, CD4/CD8 ratio, TCR diversity, antiviral T cell response, and proportion of Tregs. B cells are seemingly unaffected by this treatment; however, it is unclear whether this will result in high levels of graft-specific antibody responses, especially in the absence of humoral immunity support from CD4$^+$ T cells. To compensate, this treatment could potentially be combined with B cell-specific antibodies such as Rituximab to provide additional depletion of alloreactive B cells.

Immunosuppressants

A range of different immunosuppressants are extensively used in the treatment of immune-related disorders, including autoimmune diseases and GvHD, which could potentially be used to protect γδ T cell adoptive cell therapies from allogeneic rejection. Some of these are briefly discussed in the following section, specifically Calcineurin inhibitors (CNis), mTOR inhibitors (mTORis), and Mycophenolate mofetil, which hit specific intracellular molecular pathways associated with immune function.

CNis such as Cyclosporine and Tacrolimus are frequently used as therapeutics against acute GvHD during organ and bone marrow transplantation, as well as antiinflammatories in autoimmune disorders such as ulcerative colitis, rheumatoid arthritis, and atopic dermatitis [85]. Calcineurin is a calcium- and calmodulin-dependent serine/threonine phosphatase that dephosphorylates and activates the nuclear factor of activated T cell cytoplasmic (NFATc) transcription factor. NFATc activation in T cells results in upregulated expression of IL-2, which subsequently initiates proliferation, survival, and proinflammatory cytokine production. CNis form complexes with cytosolic proteins called immunophilins—Cyclosporine binds to cyclophilin, whereas Tacrolimus binds to FKBP-12—that bind to calcineurin and subsequently block its enzymatic function. As a result, Calcineurin can no longer dephosphorylate NFATc, thus blocking transcription of the IL-2 gene and subsequently causing widespread suppressive effects on the patient's immune response, including: (a) inhibition of $CD4^+$ and $CD8^+$ αβ T cell functions including proliferation, cytokine production and/or cytotoxicity [86,87]; (b) reduced numbers of circulating Tregs [88]; (c) suppression of Th17 function [89]; and (d) indirect suppression of B cell antibody responses by suppressing $CD4^+$ αβ T cell help [90]. Indeed, CNis have been shown to inhibit antitumor $CD8^+$ T cell responses in mouse models [91] and reduce IPP-induced cytokine production, degranulation, and cytotoxicity in peripheral blood Vδ2 cells [92]. NFATc-independent mechanisms of action for CNis have also been suggested [93]; for example, disruption of glucose metabolism by regulating the production of glucose transporters, glycolytic enzymes, and transcriptional regulators of glycolysis [94]. Multiple side effects of CNis have been described, including nephrotoxicity, hypertension, and induction of diabetes mellitus and dyslipidaemia [95]. In addition, transplant patients receiving CNis often display reduced Tregs [96], resulting in a Treg imbalance that can potentially lead to immune-related diseases and tissue damage [97].

A similar class of drugs to the CNis are the so-called mTORis. mTOR is a broadly expressed serine/threonine protein kinase downstream of the PI3K and protein kinase B pathway that regulates key cellular processes such as metabolism, proliferation, and survival [98]. Dysregulation of this enzyme is often associated with malignant transformation, and thus mTORis are frequently used to treat malignancies, particularly breast and kidney cancer [99,100]. mTORis are similar to CNis in that they display marked immunosuppressive properties; however, instead of dampening cytokine production, they preferentially inhibit T cell proliferation, and thus can be used to prevent rejection during solid organ transplantation and as a GvHD prophylaxis in HSCT [101]. Sirolimus, for example, formerly known as Rapamycin, is a fermentation product derived from *Streptomyces hygroscopicus* that shares the same target as the CNi Tacrolimus, FKBP-12. Complexes of rapamycin-FKBP-12, in contrast to Tacrolimus-FKBP-12 complexes which target Calcineurin, target mTOR and destabilize the mTOR-Raptor complex, resulting in a G1 cell cycle arrest [102].

Mycophenolate mofetil is also used to suppress graft rejection and GvHD during solid organ transplantation as well as treat autoimmune disorders such as autoimmune hepatitis, lupus-associated nephritis, and dermatitis [103]. It is a pro-drug that is converted into the active metabolite mycophenolic acid (MPA), which inhibits a specific enzyme of the de novo purine synthesis pathway called inosine 5′-monophosphate dehydrogenase enzyme (IMPDH) [104]. Blocking IMPDH prevents cells from synthesizing new purine nucleotides, and thus disrupts their ability to make RNA, DNA, and protein. Whereas most cells acquire purines via both salvaging and de novo synthesis, lymphocytes tend not to salvage their purines and thus are heavily dependent on de novo synthesis for carrying out their effector functions. Accordingly, Mycophenolate mofetil causes preferential suppression of T and B cell immunity [105].

Despite the lack of empirical evidence for their use in prolonging the persistence of allogeneic cell therapies, the different immunosuppressants mentioned here illustrate that there are potentially other unexplored strategies for evading graft rejection, which could prove to be useful persistence enhancers for future allogeneic γδ T cell therapies. However, what remains to be seen is whether their immunosuppressive effects are long-lasting enough for them to be delivered as a bolus prior to cell therapy infusion or whether they need to be combined with resistance engineering of the cell therapy in order to permit concomitant delivery. This concept has certainly been demonstrated for CNis by Brewin et al. in the context of Epstein Barr virus (EBV)-driven posttransplantation lymphoproliferative disease (PTLD), a complication that is associated with stem-cell or solid organ transplantation that can be effectively treated with adoptive cell therapy of EBV-specific CTLs [106]. The authors generated EBV-specific CTLs that were transduced to stably express two calcineurin mutants CNa12 and CNa22, which conferred resistance to Cyclosporin A and Tacrolimus, respectively. The phenotype and cytotoxicity of EBV-CTLs were unaffected by the expression of the mutants, and importantly, they retained cytokine production and proliferation against EBV targets in the presence of therapeutic levels of the drugs. Similarly, Amini et al. demonstrated that CRISPR-Cas9-based knock-out of FKBP-12 could be used to create universal adoptive antiviral T-cell therapies to treat Cytomegalovirus (CMV) and other viral infections in transplant patients with the added option of using alternative CNis as a safety switch [107].

See Fig. 2 for a summary of all the preconditioning agents discussed herein and also the genetic engineering approaches that can be used to confer resistance to the cell therapy.

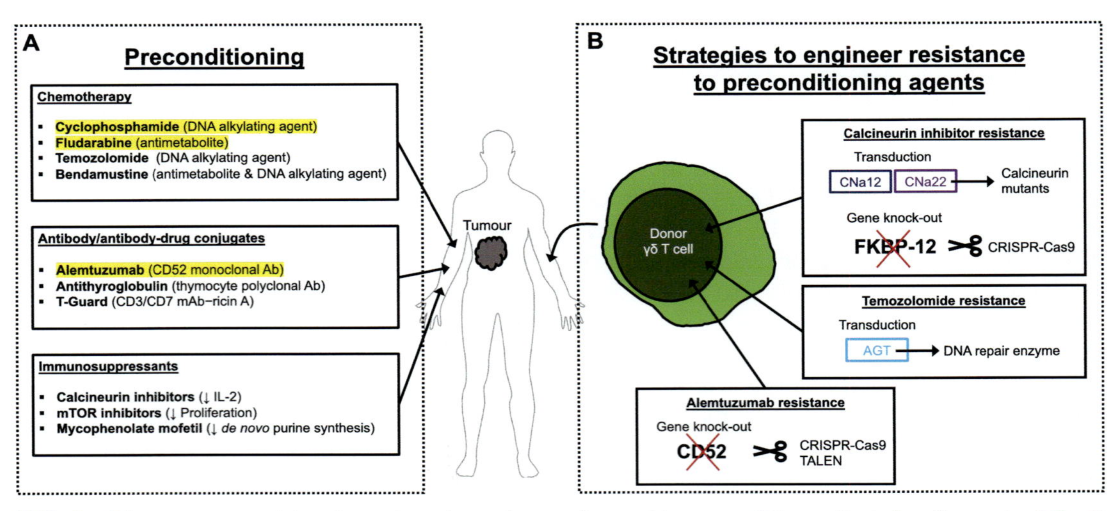

FIG. 2 Diagram summarizing the various drugs that can be used to precondition patients for allogeneic γδ T cell therapies and the complimentary genetic modifications that can be made to the cell therapy itself in order to confer resistance to the preconditioning. (A) The different agents that can potentially be used to suppress the patient's immune system in order to delay graft rejection of a donor-derived γδ T cell therapy. The three drugs that are typically used as preconditioning agents for autologous and allogeneic cell therapies are highlighted in *yellow*. (B) The allogeneic γδ T cell therapy itself can potentially be protected from certain preconditioning agents via genetic engineering. Resistance to calcineurin inhibitors or Temozolomide, for example, can be achieved through stable transduction of genes encoding either mutant calcineurins or the DNA repair enzyme AGT, respectively. Also, CRISPR-Cas9 or TALEN knock-out of CD52 can potentially confer resistance to Alemtuzumab.

Stealth engineering

An alternative approach to patient preconditioning is to genetically modify the cell therapy itself in such a way that renders it less susceptible to rejection by allogeneic immune cells. Whereas in the previous section, we touched upon how cell therapies can be genetically modified to protect them against specific preconditioning agents, such as knocking out CD52 to protect against Alemtuzumab, in this next section, we will discuss the potential genetic modifications—collectively referred to here as stealth engineering—that can be used to prevent recognition and rejection by allogeneic T cells, B cells, and NK cells. Stealth engineering approaches predominantly include knocking out the proteins responsible for generating antigraft immunity, thus rendering the cell therapy inconspicuous to a foreign immune system and enabling it to carry out its program of antitumor effector responses covertly. In addition, transgenes can be inserted into a cell's genome, which encode proteins capable of counteracting and protecting them against cell-mediated cytotoxicity. These could potentially be used to enhance the persistence of an off-the-shelf γδ T cell therapy either as a stand-alone intervention or in combination with patient preconditioning.

Studies on stealth engineering in γδ T cells are yet to be published. Nonetheless, the proof-of-concept that this approach can protect an allogeneic cell therapy against rejection without interfering with their antitumor function has already undergone preclinical evaluation for other cell types and clinical trials are underway. The current frontrunner in the field of stealth engineering is the use of CRISPR-Cas9 or TALEN to knock-out key components in the HLA class I and II gene locus. HLA class I, which is constitutively expressed in T cells, is essentially a trimeric molecule composed of a peptide, a class I heavy chain, and a B2M light chain. Ablation of the B2M light chain results in incomplete formation of this complex and thus markedly reduced expression at the cell surface [108]. HLA class II, on the other hand, is not constitutively expressed by T cells, but can be upregulated—particularly by γδ T cells—upon activation by certain physiological conditions [30]. To knock-out HLA class II, researchers have primarily targeted the gene locus for class II major histocompatibility complex transactivator (CTIIA), a transcriptional activator that functions as a master regulator of MHC class II expression on cells. Cells with a CTIIA knock-out fail to upregulate MHC class II on their surface; for example, cardiomyocytes differentiated from human induced pluripotent stem cells (hiPSCs) that had their CTIIA gene knocked out by CRISPR-Cas9 lacked discernible HLA class II compared with unmodified controls [109]. Using an alternative approach, Hu et al. have also shown that disrupting the regulatory factor X5 (RFX5) gene can generate HLA class II knock-out CAR T cells [65].

Alloreactivity of HLA single or double knock-out cells has been tested preclinically in various allogeneic settings and data suggest it can result in a marked reduction in rejection without affecting normal cell function. Kumar et al., for example, showed that HLA class I knock-out in cord blood $CD34^+$ hematopoietic cell-derived megakaryocytes using CRISPR-Cas9 to remove the B2M gene resulted in reduced susceptibility to allogeneic $CD8^+$ T cell cytotoxicity in vitro [110]. B2M and/or CTIIA knock-out has also been achieved successfully in universal CAR T cells; Ren et al., for example, generated either PSCA- or CD19-targeting CAR T cells that had undergone a triple gene knock-out of their αβTCR, HLA class I and PD-1 by using CRISPR-Cas9 to disrupt the TRAC, B2M, and PD-1 genes, respectively [111]. The TCR^{null} and HLA class I^{null} CAR T cells displayed comparable efficacy to the unmodified CAR T cell controls, but importantly, showed markedly reduced graft rejection and GvHD when coinjected into irradiated mice along with HLA mismatched allogeneic T cells. Parenthetically, knocking out PD-1—an immune checkpoint upregulated upon activation and strongly implicated in tumor suppression of cytotoxic $CD8^+$ T cells—further enhanced CAR T cell efficacy. Kagoya et al. in a similar study used CRISPR-Cas9 to target the TRAC, B2M, and CIITA genes to generate αβTCR, HLA class I, and HLA class II triple knock-out CD19-targeting CAR T cells [112]. Compared with relevant controls, the triple knock-out cells were less sensitive to rejection by allogeneic PBMCs both in vitro and in immunocompromised mouse models; however, the cells were reported to display higher levels of sensitivity to NK cell cytotoxicity.

As highlighted in the aforementioned study by Kagoya et al., although ablation of HLA class I can de-sensitize CAR T cells to allogeneic T cells, this comes at the cost of enhancing inadvertently their susceptibility to NK attack via the missing self mode of activation. Specifically, NK cells express a myriad of both activatory and inhibitory receptors, which they use to probe and interrogate the surface of any potential target cells they encounter. If an activation threshold is reached by the net balance of signals received through these receptors, NK cell cytotoxicity is triggered and the target cell is eliminated. Of the many inhibitory receptors on the surface of NK cells, HLA class I-specific killer-immunoglobulin-like receptors (KIRs) suppress NK cell cytotoxic activity upon engaging cognate HLA molecules, thus contributing to tolerance of self. Loss of classical HLA class I as a result of B2M knock-out gene editing or malignant transformation, will remove this inhibitory signal and result in a skewing of activatory signals that may trigger NK cell attack. Engineering B2M knock-out cells to express nonclassical HLA molecules such as HLA-E or HLA-G, which seemingly have a lower propensity to evoke allogeneic rejection, can deliver replacement inhibitory signals to NK cells through the receptors CD94/NKG2A and KIR2DL4, respectively [113].

Classical HLA knock-out combined with concomitant nonclassical HLA knock-in as a proof-of-concept for simultaneously protecting against allogeneic T cell and NK cell rejection, respectively, has been demonstrated by Guo et al. in 2021 [114]. In their study, the authors expressed mutant B2M-HLA-E and B2M-HLA-G fusion proteins in universal CD19-targeting CAR T cells and reported reduced allogeneic T cell and NK cell recognition in vitro. Li et al. corroborated these findings with similar in vitro experiments, but also went on to demonstrate that HLA-E knock-in CD19-targeting universal CAR T cells were able to exert their antitumor effect in NSG mice challenged with NALM-6 cells without being rejected by a coinjection of donor-mismatched NK cells [115]. Furthermore, Jo et al. used TALEN to disrupt the TRAC and B2M genes in CD22- or CD123-targeting CAR T cells and combined this with an HLA-E knock-in [116]. In this study, the gene-edited cells were able to evade NK cell and allogeneic T cell attack, thus resulting in better persistence and efficacy in vitro and in vivo. Similarly, Hoerster et al. ablated HLA class I in primary NK cells using CRISPR-Cas9 targeting of the B2M gene while coexpressing HLA-E [117]. Cells devoid of B2M were phenotypically and functionally comparable to unmodified counterparts; however, they were resistant to fratricide, did not stimulate allogeneic T cells in vitro and did not reactivate primed and expanded allogeneic T cells. Consistent with this, B2M depletion and concomitant insertion of HLA-G in umbilical cord mesenchymal stem cells caused no significant difference in phenotype and function compared with unmodified cells, but gained marked protection against allogeneic T cells and NK cells [118]. Engineered resistance to NK cells is not limited to nonclassical HLA knock-in; there is also the potential to use other mechanisms of NK cell inhibition; for example, the previously mentioned study by Hu et al. used overexpression of E-cadherin—a regulator of NK cell activity that signals through KLRG1—in CAR T cells via a gene construct encoding EC1-EC2 extracellular and transmembrane domains of E-cadherin fused to a CD28 intracellular domain [65].

Early-phase clinical trials are underway to test the concept of HLA class I knock-out universal CAR T cells. CRISPR Therapeutics, for example, are running multiple trials in this field, including: (a) the CARBON phase I trial NCT04035434 in relapsed or refractory B-cell malignancies to test CTX110, a CD19-targeting CAR T cell with TRAC and B2M knock-out using CRISPR-Cas9; (b) the phase 1 trial NCT04244656 to test CTX120, a BCMA-targeting allogeneic CAR T cell with CRISPR-Cas9 B2M and TRAC knock-out in relapsed or refractory multiple myeloma; (c) the COBALT-LYM phase 1 trial NCT04502446 in relapsed or refractory T or B cell malignancies for CTX130, a CD70-targeting CAR T cell with CRISPR-Cas9 knock-out of B2M, TRAC and CD70 (the ligand for CD27); and (d) the COBALT-RCC phase I trial NCT04438083 to test CTX130 in the context of a solid malignancy of RCC.

Despite these promising studies on the utility of HLA knock-out/knock-in as a stealth engineering strategy, we are yet to see any published data demonstrating the feasibility of this approach in γδ T cells. Furthermore, the use of immunocompromised mouse models does not address the potential impact of infusing large numbers of cells that highly express NK cell inhibitory molecules; in particular, whether endogenous NK cell effector responses will be switched off and the resulting impact this may have on overall antitumor immunity as well as other NK cell-related immunity. Also, we do not know whether γδ T cells engineered with NK cell inhibitory molecules will be self-suppressing; γδ T cells have certainly been shown to express various NK cell-associated inhibitory receptors on their surface, particularly NKG2A [119]. Stealth engineering will also add further tiers of complexity to an already complicated manufacturing process, which will create additional technical and financial challenges.

In summary, HLA knock-out in combination with some form of NK cell inhibitor is just some of the many potential stealth engineering strategies for enhancing the persistence of an allogeneic cell therapy (see Fig. 3). Although research in this field is relatively new, there is huge scope for the development of alternative approaches; for example, potential ideas that have yet to be explored include engineering with antiapoptotic signaling molecules, cytotoxicity blockers, and checkpoint inhibitors that counteract the cytotoxic effector responses of alloreactive T and NK cells.

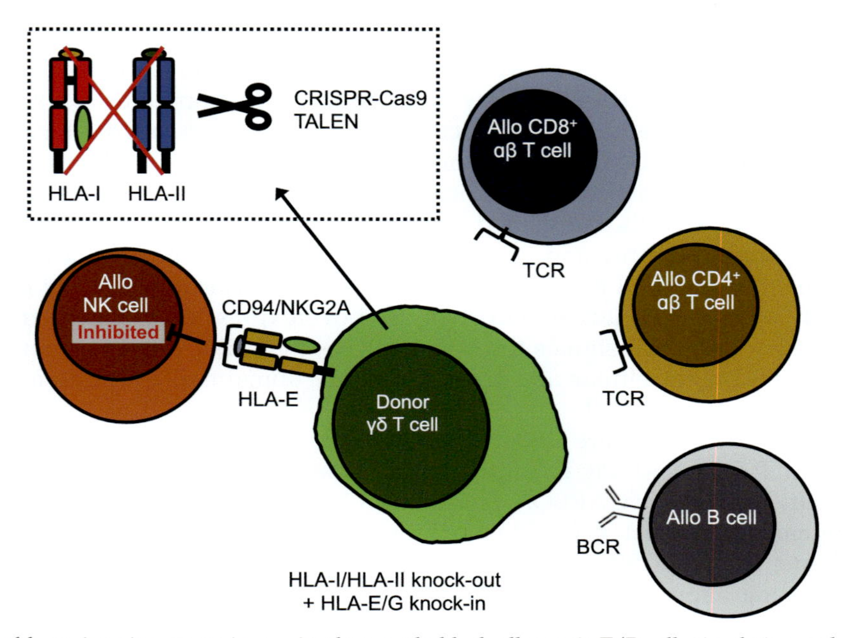

FIG. 3 Stealth engineering strategies to simultaneously block allogeneic T/B cell stimulation and missing self-recognition by NK cells. Gene editing tools such as CRISPR-Cas9 and TALEN can be used to knock-out the B2M and CTIIA genes to prevent an allogeneic γδ T cell therapy from expressing HLA class I and II, respectively, and thus prevent them from inducing allogeneic immune responses. Concomitant knock-in of nonclassical HLA molecules that do not cause graft rejection, but do provide inhibitory signals to NK cells, such as HLA-E, which signals through the NK inhibitory receptor NKG2A (CD94), will counteract any potential increase in NK cell sensitivity via the missing self mechanism of activation.

Concluding remarks

In this review, we have highlighted the exciting potential of using γδ T cells as universal cell therapies for the treatment of cancer and emphasized the importance of achieving good persistence in vivo, particularly in the context of targeting solid tumors. Furthermore, we have discussed how cell therapies created from donor-derived γδ T cells are likely to be rejected based on their high expression of mismatched HLA class I and II if suitable steps to prevent graft rejection are not taken. Hence, we have examined some of the main approaches that can be used to prevent rejection by allogeneic immune cells, and provided discussion on their limitations and potential side-effects for future consideration. Ultimately, suppressing antigraft immunity without negatively impacting the patient or cell therapy is not easy, and confounded further by the lack of appropriate in vivo models for thorough preclinical testing. This last point is particularly pertinent to the γδ T cell field, which already suffers from limited immunocompetent in vivo modeling due to the disparate γδ T cell repertoires of mice and humans. Although humanized mouse models may offer a partial solution, the cost of sufficiently powered studies using these mice is prohibitive, and whether or not they are truly representative of a human immune system remains open to debate.

From the range of different clinical trials discussed herein, patient preconditioning appears to be a staple for both autologous and allogeneic cell therapies that will likely play an important role in future therapies composed of donor-derived γδ T cells. An important question that remains open is whether or not preconditioning without additional stealth engineering is enough to support the levels of allogeneic cell therapy persistence required to generate durable clinical responses. Indeed, within the NK cell field, Liu et al. have reported low-level persistence and good clinical responses for donor-derived HLA mismatched CAR NK cells against hematological tumors using Flu/Cy preconditioning alone [120]. However, we do not yet know whether this level of persistence is enough for generating durable efficacy in other tumor types, or if stealth engineering would have improved survival and efficacy even further. A consensus on precisely how long allogeneic cells can survive in a preconditioned patient and how much persistence is required for maximum efficacy has yet to be reached, which is not surprising given the limited number of peer-reviewed studies in the literature and the huge variation between them in terms of their dosing regimens, disease indications, cell therapies, and immune monitoring. Moreover, most of the published data has focused on hematological malignancies and we have yet to overcome the technical challenges involved in tracking the persistence of allogeneic cell therapies within solid tumors. Nonetheless, it is interesting that immune reconstitution across the various preconditioning drug regimens covered herein seems to be consistently slower for T cells than it is for NK cells and myeloid cells. Due to the inherent antitumor properties of both NK cells and certain myeloid cell subsets, and their potential utility in generating concerted antitumor responses during adoptive cell therapy, γδ T cells engineered to specifically recruit and exploit these early recovering immune populations before they are rejected by alloreactive T cells could potentially be beneficial, especially where the extra help from an orchestrated immune response may be essential for achieving clinical efficacy; in the targeting of solid tumors, for example. It is important to note, however, that most of the longitudinal data we have on immune recovery postpreconditioning tracks systemic changes in immune composition within the periphery. We have little empirical evidence for the local effects of preconditioning on the immune compartment within nonhematological human tumors and thus a very limited understanding of the immune landscape with which tumor-infiltrating allogeneic γδ T cells will interact.

Although preconditioning and/or stealth engineering can potentially create a rejection-free therapeutic window long enough for allogeneic γδ T cell therapies to mount effective antitumor immunity, the intrinsic survival and persistence of the γδ T cells themselves—irrespective of any antigraft immunity—becomes a particularly important rate-limiting factor for overall efficacy and thus must be adequately bolstered. An allogeneic γδ T cell therapy must, therefore, be well equipped to survive and self-propagate, particularly within the highly immunosuppressive adverse conditions that are often associated with the microenvironment of solid tumors, such as hypoxia, nutrient deprivation, immune checkpoints, and antiinflammatory cytokines. This may require additional genetic modifications; for instance, gene editing to remove receptors such as PD-1 from the cell surface of the γδ T cells, or viral transduction to insert transgenes that drive survival, such as mitogenic cytokines like IL-15 [121] and/or antiexhaustion transcription factors like c-jun [122]. γδ T cells have certainly been reported to show both sensitivity to immune checkpoints and dependence on exogenous cytokine support for proliferation and survival in vitro [123]. Incorporation of persistence enhancing genetic modifications may prove to be essential for ensuring allogeneic γδ T cells persist long enough in vivo in order to exploit fully the rejection-free window created by preconditioning and/or stealth engineering. Indeed, a number of recent allogeneic NK cell clinical trials have yielded disappointing results in terms of persistence and efficacy, despite using preconditioning to delay rejection, supporting the notion that other persistence enhancing strategies may need to be employed on top of preconditioning and stealth engineering [124].

As ongoing and future clinical trials on allogeneic γδ T cells begin to publish their data, we will acquire greater insight into the utility of these cells as a therapy for cancer, and importantly, the effectiveness of preconditioning and stealth engineering as preventative steps against allogeneic rejection for better persistence and efficacy.

References

[1] Mantesso S, Geerts D, Spanholtz J, Kucerova L. Genetic engineering of natural killer cells for enhanced antitumor function. Front Immunol 2020;11, 607131. https://doi.org/10.3389/fimmu.2020.607131.

[2] Hadiloo K, Tahmasebi S, Esmaeilzadeh A. CAR-NKT cell therapy: a new promising paradigm of cancer immunotherapy. Cancer Cell Int 2023;23:86. https://doi.org/10.1186/s12935-023-02923-9.

[3] Fisher J, Anderson J. Engineering approaches in human gamma delta T cells for cancer immunotherapy. Front Immunol 2018;9:1409. https://doi.org/10.3389/fimmu.2018.01409.

[4] Unver N. Sophisticated genetically engineered macrophages, CAR-Macs, in hitting the bull's eye for solid cancer immunotherapy approaches. Clin Exp Med 2023. https://doi.org/10.1007/s10238-023-01106-0.

[5] Chang Y, et al. CAR-neutrophil mediated delivery of tumor-microenvironment responsive nanodrugs for glioblastoma chemo-immunotherapy. Nat Commun 2023;14:2266. https://doi.org/10.1038/s41467-023-37872-4.

[6] Dogan M, et al. Engineering human MAIT cells with chimeric antigen receptors for cancer immunotherapy. J Immunol 2022;209:1523–31. https://doi.org/10.4049/jimmunol.2100856.

[7] Ueda T, et al. Non-clinical efficacy, safety and stable clinical cell processing of induced pluripotent stem cell-derived anti-glypican-3 chimeric antigen receptor-expressing natural killer/innate lymphoid cells. Cancer Sci 2020;111:1478–90. https://doi.org/10.1111/cas.14374.

[8] Qin S, et al. Novel immune checkpoint targets: moving beyond PD-1 and CTLA-4. Mol Cancer 2019;18:155. https://doi.org/10.1186/s12943-019-1091-2.

[9] Gumber D, Wang LD. Improving CAR-T immunotherapy: overcoming the challenges of T cell exhaustion. EBioMedicine 2022;77, 103941. https://doi.org/10.1016/j.ebiom.2022.103941.

[10] Sterner RC, Sterner RM. CAR-T cell therapy: current limitations and potential strategies. Blood Cancer J 2021;11:69. https://doi.org/10.1038/s41408-021-00459-7.

[11] Jin X, et al. HLA-matched and HLA-haploidentical allogeneic CD19-directed chimeric antigen receptor T-cell infusions are feasible in relapsed or refractory B-cell acute lymphoblastic leukemia before hematopoietic stem cell transplantation. Leukemia 2020;34:909–13. https://doi.org/10.1038/s41375-019-0610-x.

[12] Caldwell KJ, Gottschalk S, Talleur AC. Allogeneic CAR cell therapy-more than a pipe dream. Front Immunol 2020;11, 618427. https://doi.org/10.3389/fimmu.2020.618427.

[13] Handgretinger R, Schilbach K. The potential role of gammadelta T cells after allogeneic HCT for leukemia. Blood 2018;131:1063–72. https://doi.org/10.1182/blood-2017-08-752162.

[14] Gentles AJ, et al. The prognostic landscape of genes and infiltrating immune cells across human cancers. Nat Med 2015;21:938–45. https://doi.org/10.1038/nm.3909.

[15] Kabelitz D, Serrano R, Kouakanou L, Peters C, Kalyan S. Cancer immunotherapy with gammadelta T cells: many paths ahead of us. Cell Mol Immunol 2020;17:925–39. https://doi.org/10.1038/s41423-020-0504-x.

[16] Bennouna J, et al. Phase-I study of Innacell gammadelta, an autologous cell-therapy product highly enriched in gamma9delta2 T lymphocytes, in combination with IL-2, in patients with metastatic renal cell carcinoma. Cancer Immunol Immunother 2008;57:1599–609. https://doi.org/10.1007/s00262-008-0491-8.

[17] Sakamoto M, et al. Adoptive immunotherapy for advanced non-small cell lung cancer using zoledronate-expanded gammadeltaTcells: a phase I clinical study. J Immunother 2011;34:202–11. https://doi.org/10.1097/CJI.0b013e318207ecfb.

[18] Noguchi A, et al. Zoledronate-activated Vgamma9gammadelta T cell-based immunotherapy is feasible and restores the impairment of gammadelta T cells in patients with solid tumors. Cytotherapy 2011;13:92–7. https://doi.org/10.3109/14653249.2010.515581.
[19] Kobayashi H, Tanaka Y, Yagi J, Minato N, Tanabe K. Phase I/II study of adoptive transfer of gammadelta T cells in combination with zoledronic acid and IL-2 to patients with advanced renal cell carcinoma. Cancer Immunol Immunother 2011;60:1075–84. https://doi.org/10.1007/s00262-011-1021-7.
[20] Kobayashi H, et al. Safety profile and anti-tumor effects of adoptive immunotherapy using gamma-delta T cells against advanced renal cell carcinoma: a pilot study. Cancer Immunol Immunother 2007;56:469–76. https://doi.org/10.1007/s00262-006-0199-6.
[21] Abe Y, et al. Clinical and immunological evaluation of zoledronate-activated Vgamma9gammadelta T-cell-based immunotherapy for patients with multiple myeloma. Exp Hematol 2009;37:956–68. https://doi.org/10.1016/j.exphem.2009.04.008.
[22] Nakajima J, et al. A phase I study of adoptive immunotherapy for recurrent non-small-cell lung cancer patients with autologous gammadelta T cells. Eur J Cardiothorac Surg 2010;37:1191–7. https://doi.org/10.1016/j.ejcts.2009.11.051.
[23] Izumi T, et al. Ex vivo characterization of gammadelta T-cell repertoire in patients after adoptive transfer of Vgamma9Vdelta2 T cells expressing the interleukin-2 receptor beta-chain and the common gamma-chain. Cytotherapy 2013;15:481–91. https://doi.org/10.1016/j.jcyt.2012.12.004.
[24] Nicol AJ, et al. Clinical evaluation of autologous gamma delta T cell-based immunotherapy for metastatic solid tumours. Br J Cancer 2011;105:778–86. https://doi.org/10.1038/bjc.2011.293.
[25] Wada I, et al. Intraperitoneal injection of in vitro expanded Vgamma9Vdelta2 T cells together with zoledronate for the treatment of malignant ascites due to gastric cancer. Cancer Med 2014;3:362–75. https://doi.org/10.1002/cam4.196.
[26] Kakimi K, et al. Adoptive transfer of zoledronate-expanded autologous Vgamma9Vdelta2 T-cells in patients with treatment-refractory non-small-cell lung cancer: a multicenter, open-label, single-arm, phase 2 study. J Immunother Cancer 2020;8. https://doi.org/10.1136/jitc-2020-001185.
[27] Almeida AR, et al. Delta one T cells for immunotherapy of chronic lymphocytic leukemia: clinical-grade expansion/differentiation and preclinical proof of concept. Clin Cancer Res 2016;22:5795–804. https://doi.org/10.1158/1078-0432.CCR-16-0597.
[28] Ferry GM, et al. A simple and robust single-step method for CAR-Vdelta1 gammadeltaT cell expansion and transduction for cancer immunotherapy. Front Immunol 2022;13, 863155. https://doi.org/10.3389/fimmu.2022.863155.
[29] Siegers GM, Lamb Jr LS. Cytotoxic and regulatory properties of circulating Vdelta1+ gammadelta T cells: a new player on the cell therapy field? Mol Ther 2014;22:1416–22. https://doi.org/10.1038/mt.2014.104.
[30] Brandes M, Willimann K, Moser B. Professional antigen-presentation function by human gammadelta T cells. Science 2005;309:264–8. https://doi.org/10.1126/science.1110267.
[31] Karahan GE, Claas FHJ, Heidt S. Pre-existing Alloreactive T and B cells and their possible relevance for pre-transplant risk estimation in kidney transplant recipients. Front Med (Lausanne) 2020;7:340. https://doi.org/10.3389/fmed.2020.00340.
[32] Li Q, Lan P. Activation of immune signals during organ transplantation. Signal Transduct Target Ther 2023;8:110. https://doi.org/10.1038/s41392-023-01377-9.
[33] Shiqi L, et al. Durable remission related to CAR-T persistence in R/R B-ALL and long-term persistence potential of prime CAR-T. Mol Ther Oncolytics 2023;29:107–17. https://doi.org/10.1016/j.omto.2023.04.003.
[34] Brudno JN, et al. Allogeneic T cells that express an anti-CD19 chimeric antigen receptor induce remissions of B-cell malignancies that progress after allogeneic hematopoietic stem-cell transplantation without causing graft-versus-host disease. J Clin Oncol 2016;34:1112–21. https://doi.org/10.1200/JCO.2015.64.5929.
[35] Muranski P, et al. Increased intensity lymphodepletion and adoptive immunotherapy—how far can we go? Nat Clin Pract Oncol 2006;3:668–81. https://doi.org/10.1038/ncponc0666.
[36] Suryadevara CM, et al. Temozolomide lymphodepletion enhances CAR abundance and correlates with antitumor efficacy against established glioblastoma. Onco Targets Ther 2018;7, e1434464. https://doi.org/10.1080/2162402X.2018.1434464.
[37] Murad JP, et al. Pre-conditioning modifies the TME to enhance solid tumor CAR T cell efficacy and endogenous protective immunity. Mol Ther 2021;29:2335–49. https://doi.org/10.1016/j.ymthe.2021.02.024.

[38] Klebanoff CA, Khong HT, Antony PA, Palmer DC, Restifo NP. Sinks, suppressors and antigen presenters: how lymphodepletion enhances T cell-mediated tumor immunotherapy. Trends Immunol 2005;26:111–7. https://doi.org/10.1016/j.it.2004.12.003.
[39] Emadi A, Jones RJ, Brodsky RA. Cyclophosphamide and cancer: golden anniversary. Nat Rev Clin Oncol 2009;6:638–47. https://doi.org/10.1038/nrclinonc.2009.146.
[40] Singh RK, Kumar S, Prasad DN, Bhardwaj TR. Therapeutic journery of nitrogen mustard as alkylating anticancer agents: historic to future perspectives. Eur J Med Chem 2018;151:401–33. https://doi.org/10.1016/j.ejmech.2018.04.001.
[41] Zhao J, et al. Selective depletion of CD4+CD25+Foxp3+ regulatory T cells by low-dose cyclophosphamide is explained by reduced intracellular ATP levels. Cancer Res 2010;70:4850–8. https://doi.org/10.1158/0008-5472.CAN-10-0283.
[42] Zhong H, et al. Low dose cyclophosphamide modulates tumor microenvironment by TGF-beta signaling pathway. Int J Mol Sci 2020;21. https://doi.org/10.3390/ijms21030957.
[43] Webb ER, et al. Cyclophosphamide depletes tumor infiltrating T regulatory cells and combined with anti-PD-1 therapy improves survival in murine neuroblastoma. iScience 2022;25, 104995. https://doi.org/10.1016/j.isci.2022.104995.
[44] Scurr M, et al. Low-dose cyclophosphamide induces antitumor T-cell responses, which associate with survival in metastatic colorectal cancer. Clin Cancer Res 2017;23:6771–80. https://doi.org/10.1158/1078-0432.CCR-17-0895.
[45] Turtle CJ, Riddell SR, Maloney DG. CD19-targeted chimeric antigen receptor-modified T-cell immunotherapy for B-cell malignancies. Clin Pharmacol Ther 2016;100:252–8. https://doi.org/10.1002/cpt.392.
[46] Turtle CJ, et al. Immunotherapy of non-Hodgkin's lymphoma with a defined ratio of CD8+ and CD4+ CD19-specific chimeric antigen receptor-modified T cells. Sci Transl Med 2016;8:355ra116. https://doi.org/10.1126/scitranslmed.aaf8621.
[47] Galmarini CM, Mackey JR, Dumontet C. Nucleoside analogues: mechanisms of drug resistance and reversal strategies. Leukemia 2001;15:875–90. https://doi.org/10.1038/sj.leu.2402114.
[48] Khan Y, Lyou Y, El-Masry M, O'Brien S. Reassessing the role of chemoimmunotherapy in chronic lymphocytic leukemia. Expert Rev Hematol 2020;13:31–8. https://doi.org/10.1080/17474086.2020.1697226.
[49] Wallen H, et al. Fludarabine modulates immune response and extends in vivo survival of adoptively transferred CD8 T cells in patients with metastatic melanoma. PLoS One 2009;4, e4749. https://doi.org/10.1371/journal.pone.0004749.
[50] Yu L, et al. GD2-specific chimeric antigen receptor-modified T cells for the treatment of refractory and/or recurrent neuroblastoma in pediatric patients. J Cancer Res Clin Oncol 2022;148:2643–52. https://doi.org/10.1007/s00432-021-03839-5.
[51] Kekre N, et al. CLIC-01: manufacture and distribution of non-cryopreserved CAR-T cells for patients with CD19 positive hematologic malignancies. Front Immunol 2022;13:1074740. https://doi.org/10.3389/fimmu.2022.1074740.
[52] Mackensen A, et al. Anti-CD19 CAR T cell therapy for refractory systemic lupus erythematosus. Nat Med 2022;28:2124–32. https://doi.org/10.1038/s41591-022-02017-5.
[53] Heczey A, et al. Anti-GD2 CAR-NKT cells in patients with relapsed or refractory neuroblastoma: an interim analysis. Nat Med 2020;26:1686–90. https://doi.org/10.1038/s41591-020-1074-2.
[54] Heczey A, et al. Anti-GD2 CAR-NKT cells in relapsed or refractory neuroblastoma: updated phase 1 trial interim results. Nat Med 2023;29:1379–88. https://doi.org/10.1038/s41591-023-02363-y.
[55] Ramos CA, et al. In vivo fate and activity of second- versus third-generation CD19-specific CAR-T cells in B cell non-Hodgkin's lymphomas. Mol Ther 2018;26:2727–37. https://doi.org/10.1016/j.ymthe.2018.09.009.
[56] Cui J, et al. Combined cellular immunotherapy and chemotherapy improves clinical outcome in patients with gastric carcinoma. Cytotherapy 2015;17:979–88. https://doi.org/10.1016/j.jcyt.2015.03.605.
[57] Aoki T, et al. Adjuvant combination therapy with gemcitabine and autologous gammadelta T-cell transfer in patients with curatively resected pancreatic cancer. Cytotherapy 2017;19:473–85. https://doi.org/10.1016/j.jcyt.2017.01.002.
[58] Maziarz RT, Diaz A, Miklos DB, Shah NN. Perspective: an international fludarabine shortage: supply chain issues impacting transplantation and immune effector cell therapy delivery. Transplant Cell Ther 2022;28:723–6. https://doi.org/10.1016/j.jtct.2022.08.002.

[59] Green S, Schultz L. Rational alternatives to fludarabine and cyclophosphamide-based pre-CAR lymphodepleting regimens in the pediatric and young adult B-ALL setting. Curr Oncol Rep 2023;25:841–6. https://doi.org/10.1007/s11912-023-01404-6.
[60] Schuster SJ, et al. Tisagenlecleucel in adult relapsed or refractory diffuse large B-cell lymphoma. N Engl J Med 2019;380:45–56. https://doi.org/10.1056/NEJMoa1804980.
[61] Ghilardi G, et al. Bendamustine is safe and effective for lymphodepletion before tisagenlecleucel in patients with refractory or relapsed large B-cell lymphomas. Ann Oncol 2022;33:916–28. https://doi.org/10.1016/j.annonc.2022.05.521.
[62] Ramos CA, et al. Anti-CD30 CAR-T cell therapy in relapsed and refractory Hodgkin lymphoma. J Clin Oncol 2020;38:3794–804. https://doi.org/10.1200/JCO.20.01342.
[63] Otegbeye F, et al. A phase I study to determine the maximum tolerated dose of ex vivo expanded natural killer cells derived from unrelated, HLA-disparate adult donors. Transplant Cell Ther 2022;28:250e251–8. https://doi.org/10.1016/j.jtct.2022.02.008.
[64] Yoon DH, et al. Phase I study: safety and efficacy of an ex vivo-expanded allogeneic natural killer cell (MG4101) with rituximab for relapsed/refractory B cell non-Hodgkin lymphoma. Transplant Cell Ther 2023;29:253e251–9. https://doi.org/10.1016/j.jtct.2022.12.025.
[65] Hu Y, et al. Genetically modified CD7-targeting allogeneic CAR-T cell therapy with enhanced efficacy for relapsed/refractory CD7-positive hematological malignancies: a phase I clinical study. Cell Res 2022;32:995–1007. https://doi.org/10.1038/s41422-022-00721-y.
[66] Wilhelm M, et al. Successful adoptive transfer and in vivo expansion of haploidentical gammadelta T cells. J Transl Med 2014;12:45. https://doi.org/10.1186/1479-5876-12-45.
[67] Vydra J, et al. A phase I trial of allogeneic gammadelta T lymphocytes from haploidentical donors in patients with refractory or relapsed acute myeloid leukemia. Clin Lymphoma Myeloma Leuk 2023;23:e232–9. https://doi.org/10.1016/j.clml.2023.02.003.
[68] Benjamin R, et al. UCART19, a first-in-class allogeneic anti-CD19 chimeric antigen receptor T-cell therapy for adults with relapsed or refractory B-cell acute lymphoblastic leukaemia (CALM): a phase 1, dose-escalation trial. Lancet Haematol 2022;9:e833–43. https://doi.org/10.1016/S2352-3026(22)00245-9.
[69] Ottaviano G, et al. Phase 1 clinical trial of CRISPR-engineered CAR19 universal T cells for treatment of children with refractory B cell leukemia. Sci Transl Med 2022;**14**:eabq3010. https://doi.org/10.1126/scitranslmed.abq3010.
[70] Mailankody S, et al. Allogeneic BCMA-targeting CAR T cells in relapsed/refractory multiple myeloma: phase 1 UNIVERSAL trial interim results. Nat Med 2023;29:422–9. https://doi.org/10.1038/s41591-022-02182-7.
[71] Lamb LS, et al. A combined treatment regimen of MGMT-modified gammadelta T cells and temozolomide chemotherapy is effective against primary high grade gliomas. Sci Rep 2021;11:21133. https://doi.org/10.1038/s41598-021-00536-8.
[72] Suzuki M, Kato C, Kato A. Therapeutic antibodies: their mechanisms of action and the pathological findings they induce in toxicity studies. J Toxicol Pathol 2015;28:133–9. https://doi.org/10.1293/tox.2015-0031.
[73] Zhao Y, et al. The immunological function of CD52 and its targeting in organ transplantation. Inflamm Res 2017;66:571–8. https://doi.org/10.1007/s00011-017-1032-8.
[74] van der Zwan M, Baan CC, van Gelder T, Hesselink DA. Review of the clinical pharmacokinetics and pharmacodynamics of alemtuzumab and its use in kidney transplantation. Clin Pharmacokinet 2018;57:191–207. https://doi.org/10.1007/s40262-017-0573-x.
[75] Amini L, et al. Preparing for CAR T cell therapy: patient selection, bridging therapies and lymphodepletion. Nat Rev Clin Oncol 2022;19:342–55. https://doi.org/10.1038/s41571-022-00607-3.
[76] Zhang X, et al. Differential reconstitution of T cell subsets following immunodepleting treatment with alemtuzumab (anti-CD52 monoclonal antibody) in patients with relapsing-remitting multiple sclerosis. J Immunol 2013;191:5867–74. https://doi.org/10.4049/jimmunol.1301926.
[77] Chakupurakal G, Freudenberger P, Skoetz N, Ahr H, Theurich S. Polyclonal anti-thymocyte globulins for the prophylaxis of graft-versus-host disease after allogeneic stem cell or bone marrow transplantation in adults. Cochrane Database Syst Rev 2023;**6**, CD009159. https://doi.org/10.1002/14651858.CD009159.pub3.
[78] Gharekhani A, Entezari-Maleki T, Dashti-Khavidaki S, Khalili H. A review on comparing two commonly used rabbit anti-thymocyte globulins as induction therapy in solid organ transplantation. Expert Opin Biol Ther 2013;13:1299–313. https://doi.org/10.1517/14712598.2013.822064.
[79] Mohty M. Mechanisms of action of antithymocyte globulin: T-cell depletion and beyond. Leukemia 2007;21:1387–94. https://doi.org/10.1038/sj.leu.2404683.

[80] Wang Y, et al. Specificity of mouse and human Fcgamma receptors and their polymorphic variants for IgG subclasses of different species. Eur J Immunol 2022;52:753–9. https://doi.org/10.1002/eji.202149766.
[81] Waller EK, et al. Pharmacokinetics and pharmacodynamics of anti-thymocyte globulin in recipients of partially HLA-matched blood hematopoietic progenitor cell transplantation. Biol Blood Marrow Transplant 2003;9:460–71. https://doi.org/10.1016/s1083-8791(03)00127-7.
[82] Feng X, et al. In vivo effects of horse and rabbit antithymocyte globulin in patients with severe aplastic anemia. Haematologica 2014;99:1433–40. https://doi.org/10.3324/haematol.2014.106542.
[83] Van Der Windt DJ, et al. Investigation of lymphocyte depletion and repopulation using alemtuzumab (Campath-1H) in cynomolgus monkeys. Am J Transplant 2010;10:773–83. https://doi.org/10.1111/j.1600-6143.2010.03050.x.
[84] Groth C, et al. Phase I/II trial of a combination of anti-CD3/CD7 immunotoxins for steroid-refractory acute graft-versus-host disease. Biol Blood Marrow Transplant 2019;25:712–9. https://doi.org/10.1016/j.bbmt.2018.10.020.
[85] Sieber M, Baumgrass R. Novel inhibitors of the calcineurin/NFATc hub—alternatives to CsA and FK506? Cell Commun Signal 2009;7:25. https://doi.org/10.1186/1478-811X-7-25.
[86] Tsuda K, et al. Calcineurin inhibitors suppress cytokine production from memory T cells and differentiation of naive T cells into cytokine-producing mature T cells. PLoS One 2012;7, e31465. https://doi.org/10.1371/journal.pone.0031465.
[87] Vafadari R, Kraaijeveld R, Weimar W, Baan CC. Tacrolimus inhibits NF-kappaB activation in peripheral human T cells. PLoS One 2013;8, e60784. https://doi.org/10.1371/journal.pone.0060784.
[88] Segundo DS, et al. Calcineurin inhibitors, but not rapamycin, reduce percentages of CD4+CD25+FOXP3+ regulatory T cells in renal transplant recipients. Transplantation 2006;82:550–7. https://doi.org/10.1097/01.tp.0000229473.95202.50.
[89] Zhang XJ, et al. Effects and mechanisms of tacrolimus on development of murine Th17 cells. Transplant Proc 2010;42:3779–83. https://doi.org/10.1016/j.transproceed.2010.08.033.
[90] Heidt S, et al. Calcineurin inhibitors affect B cell antibody responses indirectly by interfering with T cell help. Clin Exp Immunol 2010;159:199–207. https://doi.org/10.1111/j.1365-2249.2009.04051.x.
[91] Rovira J, et al. Cyclosporine A inhibits the T-bet-dependent antitumor response of CD8(+) T cells. Am J Transplant 2016;16:1139–47. https://doi.org/10.1111/ajt.13597.
[92] Li H, David Pauza C. Interplay of T-cell receptor and interleukin-2 signalling in Vgamma2Vdelta2 T-cell cytotoxicity. Immunology 2011;132:96–103. https://doi.org/10.1111/j.1365-2567.2010.03343.x.
[93] Otsuka S, et al. Calcineurin inhibitors suppress acute graft-versus-host disease via NFAT-independent inhibition of T cell receptor signaling. J Clin Invest 2021;131. https://doi.org/10.1172/JCI147683.
[94] Riella LV, Alegre ML. Novel role of calcineurin inhibitors in curbing T cells' sweet tooth. Am J Transplant 2018;18:3. https://doi.org/10.1111/ajt.14606.
[95] Azzi JR, Sayegh MH, Mallat SG. Calcineurin inhibitors: 40 years later, can't live without. J Immunol 2013;191:5785–91. https://doi.org/10.4049/jimmunol.1390055.
[96] Akimova T, et al. Differing effects of rapamycin or calcineurin inhibitor on T-regulatory cells in pediatric liver and kidney transplant recipients. Am J Transplant 2012;12:3449–61. https://doi.org/10.1111/j.1600-6143.2012.04269.x.
[97] Whitehouse G, et al. IL-2 therapy restores regulatory T-cell dysfunction induced by calcineurin inhibitors. Proc Natl Acad Sci USA 2017;114:7083–8. https://doi.org/10.1073/pnas.1620835114.
[98] Glaviano A, et al. PI3K/AKT/mTOR signaling transduction pathway and targeted therapies in cancer. Mol Cancer 2023;22:138. https://doi.org/10.1186/s12943-023-01827-6.
[99] Davoodi-Moghaddam Z, Jafari-Raddani F, Delshad M, Pourbagheri-Sigaroodi A, Bashash D. Inhibitors of the PI3K/AKT/mTOR pathway in human malignancies; trend of current clinical trials. J Cancer Res Clin Oncol 2023. https://doi.org/10.1007/s00432-023-05277-x.
[100] Occhiuzzi MA, et al. Recent advances in PI3K/PKB/mTOR inhibitors as new anticancer agents. Eur J Med Chem 2023;246, 114971. https://doi.org/10.1016/j.ejmech.2022.114971.
[101] Maenaka A, Kinoshita K, Hara H, Cooper DKC. The case for the therapeutic use of mechanistic/mammalian target of rapamycin (mTOR) inhibitors in xenotransplantation. Xenotransplantation 2023;30, e12802. https://doi.org/10.1111/xen.12802.
[102] Jozwiak J, Jozwiak S, Oldak M. Molecular activity of sirolimus and its possible application in tuberous sclerosis treatment. Med Res Rev 2006;26:160–80. https://doi.org/10.1002/med.20049.

[103] Bhat R, Tonutti A, Timilsina S, Selmi C, Gershwin ME. Perspectives on mycophenolate mofetil in the management of autoimmunity. Clin Rev Allergy Immunol 2023;65:86–100. https://doi.org/10.1007/s12016-023-08963-3.

[104] Allison AC. Mechanisms of action of mycophenolate mofetil. Lupus 2005;14(Suppl. 1):s2–8. https://doi.org/10.1191/0961203305lu2109oa.

[105] Taylor DO, Ensley RD, Olsen SL, Dunn D, Renlund DG. Mycophenolate mofetil (RS-61443): preclinical, clinical, and three-year experience in heart transplantation. J Heart Lung Transplant 1994;13:571–82.

[106] Brewin J, et al. Generation of EBV-specific cytotoxic T cells that are resistant to calcineurin inhibitors for the treatment of posttransplantation lymphoproliferative disease. Blood 2009;114:4792–803. https://doi.org/10.1182/blood-2009-07-228387.

[107] Amini L, et al. CRISPR-Cas9-edited tacrolimus-resistant antiviral T cells for advanced adoptive immunotherapy in transplant recipients. Mol Ther 2021;29:32–46. https://doi.org/10.1016/j.ymthe.2020.09.011.

[108] Thongsin N, Wattanapanitch M. CRISPR/Cas9 ribonucleoprotein complex-mediated efficient B2M knockout in human induced pluripotent stem cells (iPSCs). Methods Mol Biol 2022;2454:607–24. https://doi.org/10.1007/7651_2021_352.

[109] Mattapally S, et al. Human leukocyte antigen class I and II knockout human induced pluripotent stem cell-derived cells: universal donor for cell therapy. J Am Heart Assoc 2018;7, e010239. https://doi.org/10.1161/JAHA.118.010239.

[110] Kumar B, et al. Engineered cord blood megakaryocytes evade killing by allogeneic T-cells for refractory thrombocytopenia. Front Immunol 2022;13:1018047. https://doi.org/10.3389/fimmu.2022.1018047.

[111] Ren J, et al. Multiplex genome editing to generate universal CAR T cells resistant to PD1 inhibition. Clin Cancer Res 2017;23:2255–66. https://doi.org/10.1158/1078-0432.CCR-16-1300.

[112] Kagoya Y, et al. Genetic ablation of HLA class I, class II, and the T-cell receptor enables allogeneic T cells to be used for adoptive T-cell therapy. Cancer Immunol Res 2020;8:926–36. https://doi.org/10.1158/2326-6066.CIR-18-0508.

[113] Gornalusse GG, et al. HLA-E-expressing pluripotent stem cells escape allogeneic responses and lysis by NK cells. Nat Biotechnol 2017;35:765–72. https://doi.org/10.1038/nbt.3860.

[114] Guo Y, et al. Mutant B2M-HLA-E and B2M-HLA-G fusion proteins protects universal chimeric antigen receptor-modified T cells from allogeneic NK cell-mediated lysis. Eur J Immunol 2021;51:2513–21. https://doi.org/10.1002/eji.202049107.

[115] Li W, et al. Simultaneous editing of TCR, HLA-I/II and HLA-E resulted in enhanced universal CAR-T resistance to allo-rejection. Front Immunol 2022;13:1052717. https://doi.org/10.3389/fimmu.2022.1052717.

[116] Jo S, et al. Endowing universal CAR T-cell with immune-evasive properties using TALEN-gene editing. Nat Commun 2022;13:3453. https://doi.org/10.1038/s41467-022-30896-2.

[117] Hoerster K, et al. HLA class I knockout converts allogeneic primary NK cells into suitable effectors for "off-the-shelf" immunotherapy. Front Immunol 2020;11, 586168. https://doi.org/10.3389/fimmu.2020.586168.

[118] Meshitsuka S, et al. CRISPR/Cas9 and AAV mediated insertion of beta2 microglobulin-HLA-G fusion gene protects mesenchymal stromal cells from allogeneic rejection and potentiates the use for off-the-shelf cell therapy. Regen Ther 2022;21:442–52. https://doi.org/10.1016/j.reth.2022.09.009.

[119] Cazzetta V, et al. NKG2A expression identifies a subset of human Vdelta2 T cells exerting the highest antitumor effector functions. Cell Rep 2021;37, 109871. https://doi.org/10.1016/j.celrep.2021.109871.

[120] Liu E, et al. Use of CAR-transduced natural killer cells in CD19-positive lymphoid tumors. N Engl J Med 2020;382:545–53. https://doi.org/10.1056/NEJMoa1910607.

[121] Makkouk A, et al. Off-the-shelf Vdelta1 gamma delta T cells engineered with glypican-3 (GPC-3)-specific chimeric antigen receptor (CAR) and soluble IL-15 display robust antitumor efficacy against hepatocellular carcinoma. J Immunother Cancer 2021;9. https://doi.org/10.1136/jitc-2021-003441.

[122] Lynn RC, et al. c-Jun overexpression in CAR T cells induces exhaustion resistance. Nature 2019;576:293–300. https://doi.org/10.1038/s41586-019-1805-z.

[123] Gao Z, et al. Gamma delta T-cell-based immune checkpoint therapy: attractive candidate for antitumor treatment. Mol Cancer 2023;22:31. https://doi.org/10.1186/s12943-023-01722-0.

[124] Lian G, Mak TS, Yu X, Lan HY. Challenges and recent advances in NK cell-targeted immunotherapies in solid tumors. Int J Mol Sci 2021;23. https://doi.org/10.3390/ijms23010164.

Index

Note: Page numbers followed by *f* indicate figures and *t* indicate tables.